Lean Six Sigma
Process Improvement Tools and Techniques

Donna C. S. Summers
University of Dayton

Prentice Hall

Boston Columbus Indianapolis New York San Francisco Upper Saddle River Amsterdam
Cape Town Dubai London Madrid Milan Munich Paris Montreal Toronto Delhi
Mexico City Sao Paulo Sydney Hong Kong Seoul Singapore Taipei Tokyo

Editorial Director : Vernon R. Anthony
Acquisitions Editor: David Ploskonka
Editorial Assistant: Nancy Kesterson
Director of Marketing: David Gesell
Marketing Manager: Kara Clarke
Senior Marketing Coordinator: Alicia Wozniak
Marketing Assistant: Les Roberts
Senior Project Manager: Maren L. Miller
Senior Managing Editor: JoEllen Gohr
Associate Managing Editor: Alexandrina Benedicto Wolf
Senior Operations Supervisor: Pat Tonneman
Operations Specialist: Laura Weaver
Senior Art Director: Diane Y. Ernsberger
Cover Designer: Ali Mohrman
AV Project Manager: Janet Portisch
Full-Service Project Management: Kelly Keeler, PreMediaGlobal
Composition: PreMediaGlobal
Printer/Binder: Edwards Brothers
Cover Printer: Coral Graphics
Text Font: Minion-Regular

Credits and acknowledgments borrowed from other sources and reproduced, with permission, in this textbook appear on the appropriate page within the text. Unless otherwise stated, all artwork has been provided by the author.

Copyright © 2011 Pearson Education, Inc., publishing as Prentice Hall, One Lake Street, Upper Saddle River, New Jersey, 07458. All rights reserved. Manufactured in the United States of America. This publication is protected by Copyright, and permission should be obtained from the publisher prior to any prohibited reproduction, storage in a retrieval system, or transmission in any form or by any means, electronic, mechanical, photocopying, recording, or likewise. To obtain permission(s) to use material from this work, please submit a written request to Pearson Education, Inc., Permissions Department, One Lake Street, Upper Saddle River, New Jersey, 07458.

Many of the designations by manufacturers and seller to distinguish their products are claimed as trademarks. Where those designations appear in this book, and the publisher was aware of a trademark claim, the designations have been printed in initial caps or all caps.

Library of Congress Cataloging-in-Publication Data

Summers, Donna C. S.
  Lean six sigma : process improvement tools and techniques / Donna C. S. Summers.
     p. cm.
  Includes bibliographical references and index.
  ISBN 978-0-13-512510-6
1. Six sigma (Quality control standard)   2. Total quality management.   3. Production management—Quality control.
I. Title.
TS156.S847 2011
658.5—dc22

2010009433

10 9 8 7 6 5 4 3 2 1

ISBN 10:   0-13-512510-3
ISBN 13: 978-0-13-512510-6

Wherever you go, no matter the weather,
always bring your own sunshine.
Karl,
You are my sunshine.

PREFACE

Competitive global economy. Words that challenge organizational leaders on a daily basis. *Quality, productivity,* and *cost.* These three words give effective leaders a long to-do list that includes:

- Enhancing customer satisfaction;
- Maintaining a safe, clean, and organized workplace;
- Utilizing an effective logistics, scheduling, and shipping system;
- Deploying facilities, raw materials, and technology effectively;
- Managing incoming, work-in-process, and finished goods inventory levels; and
- Employing and motivating talented people; and maintaining equipment and facilities.

All of provides a quality product or service that customers desire. A leader's ability to manage in a complex environment is crucial to an organization's success in today's economy. Is there any organization in the world that would not want to optimize each of these three areas?

This text is designed to present both the principles of lean and Six Sigma and the tools and techniques that enable optimal performance. Just as quality, productivity, and cost are interrelated concepts, so are the tools and techniques of lean and Six Sigma. This text presents many examples that show how interrelated tools and techniques can be used to solve problems. To add to the reader's depth of understanding about how these tools and techniques came into use, the text provides the historical background about the people who popularized them. The examples come from a wide variety of venues, including service industries, manufacturing companies, and government. All of the examples in this text have been gathered from real life.

In keeping with the cost-productivity-quality theme, this text is divided into four areas:

Process Improvement Essentials
Costs: Defining Opportunities for Process Improvement
Productivity: Process Improvement Opportunities
Quality: Variation Reduction Opportunities

The order in which the concepts of quality, costs, and productivity are presented has nothing to do with their relative importance to one another. All three are equally important and very difficult for all but the best of companies to balance. The gear figure that opens each chapter was chosen to show that all must turn together in order to produce a final product or service.

Process Improvement Essentials is designed to introduce the essentials that must be present in any organization before making a commitment to a particular improvement methodology. Leaders must understand what they are getting themselves into. Customers must be the focus of any improvement initiative. Strategic plans guide activities. People make it happen.

Costs: Defining Opportunities for Process Improvement follows the first section because improvement activities must focus on what is important to the customer and the company. Costs of quality and measures of performance provide information that sets the direction for the improvement activities. Project management and problem solving processes like define, measure, analyze, improve, control (DMAIC) establish the process for making improvements.

Productivity: Process Improvement Opportunities provides many of the tools and techniques associated with lean. Examples from industry describe how these tools and techniques have been used to solve process problems. Topic coverage is broad, beginning with value-added process mapping and continuing through productive maintenance and supply chain management.

Quality: Variation Reduction Opportunities places a greater focus on Six Sigma statistical tools and techniques. Popular techniques such as variable control charts and process capability are covered. Introductions to reliability concepts and design of experiments occur in later chapters.

The tools and techniques chosen for coverage in this text are the result of careful consultation with industry professionals and the American Society for Quality website (particularly its certification section). The tools and techniques most popularly used by practicing professionals have been selected.

ONLINE INSTRUCTOR'S MANUAL

To access supplementary materials online, instructors need to request an instructor access code. Go to **www.pearsonhighered.com/irc**, where you can register for an instructor access code. Within 48 hours after registering, you will receive a confirming e-mail, including an instructor access code. Once you have received your code, go to the site and log on for full instructions on downloading the materials you wish to use.

ACKNOWLEDGMENTS

I would like to express a special thanks to Karl Summers, Murray Therrill, and Nancy Kesterson. Without their generous support, this book would not have been possible.

CONTENTS

I PROCESS IMPROVEMENT ESSENTIALS

1 Lean Six Sigma Origins 1

Introduction 1 / Lean Origins 2 / Six Sigma Origins 3 / Lean Six Sigma 7 / Summary 9 / Take Away Tips 9 / Chapter Questions 10

2 Leadership for Process Improvement 11

Leadership Traits 11 / Creating a Leas Six Sigma Culture 12 / Managing by Fact and With a Knowledge of Variation 16 / Leadership Styles 17 / Summary 18 / Take Away Tips 18 / Chapter Questions 18

3 Strategic Planning 19

Strategic Plan Creation 19 / Strategic Plan Deployment 21 / Summary 23 / Take Away Tips 23 / Chapter Questions 23

4 Creating a Customer Focus 24

Customer Value Perceptions 24 / Quality Function Deployment 26 / Summary 36 / Take Away Tips 36 / Chapter Questions 36

5 Process Improvement Teams 37

Process Improvement Teams 37 / Education and Training 39 / Enhancing Team Performance 40 / Lean Six Sigma Teams 42 / Summary 44 / Take Away Tips 44 / Chapter Questions 44

II COSTS: DEFINING OPPORTUNITIES FOR PROCESS IMPROVEMENT

6 Costs of Quality 45

Costs of Quality 45 / Defining Quality Costs 47 / Types of Quality Costs 49 / Quality-Cost Measurement System 52 / Utilizing Quality Costs for Decision Making 56 / Summary 57 / Take Away Tips 57 / Chapter Questions 57

7 Process Performance Measures 59

Defining Process Measures 59 / Measures of Performance 60 / Measures of Performance and Gap Analysis 64 / Summary 65 / Take Away Tips 67 / Chapter Questions 67

8 Project Management 68

Lean Six Sigma Projects 68 / Project Characteristics 69 / Project Selection 69 / Project Proposals 70 / Project Goals and Objectives 70 / Project Plans 72 / Project Schedules 73 / Project Budgets 76 / Project Contingency Plans and Change Control Systems 77 / Project Control 78 / Project Managers 78 / Summary 79 / Take Away Tips 79 / Chapter Questions 79

9 Problem Solving Using Define, Measure, Analyze, Improve, Control 81

Problem Solving 81 / Define 82 / Measure 82 / Analyze 87 / Improve 94 / Control 94 / Summary 96 / Take Away Tips 96 / Chapter Questions 97

III PRODUCTIVITY: PROCESS IMPROVEMENT OPPORTUNITIES

10 Value-Added Process Mapping 98

Defining Processes 98 / Identifying Key Processes 99 / Creating Value and Generating Customer Satisfaction 99 / Process Improvement 100 / Process Mapping 101 / Value Stream Process Mapping 108 / Separating the Non-Value-Added Activities from the Value-Added Activities 112 / Summary 113 / Take Away Tips 116 / Chapter Problems 116

vii

11 Just-in-Time and Kanban 118

Just-in-Time 118 / Kanban 120 / Jidoka (Autonomation) 122 / Summary 123 / Take Away Tips 123 / Chapter Problems 123

12 5S 124

5S 124 / Seiri 125 / Seiton 126 / Seiso 128 / Seiketsu 129 / Shitsuke 129 / Summary 131 / Take Away Tips 131 / Chapter Problems 131

13 Kaizen and Error Proofing 132

Kaizen 132 / Error Proofing (Poka-yoke) 142 / Summary 144 / Take Away Tips 145 / Chapter Problems 145

14 Work Optimization 146

Line Balancing 148 / Setup Time Reduction 149 / Single Piece Flow (Reduced Batch Sizes) 151 / Schedule Leveling 152 / Standardized Work 152 / Visual Management 153 / Summary 155 / Take Away Tips 155 / Chapter Questions 155

15 Productive Maintenance 156

Productive Maintenance 156 / Elimination of Equipment Losses 157 / Preventive Maintenance 160 / Predictive Maintenance 162 / Autonomous Maintenance 162 / Summary 162 / Take Away Tips 162 / Chapter Problems 164

16 Supply Chain Management 165

Supply Chains as a Process 165 / Supply Chain Management Benefits 166 / Supply Chain Management Elements 166 / Management 167 / Supply Chain Management Challenges 174 / Summary 174 / Take Away Tips 175 / Chapter Problems 175

IV QUALITY: VARIATION REDUCTION OPPORTUNITIES

17 Statistics 176

Populations Versus Samples 177 / Data Collection 178 / Accuracy, Precision, and Measurement Error 180 / Data Analysis: Graphical 182 / Data Analysis: Analytical 187 / Central Limit Theorem 193 / Normal Frequency Distribution 195 / Confidence Intervals 198 / Summary 200 / Take Away Tips 200 / Formulas 200 / Chapter Questions 201

18 Variable Control Charts 203

Variable Control Charts 203 / Dr. Walter Shewhart 203 / Control Chart Interpretation 213 / Summary 221 / Take Away Tips 221 / Formulas 222 / Chapter Questions 222

19 Process Capability 224

Individual Values Compared with Averages 225 / Estimation of Population Sigma from Sample Data 226 / Control Limits Versus Specification Limits 227 / The 6σ Spread Versus Specification Limits 228 / Calculating Process Capability Indices 230 / Calculating 6σ 230 / Summary 234 / Take Away Tips 234 / Formulas 235 / Chapter Questions 235

20 Reliability 236

Reliability Concepts 236 / Reliability Programs 237 / Product Life Cycle Curve 239 / Measures of Reliability 240 / Summary 244 / Take Away Tips 244 / Formulas 245 / Chapter Questions 245

21 Design of Experiments 247

Design of Experiments 247 / Trial and Error Experiments 247 / Definitions 248 / Conducting an Experiment: Steps in Planned Experimentation 250 / Experiment Designs 251 / Single Factor Experiments 251 / Hypotheses and Experiment Errors 255 / Experiment Analysis Methods 257 / Summary 261 / Take Away Tips 261 / Chapter Questions 261

22 Failure Modes and Effects Analysis 263

Failure Modes and Effects Analysis 263 / Systems FMEA 263 / Process FMEA 264 / Design FMEA 264 / Creating an FMEA 264 / Summary 271 / Take Away Tips 271 / Chapter Questions 271

23 Lean Six Sigma 272

Lean 272 / Six Sigma 273

Appendix 1   Normal Curve Areas P(Z ≤ Z_0)   275

Appendix 2   Factors for Computing Central Lines and 3σ, Control Limits for $\bar{X}$, s, and R Charts   277

Appendix 3   Values of t Distribution   278

Answers to Selected Problems   280

Bibliography   281

Index   289

LEAN AND SIX SIGMA ORIGINS

The longer an article is in the process of manufacture and the more it is moved about, the greater is its ultimate cost.

Henry Ford

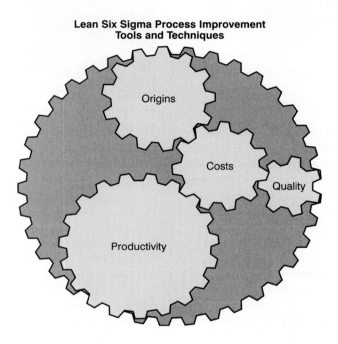

 LEARNING OPPORTUNITIES

1. To introduce the concept of lean
2. To introduce the concept of Six Sigma
3. To show the relationship between lean manufacturing and Six Sigma
4. To familiarize the reader with the benefits of implementing lean
5. To familiarize the reader with the benefits of implementing Six Sigma

INTRODUCTION

Competitive global economy are words that challenge organizational leaders on a daily basis. *Quality, productivity*, and *cost*: These three words give effective leaders a long to-do list that includes enhancing customer satisfaction; maintaining a safe, clean, and organized workplace; utilizing an effective logistics, scheduling, and shipping system; deploying facilities, raw materials, and technology effectively; managing incoming, work-in-process, and finished goods inventory levels; employing and motivating talented people; and maintaining equipment and facilities—all while providing a quality product or service that customers desire. A leader's ability to manage in a complex environment is crucial to an organization's success in today's economy. Quality, productivity, and cost. Is there any organization in the world that would not want to optimize each of these three areas?

The methodologies of lean and Six Sigma provide the tools and techniques that enable optimal performance. Just as quality, productivity, and cost are interrelated concepts, so are the tools and techniques of lean and Six Sigma. Launching a lean Six Sigma business model begins with an understanding of the origins and motivations behind these management philosophies. Originally lean and Six Sigma were two separate approaches to continuous improvement, but effective organizations have combined them in order to benefit from the strengths of both. They use lean tools and techniques to standardize work and

CHAPTER ONE

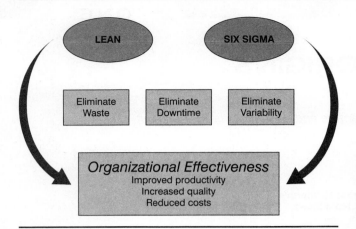

FIGURE 1.1 Lean Six Sigma Process Improvement

FIGURE 1.2 Henry Ford's Assembly Line
(**Source:** Ford Motor Company.)

remove waste and non-value-added activities, and they use Six Sigma tools and techniques to attack the variation present in processes. This chapter introduces the concepts of lean and Six Sigma. Future chapters explain tools and techniques that can be used to optimize cost, productivity, and quality (Figure 1.1).

LEAN ORIGINS

Henry Ford said, "*Time waste differs from material waste in that there can be no salvage.*" With this in mind, he focused on ways to speed up the manufacturing process without sacrificing quality. He was the first to utilize a moving assembly line on a grand scale. On Ford Motor Company's final assembly line, car bodies slid down a ramp onto waiting chassis (Figure 1.2). This moving assembly line approach, so different from previous manufacturing methods, regularly broke daily production records. Vital, adaptive organizations concentrate on three key concepts: quality, productivity, and costs. Lean tools and techniques are designed to improve process performance, increase quality and productivity, while reducing costs and enhancing profitability.

In his books *The Machine That Changed the World* (Harper Perennial, 1991) and *Lean Thinking* (Simon & Schuster, 1996), Jim Womack presents the principles and practices related to lean manufacturing. Just like Henry Ford's assembly lines, a lean system provides what is needed, in the amount needed, when it is needed. The principal focus of lean thinking concentrates on value-added process flow. For a process to have value-added flow, there must be an uninterrupted adding of value to a product or service as it is being created. Interruptions or non-value-added activities found in the process, such as downtime, rework, waiting, and inspection, must be reduced or eliminated.

Lean thinking is a mindset best described as a relentless war on waste. Companies implementing lean have reported significant reductions in cycle times, handling costs, lead times, floor space usage, inventory, and customer service activities. At the same time, they see a significant improvement in quality, inventory turns, profit margins, and customer responsiveness.

Lean focuses on eliminating wasted time, effort, and material. Time—such as time wasted waiting for value to be added, or time wasted waiting in inventory for a customer, or time wasted waiting for the next step in the process—is particularly crucial. There are many sources of waste.

Lean at Starbucks

LEAN SIX SIGMA TOOLS at WORK

On Tuesday, August 4, 2009, the *Wall Street Journal* ran a front-page article entitled "Latest Starbucks Buzzword: 'Lean' Japanese Techniques." Mr. Heydon, Starbucks's lean manager, is quoted as saying that "reducing waste will free up time for baristas—or partners, as the company calls them—to interact with customers and improve the Starbucks experience. 'Motion and work are two different things. Thirty percent of the partners' time is motion; the walking, reaching, bending,' he says. He wants to lower that." Other food service providers, including Dunkin' Donuts, McDonald's, and Burger King use lean tools and techniques to improve organization, work flow, and processes. Mr. Heydon and his lean teams study the way people work and seek ways to improve the effectiveness of their activities. Their approach to lean management includes a strong focus on customer needs, wants, and expectations. One example includes the realization that customers enjoy the sounds made while grinding coffee as well as the smell of freshly ground coffee. These sounds and smells add to their impression of overall coffee quality. Starbucks is also concerned about employees, making sure that the process improvements provide a better working environment for their baristas.

> **Lean Airlines** — **LEAN SIX SIGMA TOOLS at WORK**
>
> Global competition. Increasing fuel costs. Labor shortages. Facing such challenges, organizations are racing to improve their methods and processes. Airlines implement quality and lean tools to make improvements in the way they do business. To find the most efficient and consistent way to safely "turn a plane"—bring it in, unload it, clean it, load it, and push it out again—United Airlines applies lean principles of orderliness, communication, and standardization of the tasks involved ("Racing to Improve," by S. Carey. *Wall Street Journal*, March 24, 2006). United used ideas from pit crews in automobile racing to find ways to lean the process by removing wasteful steps and activities. When redesigning the process, improvement teams attacked the seven forms of waste. They removed unnecessary movements of people and activities and made sure that the tools were always on hand so people were not kept waiting. Their goal? A 36-minute turn time for a plane. There are a lot of maneuvers that need to be choreographed for the teams to meet this goal: the ramp crew needs to work with baggage carts, belt loaders, baggage scanners, tow bars, and push tractors.

Tadamitsu Tsuruoka, a Honda process engineer, identified seven sources of waste:

- Overproduction waste
- Idle time waste (waiting time/queue time)
- Delivery waste (transport/conveyance waste)
- Waste in the work itself
- Inventory waste
- Wasted operator motion
- Waste of rejected parts

Intellect can also be wasted. Lean thinking tackles the causes of these wastes. Lean projects focus on ineffective processes, inadequate tools/equipment, inefficient layouts, lack of training, unqualified suppliers, lack of standardization, poor management decisions, insufficient communication, mistakes by the operator, and poor scheduling.

Lean thinking generates process improvement by following five key steps:

1. Study the process by directly observing the work activities, their connections, and flow.
2. Study the process to systematically eliminate wasteful activities, their connections, and flow.
3. Establish agreement among those affected by the process in terms of what the process needs to accomplish and how the process will accomplish it.
4. Attack and solve problems using a systematic method.
5. Integrate the above approach throughout the organization.

Lean Measures

Three key performance measures related to lean are cycle time, value-creation time, and lead time. *Cycle time* represents how often a product is completed by a process or the time it takes an operator to complete all the steps in the work cycle before repeating them. *Value-creation* time is the part of the cycle time during which the work activities actually transform the product in a way the customer is willing to pay for. *Lead time* is the time it takes for one piece to move all the way through a process from start to finish.

Lean Tools

Lean thinking essentially takes a diverse set of tools, techniques, and practices and combines them into a system. The lean thinking tools described in later chapters include the following:

- Kaizen
- Value stream process mapping
- 5S
- Kanban (pull inventory management)
- Error proofing (poka-yoke, pronounced "poke-a-yoke")
- Productive maintenance
- Setup time reduction (single minute exchange of dies; SMED)
- Reduced lot sizes (single piece flow)
- Line balancing
- Schedule leveling
- Standardized work
- Visual management

As future examples will show, lean improvement projects include the following steps:

1. Practice the five Ss.
2. Develop a continuous flow that operates based on takt time.
3. Establish a pull system to control production.
4. Introduce line balancing and level scheduling.
5. Practice kaizen to continually eliminate waste, reduce batch sizes, and create continuous flow.

SIX SIGMA ORIGINS

Six Sigma is a methodology that blends together many of the key elements of past quality initiatives while adding its own special approach to business management. Essentially, *Six Sigma* is about results; enhancing profitability and reducing costs through improved quality, productivity, and efficiency. Six Sigma emphasizes the reduction of variation, a focus on doing the right things right, combining of customer knowledge with core process improvement efforts, and a subsequent improvement in company sales and revenue growth.

> **Motorola and Six Sigma** — LEAN SIX SIGMA TOOLS at WORK
>
> In the 1970s, Motorola learned about quality the hard way—by being consistently beaten in the market. When a Japanese firm took over a Motorola factory that manufactured television sets in the United States, it promptly set about making drastic operational changes. Under Japanese management, the factory was soon producing TV sets with one-twentieth the number of defects they had produced under previous management. Bob Galvin, then CEO of Motorola, started the company on the quality path and became a business icon largely as a result of what he accomplished in quality at Motorola. In accepting the first ever Malcolm Baldrige National Quality Award at the White House in 1988, Bob Galvin briefly described the company's turnaround. He said it involved something called Six Sigma.
>
> *Source:* Adapted from T. Pyzdek, "Six Sigma Is Primarily a Management Program," *Quality Digest* (June 1999).

The Six Sigma methodology encourages companies to improve their business processes by taking a customer focus.

The Six Sigma concept was conceived by Bill Smith, a reliability engineer for Motorola Corporation. His research lead him to believe that the increasing complexity of systems and products used by consumers created higher than desired system failure rates. His reliability studies showed that to increase system reliability and reduce failure rates, the components utilized in complex systems and products have to have individual failure rates approaching zero. With this in mind, Smith took a holistic view of reliability and quality and developed a strategy for improving both. Smith worked with others to develop the Six Sigma Breakthrough Strategy, which is essentially a highly focused system of problem solving. Six Sigma's goal is to reach 3.4 defects per million opportunities over the long term.

Dr. W. Edwards Deming, considered the founder of the quality management movement, stated that the term *quality* means "*non-faulty systems.*" To create non-faulty systems, variation must be removed. Several options exist to reduce variation. Improvements can be made to the product, the raw materials going into the product, the process equipment or machines producing the product, the humans in the system, or the process procedures. A reduction of variation related to any of these can result in less waste, higher yields, shorter cycle times, improved overall equipment effectiveness, and so on. Consider this list again. Where is most of the variation coming from? In the list of five, only one is the human in the system. Unfortunately, the human in the system is often blamed for causing most of the variation. However, the humans in the systems are using the raw material supplied to them while running machines or equipment designed, set-up, and repaired by someone else, using work instructions created by someone else, and working in a given environment. Dr. Deming's non-faulty systems support human performance excellence. Six Sigma variation reduction efforts expand beyond the human in the system and concentrate on reducing variation from all sources.

The Six Sigma methodology is based on knowledge. Practitioners need to know statistical process control techniques (Chapters 9, 10, 18, 19), data analysis methods (Chapter 17), and project management techniques (Chapter 8). Systematic training within Six Sigma organizations must take place.

Motorola Corporation, the originators of the Six Sigma concept, started the practice of utilizing terminology from the martial art of karate to designate the knowledge, experience, and ability levels of Six Sigma project participants. Green Belts are individuals who have completed a designated number of hours of training in the Six Sigma methodology (Figure 1.3). To achieve Green Belt status, they must also complete a cost-savings project of a specified size, often $10,000, within a stipulated amount of time. Black Belts are individuals with extensive training in the Six Sigma methodology (Figure 1.4). To become a Black Belt, the individual must have completed a number of successful projects under the guidance and direction of Master Black Belts. Often companies expect the improvement projects overseen by a Black Belt to result in savings of $100,000 or more. The American

Quality Philosophies	Statistics
Performance Measures/Metrics	Data Collection: Data types and sampling techniques
Problem Solving	$\bar{X}$ and R charts
Process Mapping	Process Capability Analysis
Check Sheets	P, u, c charts
Pareto Analysis	Root Cause Analysis
Cause-and-Effect Diagram Analysis	Variation Reduction
Scatter Diagrams	Six Sigma philosophy
Frequency Diagrams	Green Belt Project
Histograms	

FIGURE 1.3 Training Typically Required for Green Belt Certification

Green Belt Requirements plus:	
Variables Control Charts	Reliability
Attribute Control Charts	ISO 9000, MBNQA
Process Capability Analysis	Voice of the Customer: Quality Function Deployment
Hypothesis Testing	Regression Analysis
Design of Experiments	Black Belt Project
Gage R&R	

FIGURE 1.4 Training Typically Required for Black Belt Certification

Lean and Six Sigma Origins   5

Responsibility	Phase
Management	Recognize
Management/Master Black Belts	Define
Black Belts/Green Belts	Measure
Black Belts/Green Belts	Analyze
Black Belts/Green Belts	Improve
Black Belts/Green Belts	Control
Management	Standardize
Management	Integrate

FIGURE 1.5 Six Sigma Responsibility Matrix

Society for Quality offers certification for a Black Belt. Master Black Belts are individuals with extensive training who have completed a large-scale improvement project, usually saving $1,000,000 or more for the company. Before designating someone a Master Black Belt, many companies will require a Master's Degree in an appropriate field from an accredited university. Master Black Belts provide training and guide trainees during their projects. Figure 1.5 shows the responsibilities of project participants.

The term *Six Sigma* describes a methodology, whereas the mathematical designation, 6σ (also referred to with lowercase letters as six sigma), is the value used to calculate process capability. Providers of products and services are very interested in whether their processes can meet the specifications as identified by the customer. The spread of a distribution of average process measurements, that is, what the process produces, can only be compared with the specifications set by the customer by using C_p, where

$$C_p = \frac{USL - LSL}{6\sigma}$$

When 6σ = USL – LSL, process capability $C_p = 1$. When this happens, the process is considered to be operating at three sigma (3σ). This means that three standard deviations added to the average value will equal the upper specification limit and three standard deviations subtracted from the average value will equal the lower specification limit (Figure 1.6). When $C_p = 1$, the process is capable of producing products that conform to specifications provided that the variation present in the process does not increase and that the average value equals the target value. In other words, the average cannot shift. That is a lot to ask from a process, so those operating processes often reduce the amount of variation present in the process so that 6σ < USL – LSL.

Some companies choose adding a design margin of 25 percent to allow for process shifts, requiring that the parts produced vary 25 percent less than the specifications allow. A 25 percent margin results in a $C_p = 1.33$. When $C_p = 1.33$, the process is considered to be operating at four sigma. Four standard deviations added to the average value will equal the upper specification limit and four standard deviations subtracted from the average value will equal the lower specification limit. This concept can be repeated for five sigma and $C_p = 1.66$.

When $C_p = 2.00$, six sigma has been achieved. Six standard deviations added to the average value will equal the upper specification limit and six standard deviations subtracted from the average value will equal the lower specification limit (Figure 1.7). Those who developed the Six Sigma methodology felt that a value of $C_p = 2.00$ provides adequate protection against the possibilities of a process mean shift or an increase in variation.

Operating at a Six Sigma level also enables a company's production to have virtually zero defects. Long-term expectations for the number of defects per million opportunities is 3.4. Compare this to a process that is operating at three sigma and centered. Such a process will have a number of defectives per million opportunities of 1,350 out on each side of the specification limits for a total of 2,700. If the process center were to shift 1.5 sigma, the total number of defects per million opportunities at the three sigma level would be 66,807. A process operating at four sigma will have 6,210 defects per million opportunities over the long term, whereas a process operating at the five sigma level will have 320 defects per million opportunities long term. Even if the cost to correct the defect is only $100, operating at the three sigma level while experiencing a process shift will cost a company $6,680,700 per million parts. Improving performance to four sigma reduces that amount to $621,000 per million parts produced. Six sigma performance costs just $340 per million parts with a yield rate of 99.9997 percent (Figure 1.8).

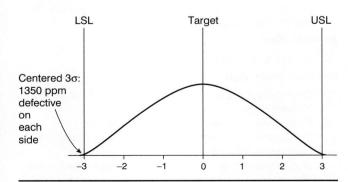

FIGURE 1.6 Three Sigma Occurs When 6σ = USL – LSL, Process Capability, $C_p = 1$

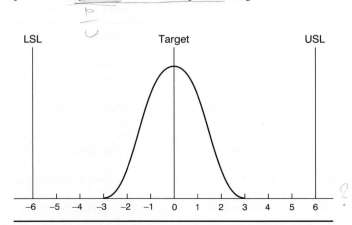

FIGURE 1.7 Six Sigma Occurs When 6σ < USL – LSL, Process Capability, $C_p = 2$

Sigma	Defects per Million Opportunities	Yield (%)
1	690,000	30.90
2	308,000	69.20
3	66,807	93.30
4	6,210	99.40
5	320	99.98
6	3.4	99.9997

FIGURE 1.8 Six Sigma Performance

Six Sigma Measures

Companies using the Six Sigma methodology see an enhanced ability to provide value for their customers. Internally, they have a better understanding of their key business processes, and these processes have undergone process flow improvements. Improved process flow means reduced cycle times, elimination of defects, and increased capacity and productivity rates. Accelerated improvement efforts reduce costs and waste while increasing product and service reliability. All of these changes result in greater value for the customer as well as improved financial performance for the company. Six Sigma places a heavy reliance on metrics and graphical methods for analysis. Key business measures include revenue dollars, labor rates, fixed and variable unit costs, gross margin rates, operating margin rates, inventory costs, general and administrative expenses, cash flow, warranty costs, product liability costs, and cost avoidance. Remember, Six Sigma's goal is to reach 3.4 defects per million opportunities over the long term. Figure 1.9 shows the impact of achieving 6σ levels of process performance. Did you know that according to data published by the American Society for Quality, accurate prescription writing, correct restaurant bill calculation, and proper airline baggage handling takes place in the 4σ range? Automotive expectations of quality exceed 5σ, as do safe aircraft carrier landings. Domestic airline fatality avoidance rates reach nearly 7σ.

99.74% Good = Three Sigma (3σ)	99.9998% Good = Six Sigma (6σ)
20,000 lost articles of mail per hour	Seven lost articles of mail per hour
Unsafe drinking water for almost 15 minutes each day	Unsafe drinking water one minute every seven months
5,000 incorrect surgical operations per week	1.7 incorrect surgical operations per week
Two short or long landings at most major airports each day	One short or long landing every five years
200,000 wrong drug prescriptions each year	68 wrong drug prescriptions each year
No electricity for almost seven hours each month	One hour without electricity every 34 years

FIGURE 1.9 Achieving 6σ Levels of Process Performance

Six Sigma Tools

The Six Sigma methodology focuses on customer knowledge. By translating customer needs, wants, and expectations into areas for improvement, the Six Sigma methodology concentrates effort on improving critical business activities. Six Sigma focuses attention on these processes to make sure that they deliver value directly to the customer.

The backbone of Six Sigma efforts are improvement projects, chosen based on their ability to contribute to the bottom line on a company's income statement by being connected to the strategic objectives and goals of the organization. Good projects improve productivity, reduce costs, and increase quality. Since the Six Sigma methodology seeks to reduce the variability present in processes, Six Sigma projects are easy to identify. Project teams seek out sources of variation, investigate production backlogs or areas in need of more capacity, and focus on customer and environmental issues. Six Sigma projects have eight essential phases:

Recognize
Define
Measure
Analyze
Improve
Control
Standardize
Integrate

This cycle is often expressed as DMAIC (define, measure, analyze, improve, and control). As Figure 1.10 shows, the generic steps for Six Sigma project implementation

---

Define, Measure, Analyze (Plan)

1. Select appropriate metrics: key process output variables (KPOVs).
2. Determine how these metrics will be tracked over time.
3. Determine current baseline performance of project/process.
4. Determine the key process input variables (KPIVs) that drive the key process output variables (KPOVs).
5. Determine what changes need to be made to the key process input variables in order to positively affect the key process output variables.

Improve (Do)

6. Make the changes.

Control (Study, Act)

7. Determine if the changes have positively affected the KPOVs.
8. If the changes made result in performance improvements, establish control of the KPIVs at the new levels. If the changes have not resulted in performance improvement, return to step 5 and make the appropriate changes.

FIGURE 1.10 Six Sigma Problem-Solving Steps

APQP	Advanced product quality planning
CTQ	Critical to quality
DFSS	Design for Six Sigma
DMAIC	Define, measure, analyze, improve, control
DPMO	Defects per million opportunities
DPU	Defect per unit
EVOP	Evolution operation
FMEA	Failure modes and effects analysis
KPIV	Key process input variable
KPOV	Key process output variable
Process Owners	The individuals ultimately responsible for the process and what it produces
Master Black Belts	Individuals with extensive training qualified to teach Black Belt training classes who have completed a large-scale improvement project; often a master's degree is required
Black Belts	Individuals with extensive training in the Six Sigma methodology who have completed a number of improvement projects of significant size
Green Belts	Individuals trained in the Six Sigma methodology who have completed an improvement project of a specified size
Reliability	Measured as mean-time-to-failure
Quality	Measured as process variability and defect rates

FIGURE 1.11 Six Sigma Acronyms and Definitions

are similar to the Plan-Do-Study-Act problem-solving cycle espoused by Dr. W. Edwards Deming and Dr. Walter Shewhart, early pioneers of the quality field. Using DMAIC as a guideline, organizations seek opportunities to enhance their ability to do business. DMAIC is covered in detail in Chapter 9. Six Sigma acronyms are provided in Figure 1.11. Process improvement of any kind leads to benefits for the company, including the reduction of variation, waste, costs, and lost opportunities. Ultimately, it is the customer who enjoys enhanced quality and reduced costs.

The tools utilized during a project include statistical process control techniques, customer input through quality function deployment (QFD), failure modes and effects analysis (FMEA), design of experiments (DOE), reliability, teamwork, and project management. Statistical and probabilistic methods, control charts, process capability analysis, process mapping, and cause-and-effect diagrams are key tools of Six Sigma.

As with any process improvement methodology, there are issues that need to be examined carefully. One criticism is that Six Sigma methodology does not offer anything new. Comparisons have been made between the qualifications for a Black Belt and those for a Certified Quality Engineer (CQE). The similarities are striking (Table 1.1). Certifications for both are available through the American Society for Quality (www.asq.org). Comparisons have also been made between Six Sigma strategies and ISO 9000, the Baldrige Award, continuous improvement, and lean strategies. Significant similarities exist (Table 1.2).

The Six Sigma methodology is often adopted by large corporations with the financial resources to focus on training and projects. Because the methodology requires training and team participation, it is perceived to be costly to implement. Six Sigma should not be limited to large firms. Its ideas and concepts are viable regardless of the size or type of industry with costs being scaled accordingly.

Another criticism is the focus on defectives per million. Can an organization really call them defectives? The term itself brings to mind product liability issues. How does a customer view a company that is focused on counting defectives? Should defect counts be seen as the focus or are companies really trying to focus on process improvement? Lean Six Sigma organizations must be prepared to face these issues.

LEAN SIX SIGMA

Effective organizations realize that devising new methods to cut production costs, improve quality, speed up assembly, and increase throughput are vital to staying competitive. The Six Sigma approach blends nicely with lean techniques, enhancing the effectiveness of both. One of the critical aspects of these two approaches is the realization that working hard to keep things simple saves money. Six Sigma strengthens company performance by concentrating on reducing process variation. Lean thinking enhances company performance by focusing on the reduction of waste. Effective process improvement efforts involve both. Lean and Six Sigma methodologies work together so that each step in the process is:

Valuable (customer focused)

Capable (Six Sigma tools and techniques)

Available (lean: total productive maintenance)

Adequate (lean tools and techniques)

Flexible (lean tools and techniques)

Combining lean and Six Sigma tools support a focus on cost, quality, and productivity. A highly cost conscious company will overlook improvements in quality and

TABLE 1.1 Body of Knowledge Comparison of CQE and Black Belt Certification

Category	ASQ Certified Quality Engineer (CQE) Certification Requirements	Black Belt Requirements
Leadership	Management and leadership in quality engineering	Enterprise-wide deployment
Business processes	Not covered	Business process management
Quality systems	Quality systems development, implementation, and verification	Not covered
Quality assurance	Planning, controlling, and assuring product and process quality	Not covered
Reliability	Reliability and risk management	Not covered
Problem solving	Problem solving and quality improvement	Define-measure-analyze improve-control
Quality Tools	Problem solving and quality improvement	DMAIC
Project Management	Not covered	Project management
Team Concepts	Not covered	Team leadership
Statistical Methods	Probability and statistics collecting and summarizing data	Probability and statistics collecting and summarizing data
Design of experiments	Designing experiments	Design of experiments
Process capability	Analyzing process capability	Analyzing process capability
Statistical process control	Statistical process control	Statistical process control
Measurement systems (metrology/calibration)	Measurement systems metrology	Measurement systems metrology
Lean manufacturing	Not covered	Lean enterprise
Other techniques	FMEA, FMECA, FTA	FMEA, QFD multivariate studies

TABLE 1.2 Comparison of ISO 9000, the Malcolm Baldrige Award Criteria, and Continuous Improvement/Quality Management

	ISO 9000	Baldrige Award	CI/QM	Lean	Six Sigma
Scope	Quality management system Continuous improvement	Quality of management	Quality management and corporate citizenship Continuous improvement	Elimination of waste and non-value-added activities	Systematic reduction of process variability
Basis for defining quality	Features and characteristics of product or service	Customer driven	Customer driven	Customer focused Value added	Defects per million opportunities
Purpose	Clear quality management system requirements for international cooperation Improved record keeping	Results-driven competitiveness through total quality management	Continuous improvement of customer service	Elimination of non-value-added activities	Improve profitability by reducing process Variation
Assessment	Requirements based	Performance based	Based on total organizational commitment to quality	Increased process speed	Defects per million opportunities
Focus	International trade Quality links between suppliers and purchasers Record keeping	Customer satisfaction Competitive comparisons	Processes needed to satisfy internal and external customers	Locating and eliminating sources of process waste	Locating and eliminating sources of process variability

Organizational Evolution

JF Inc., a job shop specializing in machining forgings into finished products, utilizes three separate inspections as its primary method of ensuring the quality of its products. The first inspection occurs following the initial machining operations (grinding, milling, and boring) and before the part is sent to a subcontractor for heat treating. After the part returns from heat treatment, key dimensions are checked at the second inspection. A final inspection is conducted before the finished part leaves the plant.

With these three inspections, discrepancies between actual part dimensions and the specifications are found only after the part had completed several machining operations. This approach results in significant scrap and rework costs. If the forgings pass incoming inspection and begin to progress through the machining operations, for a typical part, four to six operations will have been completed before any errors are caught during the in-process inspection that occurs before the part is shipped out for heat treatment. The work done after the operation where the error occurred is wasted because each subsequent machining operation is performed on a faulty part.

A project team studied the inspection method and determined that process improvements could be made to improve incoming forging consistency and reduce variation. Operators at each step in the process are now responsible for checking actual part dimensions against specifications. The team also initiated a corrective action plan that required a root cause analysis and corrective action for each nonconformance to standards. Corrective actions plans prevent future errors.

As a result of its efforts, JF Inc. has seen an improvement of one sigma, which translates to a tenfold reduction in the number of defects. At three sigma, it could expect 66,807 defects per million. At an average cost of $10/piece to fix, the costs incurred were $668,007. Now that JF Inc. is operating at four sigma, it expects 6,210 defects per million. At $10/piece to fix, its costs are $62,100. Further improvement activities are planned to reduce variation and improve performance to five sigma.

---

productivity. In the same way, a quality conscious firm fails to consider productivity and cost improvement opportunities. Productivity conscious organizations place a lower emphasis on cost and quality, losing process improvement opportunities in these areas. Combined usage of both lean and Six Sigma tools and techniques provides a balance among the concepts of cost, quality, and productivity. This balance enables an organization to optimize all three. Throughout this text, Lean Six Sigma Tools at Work examples will show how lean and Six Sigma tools can be used to enhance organizational performance.

SUMMARY

Six Sigma is a data-driven, profit-focused improvement methodology for business. Six Sigma attacks variation in a process. Lean seeks to eliminate waste and non-value-added activities. Organizations interested in increasing their customer satisfaction and overall organization success can use both the methodologies and their attendant tools and techniques to study their processes and become more proactive when dealing with issues that prevent excellent performance. The lean Six Sigma methodology not only reduces process defects and waste, but also provides a framework for overall organizational culture change.

TAKE AWAY TIPS

1. Lean thinking generates process improvement by following five key steps:
   a. Study the process by directly observing the work activities, their connections, and flow.
   b. Study the process to systematically eliminate wasteful activities, their connections, and flow.
   c. Establish agreement among those affected by the process in terms of what the process needs to accomplish and how the process will accomplish it.
   d. Attack and solve problems using a systematic method.
   e. Integrate the above approach throughout the organization.
2. Lean improvement projects include the following:
   a. Practice the five Ss.
   b. Develop a continuous flow that operates based on takt time.
   c. Establish a pull system to control production.
   d. Introduce line balancing and level scheduling.
   e. Practice kaizen to continually eliminate waste, reduce batch sizes, and create continuous flow.
3. Six Sigma is data driven and profit focused.
4. Six Sigma enhances an organization's ability to provide value for its customers.
5. Six Sigma enhances an organization's understanding of its key business processes.
6. Six Sigma improves profit performance.
7. Six Sigma requires training in statistical process control techniques, data analysis methods, project management techniques, and other areas shown in Table 1.1.
8. Each step in the process must be:
   Valuable (customer focused)
   Capable (Six Sigma tools and techniques)
   Available (total productive maintenance)
   Adequate (lean tools and techniques)
   Flexible (lean tools and techniques)
9. Six Sigma tools work to reduce variation and eliminate process defects whereas lean tools work to increase process speed.

CHAPTER QUESTIONS

1. Describe the goals of the lean methodology.
2. What are the benefits of implementing lean?
3. What do lean projects focus on? Why?
4. Why would an organization want to implement lean methodology?
5. Review the seven sources of waste provided in the chapter. Provide examples of each of these types of wastes in your organization.
6. Describe the five steps of lean process improvement in your own words.
7. Describe the cycle that lean improvement projects often follow.
8. Describe the Six Sigma methodology to someone who has not heard of it.
9. What do Six Sigma projects focus on? Why?
10. Describe the changes that occur to the spread of the process when the amount of variation in the process decreases.
11. What are the benefits of implementing the Six Sigma methodology?
12. Why would a company want to follow the Six Sigma methodology?
13. Describe what it takes to become a Green Belt.
14. What does a person need to do to become a Black Belt?
15. Describe the difference between a Black Belt and a Master Black Belt.
16. How do Green Belts, Black Belts, and Master Black Belts interact when working on a project?
17. How does Six Sigma work together with lean manufacturing concepts?

CHAPTER TWO

LEADERSHIP FOR PROCESS IMPROVEMENT

Create a constancy of purpose toward improvement of product and service, with the aim to become competitive and to stay in business and to provide jobs.

Dr. W. Edwards Deming, Out of the Crisis

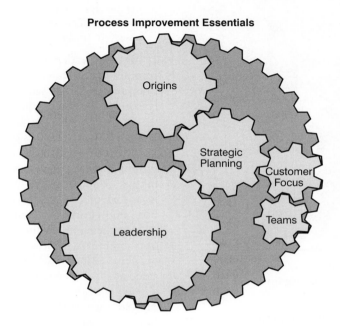

Process Improvement Essentials

LEARNING OPPORTUNITIES

1. To understand what traits define effective leadership
2. To understand how leaders guide lean Six Sigma organizations
3. To understand how to manage by fact and with a knowledge of variation
4. To become familiar with leadership styles

LEADERSHIP TRAITS

Dr. W. Edwards Deming (1900–1993) made it his mission to teach optimal management strategies and practices for effective organizations. Dr. Deming (Figure 2.1) encouraged top-level management to create an organizational culture that supports continuous improvement. He first began spreading his quality message shortly after World War II. In the face of American prosperity following the war, his message was not heeded in the United States. His work with the Bureau of the Census and other government agencies led to his eventual contacts with Japan as that nation began to rebuild. There he helped turn Japan into an industrial force to be reckoned with. His efforts resulted in his being awarded the Second Order of the Sacred Treasure from the Emperor of Japan. It was only after his early 1980s appearance on the TV program "If Japan Can, Why Can't We?" that Dr. Deming found an audience in the United States. Over time, he became one of the most influential experts on quality assurance and process improvement.

Dr. Deming, who described his work as "management for quality," felt that leaders are responsible for managing costs, quality, and productivity in the production of a product or the provision of a service. Dr. Deming considered quality and process improvement activities as the catalysts necessary to start an economic chain reaction. Improving quality leads to decreased costs, fewer mistakes, fewer delays, and better use of resources. This, in turn, leads to

FIGURE 2.1 W. Edwards Deming
(**Source:** Courtesy of the W. Edwards Deming Institute.)

improved productivity, which enables a company to capture more of the market. Higher sales enable the company to stay in business, which results in providing more jobs (Figure 2.2). In his book *Out of the Crisis* (MIT CAES, 1986), Dr. Deming describes the gains that must be made and sustained because progress cannot be made without sustaining gains. Organizational survival depends on utilizing four key methods to sustain gains: good work procedures (Chapters 10–16), sound employee training (Chapter 5), visual process management (Chapter 14), and routine interaction with leadership (Chapter 2).

Dr. Deming's teachings focus heavily on management involvement, continuous improvement, statistical analysis, goal setting, and communication. His message, in the form of 14 points, is aimed primarily at management. Dr. Deming's first point is to *create a constancy of purpose toward improvement of product and service, with the aim to become competitive and to stay in business and to provide jobs*. This point encourages leadership to accept the obligation to constantly improve the product or service through innovation, research, education, and continual improvement in all facets of the organization. A company is like an Olympic athlete who must constantly train, practice, learn, and improve in order to attain a gold medal. Without dedication, the performance of any task cannot reach its best. In *Out of the Crisis*, Dr. Deming states:

> The aim of leadership should be to improve the performance of man and machine, to improve quality, to increase output and simultaneously to bring pride of workmanship to people. Put in a negative way, the aim of leadership is not merely to find and record failures of men, but to remove the causes of failure: to help people to do a better job with less effort. (p. 248)

Because leaders set the vision for their organization, the commitment to lean Six Sigma can come only from leaders. They are the people who realize that there is a compelling reason to adopt this approach. They are the ones who lead the way in the change process by establishing the goals, objectives, and measures of performance for the organization. For this reason, leaders must learn to use their talents, training, and skill to serve two purposes: as a workforce motivator and as a decision maker (Figures 2.3 and 2.4). As a workforce motivator, leaders set performance expectations and clearly communicate them. While tracking the progress being made, they provide feedback. As befits a key decision maker, leaders also have an overall understanding of the situation. They assess the situation and analyze the associated problems. From this, they set a strategy that is aligned with the goals and objectives of the organization. This strategy guides them as they evaluate potential solutions and make effective decisions. They follow through with their decisions and deploy their strategy. Achieving these two purposes ensures that the goals of the individuals are aligned with the goals of the job, which are aligned with the goals of the department, which are aligned with the goals of the organization.

CREATING A LEAN SIX SIGMA CULTURE

Leaders of lean Six Sigma organizations create a culture focused on customers, the reduction of variation, and the elimination of waste. *A **culture** is defined as a pattern of shared beliefs and values that provides the members of an organization with rules of behavior or accepted norms for conducting operations.* In a lean Six Sigma culture, the philosophies, ideologies, values, assumptions, beliefs, expectations, attitudes, and norms are focused on creating value for their customers. Lean Six Sigma leaders apply a missionary zeal to the job of creating an organizational culture focused on creating value for their customers by improving quality, reducing costs, and increasing productivity. Effective leaders realize that they must visibly practice and support the desired culture on a daily basis. To do so, they define systems and standards that support the overall organizational quality, productivity and cost goals, and objectives. In a lean Six Sigma culture, the tools and techniques used in problem

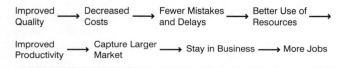

FIGURE 2.2 Dr. Deming's Economic Chain Reaction

Leadership for Process Improvement

Workforce Motivator	Decision Maker
Set Performance Expectations Set realistic goals Set realistic standards Establish checkpoints to measure progress Establish measures of performance to measure progress Encourage creativity	**Assess Situation** Provide clarification Ask questions Simplify confusing situations
Communicate Effectively Clearly communicate goals and objectives Handle resistance to change Listen and act on employee concerns Inspire cooperation and commitment Encourage discussion of ideas	**Analyze Problems** Formulate well-structured problem statements and descriptions Isolate root causes Look beyond the obvious Avoid jumping to conclusions
Provide Feedback Create feedback system Provide timely feedback Provide constructive criticism Manage conflict	**Define Strategy** Create a strategy based on situation assessment and problem analysis Clearly align strategy with goals and objectives of the organization
Track Progress Set priorities Approve solutions Encourage improvement Manage differences Provide recognition and rewards	**Evaluate Potential Solutions** Explore potential solutions Think outside the box Listen to others' ideas
	Make Effective Decisions State goals clearly Align goals with strategy Formulate alternatives Involve others in decision making
	Deploy Strategy Follow through! Take steps to make planned changes Use measures of performance to monitor progress Use checkpoints to track progress Follow up with delegated tasks

FIGURE 2.3 Leadership Roles

solving enable organizations to make and sustain gains. A lean Six Sigma culture includes maintaining a focus on continuous improvement, improving overall equipment effectiveness, and achieving and sustaining productivity, quality, and cost goals.

Effective leaders "walk the talk." In other words, their actions display to other members of the organization what is expected. As Dwight Eisenhower so aptly stated, "They never listen to what I said; they always watch what I do." In Japanese, this concept translates as *genchi genbutsu*.

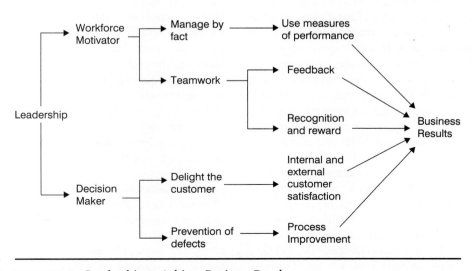

FIGURE 2.4 Leadership to Achieve Business Results

> **Leadership Contact**
>
> **LEAN SIX SIGMA TOOLS at WORK**
>
> Effective leaders understand how important it is to get to know employees. At JF, leadership makes it a point to talk personally and make eye contact with each employee on a regular basis. This means greeting employees while walking through the plant or offices, stopping to exchange a few words, and being present for both shifts. These meetings create an opportunity for interaction and enable leaders to hear employees' ideas and concerns. To get at employees' thoughts, they ask questions.
>
> The owner sees his job as making sure that work, in terms of sales, gets into the shop and that his employees are responsible for making quality product. As a leader, he wants buy-in on his vision for the organization: grow the company. To do this, his employees will have to grow themselves, obtaining new skills and performing new jobs. These one-on-one meetings provide him with the opportunity to motivate employees, set performance expectations, provide feedback and encouragement, and track progress. He also has the opportunity to clarify any questions his employees may have. Lots of changes have been taking place over the past two years, and his one-on-one meetings have reduced the resistance to change.
>
> The owner also serves as a decision maker. By walking through the plant and interacting with employees on a regular basis, he can see what is working and what is not. As he assesses the situation and analyzes problems, he can make adjustments to JF's strategy, goals, and objectives. He involves employees in designing solutions to the issues that arise. In many cases, he just asks the questions to get them started. If there are signs of frustration, he steps in and provides guidance and direction. This ensures that his employees feel good about the changes they helped put in place and the accomplishments they achieved. How does he know it's working? He has started to hear his own words coming back at him as his employees seek solutions to achieving the strategic plan.

The concept encourages leaders to go and see for themselves in order to fully understand the situation by observing what happens. Effective leaders recognize that their employees closely watch their actions. Because employees focus on the activities of their leaders, leadership involvement in key activities is crucial when encouraging people to work in a manner that creates value for their firm. If leaders only appear when there is a problem, workers come to dread their presence. However, if leaders regularly visit work areas as part of their normal workday, they will learn a lot about normal day-to-day operations. Effective leaders recognize that by actively listening to their workforce they will receive key information instead of what their employees think that management wants to hear.

Effective leaders provide guidance. These leaders make sure that their actions correspond with a written vision or mission statement. *Vision statements describe where leadership sees the organization in the future.* Like the stars used as guides by ancient mariners, vision statements provide a star to chart a course by. *Mission statements are usually more specific and provide greater detail about the firm's objectives.* Figure 2.5 provides example wordings of a variety of vision and mission statements. In lean Six Sigma organizations, vision and mission statements include creating value for the customer while meeting cost, productivity, and quality goals through the reduction of variability and elimination of waste.

If, like Dr. Deming says, employees come to work to do a good job, then why does it seem as if employees spend much of their time solving recurring problems and fighting the same fires? This is a common organizational problem that leaders of lean Six Sigma organizations try to avoid. It takes discipline to follow the Plan-Do-Study-Act (PDSA) problem-solving process popularized by Dr. Deming. Following PDSA (Chapter 9) is one way to make sure the problem is truly solved. Essentially, effective leaders practice Dr. Deming's point: *implement leadership*. Effective leaders are the first to ask:

- Does the employee know what he/she is supposed to do?
- Does the employee have the means to determine whether he/she is doing the job correctly?
- Does the employee have the authority and the means to correct the process when something is wrong?

Effective leaders realize that if any of these questions go unanswered, or if the answer to any of them is no, the fault probably lies with leadership, not with employees. Leaders of lean Six Sigma organizations communicate the values of the organization to their employees by translating the vision and mission into day-to-day activities. It is the day-to-day activities that require knowledge of lean Six Sigma tools and techniques. Effective leaders set realistic goals for their employees and give timely rewards to those who meet these goals. Leaders seek out employees' ideas and actively support the good ones. Leaders also place a high priority on expanding employee capabilities and responsibilities through education and training. Other employee-related activities of effective leadership are covered in Chapter 5.

Effective leaders have a very powerful tool that helps to link the activities of individuals within an organization with the customers served by the organization: the strategic plan. Strategic plans enable leaders to communicate effectively to all levels of the organization. Strategic plans are also the vehicle for translating the organization's lean Six Sigma vision, mission, and strategic objectives into deployable action plans. The strategic plan and its impact on creating a lean Six Sigma organization will be covered in greater detail in Chapter 3.

Hospital

As a major teaching institution, we will continue to be a leader in providing a full range of health care services. Working together with our medical staff, we will meet and exceed our patients' needs for high-quality health care given in an efficient and effective manner.

Grocery Store

The mission of our store is to maintain the highest standards of honesty, trust, and integrity toward our customers, associates, community, and suppliers. We will strive to provide quality merchandise consistent with market values.

Student Project

Our mission is to provide an interesting and accurate depiction of our researched company. Our project will describe their quality processes compared with the Malcolm Baldrige Award standards.

Pet Food Company

Our mission is to enhance the health and well-being of animals by providing quality pet foods.

Customer Service Center

Meet and exceed all our customers' needs and expectations through effective communication and intercompany cohesiveness.

Manufacturing

To produce a quality product and deliver it to the customer in a timely manner while improving quality and maintaining a safe and competitive workplace.

Manufacturing

To be a reliable supplier of the most efficiently produced, highest-quality automotive products. This will be done in a safe, clean work environment that promotes trust, involvement, and teamwork among our employees.

FIGURE 2.5 Examples of Mission Statements

Walking the Talk

LEAN SIX SIGMA TOOLS at WORK

JF leaders "walk the talk." They communicate with their employees verbally, in writing, and by their actions. Quarterly meetings are held with all employees to discuss key performance measures. Weekly data, such as sales numbers, number of returns, late shipments, internal quality issues, and number of good shafts out the door, are posted next to the employee break room. These numbers are important for employees to track because it affects their quarterly bonus. Through this bonus, JF communicates financially with its employees. Employees receive a bonus when quality parts shipped totals more than $1.5 million in a single quarter. The bonus is a percentage relationship based on their particular salary versus the total dollars spent on salaries during the quarter.

JF leaders show that they mean business by their actions, too. During the first few weeks of new ownership, employees were surprised when they were ordered to turn off their machines under unsafe conditions. In the past, production needs meant ignoring unsafe conditions and actions. JF's new management dealt with safety issues first, returning guards to machines, opening up paths to fire exits, replacing/restoring fire extinguishers, reprimanding employees for unsafe acts, improving lighting, replacing hot fuses, redesigning transformers, mopping up spills and slick spots, making the break room and restrooms more hygienic, and getting live electrical cords out of coolant pans.

They made changes on the production side, too. To improve quality, welds on shafts were no longer permitted. Fluorescent lighting was added to the existing mercury vapor lamps. This eliminated shadows on the parts and permitted employees to monitor their production more carefully, thus improving part quality. In the past, as machines wore out, employees developed all sorts of workarounds. Over time, as money became available, the new leadership replaced several machines, starting with one machine that squeaked so badly that, when operating, the sound level in the plant approached OSHA's 90-decibel maximum permissible level. Grinders did not exhibit good repeatability, so money was spent to scrape the ways, replace the bearings, and replace switches and handles. New gauges have been purchased and setup boards created. Each chip-making machine now has a vacuum that sucks chips directly into a barrel. These chips are taken to a newly acquired "puck" machine that compresses the chips into recyclable discs of metal. A general cleanup of the plant is under way, too. Based on employee suggestions, lathes, once repaired, are painted blue and white; grinders, burnt orange; and mills, gray. New machine purchases ensured that more capable machines with newer CNC controls increased productivity and agility in the plant.

The shop isn't the only part of the company that is experiencing changes. The front office is involved in the improvement process, too. Customer quoting and order confirmation is now computerized and online. Basic services, such as payroll and accounting, have been either standardized procedurally or outsourced. By shopping around, better sources were found for everything from printer paper to shipping boxes and pallets. Establishing safety policies and improving plant conditions resulted in a $50,000 per year savings in insurance costs.

The leaders at JF are intent on creating a positive organizational culture focused on their vision of "Grow the Business." Their actions have spoken loudly to employees, telling them that there is a new way of doing business, one that respects employees and customers. Over time, as JF employees watched their leaders, they became involved in helping achieve the goals and objectives set in the strategic plan. By finding more and more ways to improve their business, they have benefited in the form of a better working environment and great bonuses.

MANAGING BY FACT AND WITH A KNOWLEDGE OF VARIATION

Management has the responsibility of providing employees with non-faulty systems to work in by supplying appropriate raw material, functional equipment and machines, correct work instructions, and an appropriate work environment. These key sources of variation are under the control of leadership. Understanding that workers have little control over their work environment with its multiple sources of variation helps leaders manage by fact and with a knowledge of variation. Leaders of lean Six Sigma organizations know that people must be included in their organization's decision-making processes. They also know that sharing information is critical to making good decisions. Management by fact involves understanding the hows and whys of a situation

---

Managing by Fact and With a Knowledge of Variation

LEAN SIX SIGMA TOOLS at WORK

A sales manager has three salespersons, each covering different areas of the country. The performance of each salesperson is reviewed monthly. These reviews are often followed by praise for the highest achiever and a "pep talk" for the underachievers.

In response to upper management's request, the sales manager has graphed the performance of each salesperson versus the sales goal for the past 15 months. The resulting graphs (Figure 2.6) justified his suspicions concerning performance.

However, it is only when these graphs are combined with factual information and an understanding of variation that the true patterns of performance begin to emerge. Figure 2.7 shows how control limits were calculated based on the variation present in each individual's process. These limits show what each process is actually capable of, compared to the previous graph, which showed performance versus specifications (goals set). For more information on how these limits were calculated, see the discussion on control charts in Chapter 18.

Note the cycle present in the Great Lakes Region sales. When the variation present in the process is studied, it appears that this salesperson expends efforts to get sales, but then his performance drops off until he gets too close to the established goal. Another fact to add to this analysis is that this firm specializes in automotive components and this sales territory encompasses the lucrative territories of Michigan, Ohio, and Indiana. Making sales in this region is relatively easy.

Note the Mid-Region (Kentucky, Tennessee, North and South Carolina) performance. A large amount of variation is present in the process. This region has been experiencing much growth in the automotive field. The large amount of variation present could signify a lackadaisical attitude on the part of this salesperson or a need for more training.

Note the performance of the Southern Region salesperson. The first graph, Figure 2.6, shows that his performance is consistently below the goal. Yet, if he were managed by facts with a knowledge of variation, it would become apparent that something good is going on here. Based on this graph, this salesperson has achieved high performance on several occasions. Further investigation reveals that although his region encompasses Florida, Georgia, and Alabama, areas not prime for automotive work, he has been able to generate significant sales. These sales, in the area of medical and computer devices, are helping the manufacturing firm diversify and move away from its dependence on automotive business. This salesperson should be congratulated for his efforts.

Using control charts and managing with a knowledge of variation provides an entirely different outlook for the performance of these individuals.

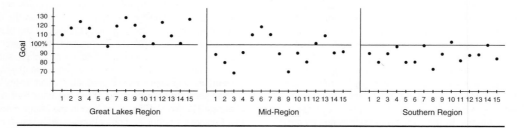

FIGURE 2.6 Performance Graphs of Each Salesperson for the Past 15 Months

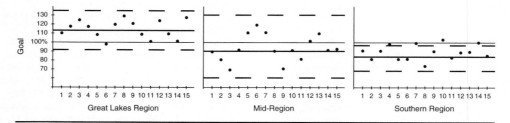

FIGURE 2.7 Performance Graphs with Control Limits

before taking action. Management by fact requires an appreciation for and an understanding of the key systems of an organization. Effective leaders realize that management of systems requires knowledge of the interrelationships between all of the components within the system and the people who work in it. Information can sometimes be misleading. When leaders manage by fact, they use objective evidence to support their decisions. Objective evidence is not biased and is expressed as simply and clearly as possible. More importantly, it is traceable back to its origin, whether that be a customer, an order number, a product code, a machine, or an employee. Effective leaders remember to ask Dr. Deming's favorite question: "How do we know?" Knowing the answer to this question verifies the source of the information and its importance to and relationship with the issue at hand. Having this knowledge means having the facts available to support a plan of action.

Dr. Walter Shewhart, the creator of modern control charts, identified two sources of variation. Controlled variation, the variation present in a process due to the very nature of the process, can be removed from the process only by changing the process. Uncontrolled variation, on the other hand, comes from sources outside the process. Normally uncontrolled variation is not part of the process and can be identified and isolated as the cause of a change in the behavior of the process. To manage with an understanding of variation means that a leader can recognize the type of variation present and respond accordingly. Dr. Deming applied Dr. Shewhart's concept to managing. He felt it would be a mistake to react to any fault, complaint, mistake, breakdown, accident, or shortage as if it came from a special cause when in fact there was nothing special at all—that is, it came from random variation due to common causes in the system. It is also a mistake to attribute to common causes any fault, complaint, mistake, breakdown, accident, or shortage when it actually came from a special cause. An understanding of variation enables leaders to make the right choices and deal appropriately with problems as they arise.

LEADERSHIP STYLES

Four primary styles of leading exist: directing, consultative, participative, and delegating (Figure 2.8). Effective leaders realize that different situations call for different leadership styles. A leader must be skilled at recognizing the need for and then putting to use the appropriate style.

The *directing* style of leadership is an autocratic style. It is normally used when the leader must make a unilateral decision that must be followed without comment or question from the rank and file. The need to use the directing style of leadership may come about because the leader has more knowledge of the situation or because the decision affects the common good of the organization. This leadership style may be something as simple as a "no horseplay will be tolerated on the job" rule or as complex as "here are the customers the organization will focus on." Leaders using this style expect to be obeyed. Those who work for them have very little, if any, input.

The *consultative* style of leadership is used when a leader is seeking input from those working under him or her. It is considered a more developing problem solving style of leading because it encourages participation. Leaders may be facing a customer issue that requires the input of a specialist, such as a chemist; or they may be facing an employee issue that requires input from all employees, such as the selection of a health insurance carrier. When using this style of leadership, the leader seeks the advice, suggestions, and input of those around him or her but still remains the final decision maker.

When using the *participative* style of leadership, a leader assigns work to the employees, provides guidance during the work process, and makes a decision based on the conclusions of the employees working on the task. Unlike the consultative style of leadership, in this situation the leader is more likely to take the word or work of the employee(s) as the final decision in the matter. This leadership style develops employees by allowing them to participate fully in key decisions.

The *delegating* style of leadership is the style in which the leader takes the smallest role. In this style of leadership, the

Participative

Provides guidance
Gets involved only when necessary
Accepts work and decisions of employees
Helps others analyze and solve problems
Recognizes employees for seeking support

Consultative

Seeks input, advice, and suggestions
Makes final decision based on employee input
Recognizes employees for their contributions

Delegating

Assigns responsibility
Assigns authority
Provides minimal input
Provides recognition
Verifies work
Recognizes employees for accepting responsibility

Directing

Engages in unilateral decision making
Expects employees to follow orders
Gives information about what to do
Gives information about how to do it
Gives information about why it should be done
Recognizes employees for following directions

FIGURE 2.8 Leadership Styles

leader essentially tells the employee or team what needs to be done, assigns the responsibility, and provides the individual or team with the authority to get the job done. The individual or team, having been given both the responsibility and the authority, completes the work with minimal input from the leader. In this style of leadership, the leader checks to verify the successful completion of the assignment and participates only if necessary.

Leaders must take care to wisely match their leadership style to the situation. Using a directing style when a delegating style is called for may leave an employee feeling stifled and asking the question, "Why was I even asked to participate?" The directing style of leadership is the most heavy-handed. Applied incorrectly, it could stifle employee creativity and motivation. Conversely, employing a delegating style instead of a consultative style style of leadership when more guidance is needed will leave the employee feeling stranded. The delegating style allows the employee the most freedom. However, if the needs of the project have not been clearly communicated, or if the employee is inadequately prepared to do the work, both the leader and the employee will be dissatisfied with the result. Leaders who try to apply the consultative or problem-solving style in all situations may discover that employees make decisions contrary to key policies and procedures. Some circumstances require that established rules be followed under a directing style of leadership. Effective leaders feel comfortable adopting each leadership style as needed in appropriate situations.

SUMMARY

Strong leadership is essential for lean Six Sigma. Leaders of lean Six Sigma organizations focus attention on cost, productivity, and quality improvements. They hold their workforce accountable for results. To be a leader of a lean Six Sigma organization, an individual must be able to guide an organization by aligning the needs, wants, and expectations of the customer with the processes and day-to-day activities of its employees. A leader must be able to react to different situations with the appropriate leadership style in order to motivate employees. Effective leaders establish trust with their employees. They provide them with clear direction and guidance. They stay close to their employees, seeking information directly from those closest to the situation. They concentrate on the essential, establishing priorities and monitoring progress. A leader must have and utilize an understanding of managing by fact with a knowledge of variation when making decisions. A leader must create a customer-focused culture that works to create value for customers each and every day in all activities. With good leadership, positive changes can occur; without it, nothing is possible.

Effective leaders share key characteristics. They are optimistic and kind, with a preference for personal contact. Although they display independent judgment, they are loyal team players, backing up their employees. Leaders display a characteristic calmness under stress. This trait enables them to face bad news squarely. They are decisive, able to combine a broad understanding of the whole picture and still see the detail. They define their jobs and the cultures of the organizations they work in.

TAKE AWAY TIPS

1. According to Dr. Deming, the "aim of leadership should be to improve the performance of man and machine, to improve quality, to increase output and simultaneously bring pride of workmanship to people" (*Crisis* by W. E. Deming, p. 248).
2. Leaders are change agents.
3. Leaders set the direction and define the culture for an organization.
4. Leaders serve as workforce motivators and as key decision makers.
5. As workforce motivators, leaders set performance expectations, communicate effectively, provide feedback, and track progress.
6. As decision makers, leaders assess situations, analyze problems, define strategies, evaluate potential solutions, make effective decisions, and deploy strategy.
7. Effective leaders translate the vision and mission into day-to-day activities.
8. The four different styles of leadership are directing, consultative, participative, and delegating.
9. Effective leaders manage by fact and with a knowledge of variation.

CHAPTER QUESTIONS

1. What is an organizational culture? What cultural aspects would you expect to see in a lean Six Sigma organization?
2. Describe each type of leadership style. Include a description of where you have seen each of these styles used.
3. What role does leadership play in running a lean Six Sigma organization?
4. What does it mean to manage by fact and with a knowledge of variation?
5. What characteristics should effective leaders have?
6. What does a leader need to do in order to be a workforce motivator?
7. What does a leader need to do in order to be a decision maker?
8. Research Dr. Deming's 14 points. Which of his points deal with leadership? Give an example of where you have seen each of these points applied (or where they need to be applied).

STRATEGIC PLANNING

"Cheshire Puss," she began, rather timidly, as she did not at all know whether it would like the name: however, it only grinned a little wider. "Come, it's pleased so far," thought Alice, and she went on.

"Would you tell me, please, which way I ought to go from here?"
"That depends a good deal on where you want to get to," said the Cat.
"I don't much care where—" said Alice.
"Then it doesn't matter which way you go," said the Cat.
"—so long as I get somewhere," Alice added as an explanation.
"Oh, you're sure to do that," said the Cat, "if you only walk long enough."

Lewis Carroll
Alice's Adventures in Wonderland, *Chapter 6*

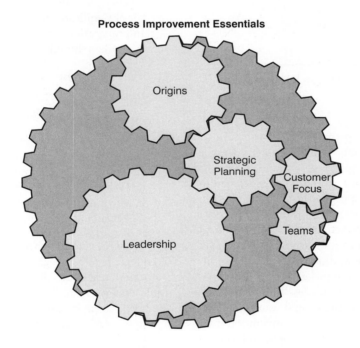

 LEARNING OPPORTUNITIES

1. To discuss the role of strategic planning in a lean Six Sigma organization
2. To present the basics of strategic plan development
3. To discuss the importance of strategic plan deployment

STRATEGIC PLAN CREATION

Strategic plans enable an organization's leadership to translate the organization's vision and mission into actionable and measurable activities. Strategic plans describe how an organization plans to grow. Lean Six Sigma organizations create strategic plans that focus on making quality, costs, and productivity improvements happen. Leaders of these organizations then ensure the deployment of these plans. Developing a strategic plan requires taking a systematic look at the organization to see how each part of it interrelates with the

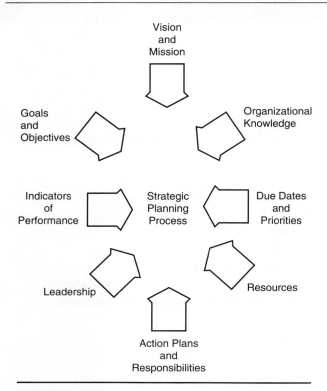

FIGURE 3.1 Elements Needed for an Effective Strategic Planning Process

whole (Figure 3.1). Strategic plans enable an organization to advance from wishful thinking—the customer is number one—to taking action—changing corporate behavior and the actions of employees in order to support a focus on the customer. In a lean Six Sigma organization, strategic plans describe how the organization's process improvement vision is translated into day-to-day activities. Without strategic plans, lean Six Sigma organizations are not able to make gains. If lean Six Sigma goals and objectives are not in the strategic plan, then people will not make the changes happen.

A strategic plan defines the business the organization intends to be in, the kind of organization it wants to be, and the kind of economic and noneconomic contribution it will make to its stakeholders, employees, customers, and community. Lean Six Sigma strategic plans describe how an organization is going to provide value for its customers through improvements in costs, quality, and productivity. The plan spells out the organization's goals and objectives and how the organization will achieve these goals and objectives. Lean Six Sigma strategic plans concentrate on the critical success factors (CSFs) for the organization, providing plans for closing the gaps between what the organization is currently capable of doing versus what it needs to be able to do (Figure 3.2). Using indicators or performance measures, the organization will monitor its progress toward meeting the short-term, mid-term, and ultimately long-term goals. These measures of performance are critical because they enable an organization to determine whether they are on target toward reaching their goals. Lean Six Sigma organizations analyze the gap between what the current performance is and what was originally planned for in the strategic plan. For a negative gap showing lower than expected performance, the organization must take corrective action to eliminate the root cause, improve performance toward the goal, and narrow the gap. For a positive gap showing better than expected performance, the organization may choose to take action to further enhance the gap. A good strategic plan also includes contingency plans in case some of the basic assumptions are in error or significant changes in the market occur.

In preparation for creating a strategic plan, the organization's leaders should determine the following:

1. The organization's business (what business are they really in?)
2. The principal findings from the internal and external assessments
   a. Strengths and weaknesses
   b. Customer information
   c. Economic environment information
   d. Competition information
   e. Government requirements
   f. Technological environment

Once these issues have been addressed, the creation of a strategic plan can begin. The strategic plan is essentially a framework that assists the organization in achieving its vision while allowing flexibility to deal with unforeseen changes in the business environment (Figure 3.3). The components of a strategic plan are as follows:

1. **Vision:** the organization's strategic direction for the foreseeable future
2. **Mission:** the translation of the organization's vision into strategic actions

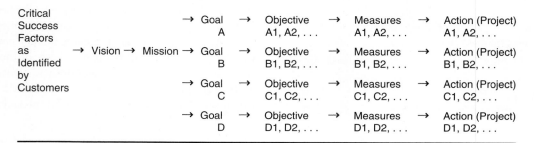

FIGURE 3.2 Strategic Planning

Preparation
1. The organization's business (what business are they really in?)
2. The principal findings from the internal and external assessments
   A. Strengths and weaknesses
   B. Customer information
   C. Economic environment information
   D. Competition information
   E. Government requirements
   F. Technological environment

Planning
1. *Vision:* the organization's strategic direction for the foreseeable future
2. *Mission:* the translation of the organization's vision into strategic actions
3. *Critical success factors as identified by customers:* the three to ten things that a company absolutely must do well if the company is going to thrive
4. *Goals, objectives, and indicators of performance:*
   A. Goal 1
      Objective 1.1
         Indicator
      Objective 1.2
         Indicator
   B. Goal 2
      Objective 2.1
         Indicator
   C. Goal 3
      ⋮
5. Contingency plans

FIGURE 3.3 Generic Strategic Plan

3. **Critical success factors as identified by customers:** the three to ten things that absolutely must be done well if the company is going to thrive. Critical success factors answer the question: *Where should the organization focus its resources in order to provide value to their customers?*

4a. **Goals:** what must be achieved in order to support the critical success factors

4b. **Objectives:** the specific and quantitative actions that must be taken in order to support the accomplishment of the goals and ultimately the mission and vision

4c. **Indicators:** the performance measures that indicate whether the organization is moving toward meeting its objectives, goals, mission, and vision

5. **Contingency plans:** the plans in place that enable an organization to remain flexible in a complex, competitive environment

STRATEGIC PLAN DEPLOYMENT

Beyond a lack of understanding, commitment, and participation by leadership, a variety of pitfalls in strategic planning and its deployment exist. It is not unusual for an unbelievable vision to be proposed, only to be followed up by an inadequate definition of operating expectations. A strategic plan that lists objective after objective suffers from a lack of prioritizing. A strategic plan that fails to clearly assign responsibility for results is weak, as is one that fails to identify and utilize performance indicators. The performance measures themselves may be a problem. Organizations do not benefit from strategic plans containing performance measures that are not connected with the activities proposed by the plan or that are vague or unclear. Strategic plans are also hampered when *what* the organization wants to accomplish is not supported by a corresponding *how* the *what* is going to be accomplished. Without the *how* and *what,* strategic plan deployment cannot happen.

To be effective, a strategic plan must be deployed. As living documents, strategic plans are not meant to sit on a shelf, only to be touched when it is time for an annual revision. Essentially, creating alignment is policy deployment. *Alignment means that if you push on one end, the other end will move in the direction you want.* Effective leaders enable members of the organization to make the transition between the strategic plan and daily business activities by translating *what* needs to be accomplished into *how* it will be accomplished. Leaders of lean Six Sigma organizations make sure that the day-to-day activities and the goals of the strategic plans of the organization are in harmony and are focused on what is critical to the success of the organization. The leaders want to ensure that if they push on the strategic plan, the actions of their employees will go in the desired direction. For this reason the strategy must be clearly

Strategic Planning

PREPARATION

The availability of inexpensive computers, printers, and copiers has increased competition in the printing and design field. Leaders of PM Printing and Design recently met to clarify portions of their strategic plan. Using the plan, they hope to communicate to their employees the importance of creating and maintaining a customer-focused process orientation as they use lean Six Sigma tools and techniques to improve the way they do business.

Based on their meeting, the leaders have determined that their best market niche, or the business that they are really in, is concept to delivery. This includes designing, reproducing, and mailing customer brochures and literature. Market research has shown that no other full-service printing and design company exists in their market area and customers are seeking an organization that can take their ideas and turn them into a finished product in their customers' mailboxes. This research has identified their strengths and weaknesses; provided customer information; defined their economic, technological, and competitive environments; as well as specified appropriate government requirements.

PLANNING

PM Printing and Design Vision
- PM Printing and Design will be recognized as the best source for printed and duplicated material through the recognition and implementation of customer-driven change in a service-focused environment.

PM Printing and Design Mission
- PM Printing and Design is a full-service design, reprographic, and mailing facility committed to serving the local community by producing the highest quality product in the most cost-effective manner.

Critical Success Factors
- Provide a full-service reprographic facility.
- Provide a full-service design process.
- Provide a full-service mailing facility including mailer creation and mailing processes.
- Employ talented designers and skilled technicians.
- Provide quality printed material in a cost-effective manner.

Goals, Objectives, and Indicators of Performance
- Goal 1: Improve customers' knowledge of our services.
    - Objective 1.1: Advertise services community wide.
        - Indicator: Number of customers
        - Measure: Number of repeat sales
    - Objective 1.2: Increase market share.
        - Measure: Number of customers
        - Measure: Number of repeat sales
        - Measure: Number of referrals
- Goal 2: Improve customers' perceptions of our services.
    - Objective 2.1: Reduce the number of customer complaints.
        - Measure: Number of complaints per month
        - Measure: Average time to resolve complaints
- Goal 3: Design and process orders rapidly while maintaining high quality.
    - Objective 3.1: Reduce order time to completion.
        - Measure: Average order turnaround time
- Goal 4: Increase customer value.
    - Objective 4.1: Improve quality and lower cost.
        - Measure: Cost per printed unit (impression)
        - Measure: Cost avoidance (work performed in-house versus contracting work)
    - Objective 4.2: Remove non-value-added activities.
        - Measure: Cost per printed unit (impression)
    - Objective 4.3: Improve first time through quality.
        - Measure: Reduction in rework/scrap
- Goal 5: Provide a desirable work environment for PM employees.
    - Objective 5.1: Expand employee opportunities for growth.
        - Measure: Progress toward cross-training goals for critical processes as identified by customers
    - Objective 5.2: Improve employee retention.
        - Measure: Length of employee service
        - Measure: Number of employees with 1+ years of service

Contingency Plan
- Maintain good relationships with other area printers in order to have a source of extra production capability if needed for a rush job or in the event of an equipment malfunction.

Though this strategic plan is still in a state of development, it does provide a concrete example of how a strategic plan is aligned throughout. This alignment can be seen by goals, supported by objectives, and monitored by performance indicators. Leaders of PM Printing and Design used the strategic plan deployment tree (Figure 3.4) as an example and created an action plan with responsibilities (Figure 3.5). The action plan assigns activities and responsibilities to specific individuals, shows the time frame for accomplishment, and establishes priorities for the objectives.

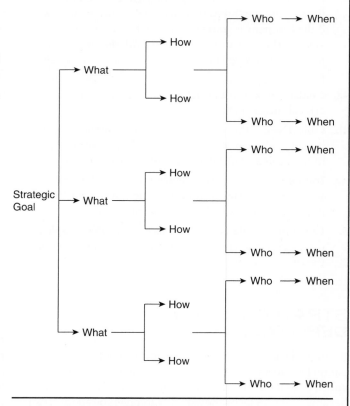

FIGURE 3.4 Strategic Plan Deployment Tree

Goal 1: Improve customers' knowledge of our services

Objective 1.1: Advertise services community wide
  Action: Contract with local radio, TV, and newspapers for advertisements
  Responsibility: QS, MF
  Due: September 15
  Priority: 1
    Indicator: Number of customers
    Indicator: Number of repeat sales

Goal 4: Increase customer value

Objective 4.1: Improve quality and lower cost
  Action: Compare and contrast reproducing machines available
  Action: Select improved machine if available
  Responsibility: RP
  Due: October 15
  Priority: 2
    Indicator: Cost per printed unit (impression)
    Indicator: Cost avoidance (work performed in-house versus contracting work)

Objective 4.2: Remove non-value-added activities
  Action: Create process maps for all critical processes
  Action: Set up one team per process to identify and remove non-value-added activities
  Responsibility: DS
  Due: December 1
  Priority: 1
    Indicator: Cost per printed unit (impression)

FIGURE 3.5 PM Printing and Design Action Plan with Responsibilities

communicated throughout the organization. Effective leaders make sure that the strategic plan contains clear objectives, provides and utilizes measures of performance, assigns responsibilities to specific individuals, and denotes timing (Figure 3.5). For effective strategic plan deployment, the organization's reward and recognition system must support plan deployment.

SUMMARY

Lean Six Sigma organizations recognize that good leadership combined with strategic planning and deployment maximize long-term organizational health. If good strategic planning is not practiced:

- Goals are not known throughout the company
- Goals change too often
- Goals are not achieved
- Goals are achieved without real improvement
- Progress is not sustained
- Organizational frustration exists
- Lean Six Sigma strategic plans emphasize creating value for customers and describe how the organization plans to close the gaps that affect the organization's ability to do so.

TAKE AWAY TIPS

1. Strategic plans align the needs, wants, and expectations of the customer with the processes and day-to-day activities of its employees.
2. Good strategic plans respond to the question: What business are we really in?
3. Strategy deployment is as crucial as strategy development.

CHAPTER QUESTIONS

1. What role does leadership play in strategic planning?
2. Why does an effective organization need a strategic plan?
3. What are the benefits of a strategic plan?
4. Describe each of the elements needed for the strategic planning process.
5. Describe the steps necessary to create a strategic plan.
6. Why is strategy deployment as important as strategic planning?
7. For the company you work for or have worked for, ask to see the strategic plan. How does it compare with what you've learned about the strategic planning process?

CREATING A CUSTOMER FOCUS

From a customer's standpoint, neither quality, cost nor schedule always comes first. When customers evaluate the products and services they receive, they make trade-offs between all three key factors in order to maximize value. The challenge that suppliers face is to provide their customers with the maximum value, which often is a balancing act between quality, cost and schedule.

The First Among Equals, Quality Digest, *June 1999*

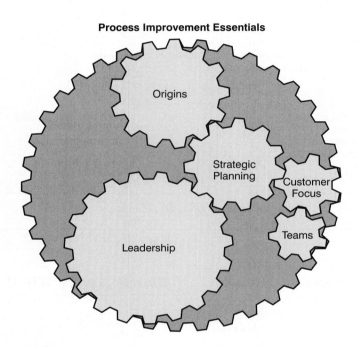

 LEARNING OPPORTUNITIES

1. To explain the difference between satisfaction and perceived value
2. To discuss how customers define value
3. To introduce the reader to quality function deployment

CUSTOMER VALUE PERCEPTIONS

The current global business environment is extremely competitive. Today's consumers are more than willing to switch from supplier to supplier in search of better service or availability or courtesy or features or for any variety of reasons.

The quote that begins this chapter discusses how customers make purchasing decisions. Successful companies seek to understand their products, services, and processes from their customer's point of view. To attract and retain customers, lean Six Sigma organizations focus on determining and providing what their customers want and value. Advertising, market positioning, product/service imaging, discounting, crisis handling, and other methods of attracting the customer's attention are not enough. Organizations utilizing the lean Six Sigma methodologies do well because they talk to customers, translate what their customers said into appropriate actions, and align their key business processes and cost, quality, and productivity improvement projects to support what their customers want.

Armand Feigenbaum (1920–) is considered to be the originator of the total quality movement. His teachings provide sound management advice beyond the field of quality. His landmark text, *Total Quality Control*, first published in 1951 and updated regularly since then, continues to influence modern industrial practices. In his original text, he predicted that quality would become a significant customer-satisfaction issue, even to the point of surpassing price in importance in the decision-making process. As lean Six Sigma organizations know, consumers have come to expect quality as an essential dimension of the product or service they are purchasing. Dr. Feigenbaum defined quality based on a customer's actual experience with the product or service, stating that quality is what the customer says it is. He formally defines quality as:

> A customer determination which is based on the customer's actual experience with the product or service, measured against his or her requirements—stated or unstated, conscious or merely sensed, technically operational or entirely subjective—always representing a moving target in a competitive market. (*Total Quality Control*, p. 7)

This definition describes what customers value. It stresses that quality is a customer determination, meaning that only a customer can decide if and how well a product or service meets his or her needs, requirements, and expectations. These needs, requirements, and expectations may be stated or unstated, conscious or merely sensed, technically operational or entirely subjective. Quality is also based on the customer's actual experience with the product or service throughout its life, from purchase to disposal. Dr. Feigenbaum's definition recognizes that quality and therefore customer value perceptions is a moving target in a competitive market.

To Dr. Feigenbaum, quality is more than a technical subject; it is an approach to doing business that makes an organization more effective. Throughout his life, he has consistently encouraged treating quality as a fundamental element of a business strategy. In his article "Changing Concepts and Management of Quality Worldwide," from the December 1997 issue of *Quality Progress*, he asserts that quality is not a factor to be managed but a method of "managing, operating, and integrating the marketing, technology, production, information, and finance areas throughout a company's quality value chain with the subsequent favorable impact on manufacturing and service effectiveness." In May 1998, in an interview with *Quality Digest*, Dr. Feigenbaum stated, "The demand for complete customer satisfaction indicates a profound social shift for both global consumers and business buyers." As he predicted, demanding customers with their shifting value perceptions have placed economic pressure on businesses. Organizations must try a more systematic process of managing. Their approach to business must recognize that sustained growth demands improvement in customer satisfaction, cost, quality, productivity, and human resource effectiveness. Many organizations are turning to lean Six Sigma to enhance their organizational effectiveness. In the February 2007 issue of *Quality Progress*, Dr. Feigenbaum reiterated that management is responsible for recognizing the evolution of the customer's definition of quality for their products and services. In July 2008, in an article in *Quality Progress*, Dr. Feigenbaum tackles globalization. He points out that globalization and outsourcing have made measuring and managing quality costs even more crucial to organizational success. As rising costs conflict with a market demand for lower prices, organizations must manage their quality costs, the costs of not doing things right the first time. The lean Six Sigma methodology is a method of managing, enabling an organization to achieve higher customer satisfaction, lower overall costs, greater productivity, increased quality, higher profits, and greater employee effectiveness and satisfaction. Due to his work in the field of quality management, in September 2009, Dr. Feigenbaum was named laureate of the National Medal of Technology and Innovation, the highest honor for technological achievement awarded by the United States government.

Lean Six Sigma organizations recognize that customers base their decisions about the quality of a product or service on the perceived value they receive. ***Value***, *a product's or service's attributed or relative worth or usefulness*, is judged by a consumer each time he or she trades something of worth (usually money) in order to acquire the product or service. More simply, value can be defined as a solution to a customer's problem or an activity that a customer is willing to pay for. Customer value judgments, because they involve actual experience, requirements, wants, needs, and expectations, are complex. Lean Six Sigma organizations realize that they are offering product or service features to their customers, but what the customers are really buying is the benefits those products or services offer. Perceived value is the customer's viewpoint of those benefits. Customer satisfaction, on the other hand, centers around how they felt the last time they bought a product or service from a company. It is a comparison between customer expectations and customer experience. Perceived value goes beyond customer satisfaction and concentrates on future transactions. Consumers' perception of the value they have received in the recent transaction will affect their future decision to purchase the same thing again. If they perceive their overall experience with the product or service as valuable, they will most likely purchase in the future; if they do not, they won't. Lean Six Sigma organizations realize that how customers perceive the value of that transaction will determine whether they will buy from the same organization the next time.

Lean Six Sigma organizations seek to understand every aspect of their customers' interaction with their company. They understand that this process begins when the customer first contacts the company and continues until the product has been consumed or the service completed. Lean Six Sigma organizations study their processes from their customers' point of view, selecting improvement projects that focus on designing, developing, and improving key customer processes, making them seamless, flawless, and easy to negotiate. Having hassle-free processes adds considerable value from the customer's viewpoint. These types of processes save money and time. Customers willingly participate in processes they can understand, which is essential in the service industries where customer input is vital to the success of the process. Information, whether about customer perceived value or customer satisfaction, is more meaningful if it is obtained on the customers' terms and from their perspective.

Organizations practicing the lean Six Sigma methodologies need an accurate understanding of the gap between their current performance and what the customer requires. Companies use information related to how their customers define quality and value to help select and refine improvements to their processes. To guide project selection and focus, lean Six Sigma organizations talk with their customers.

QUALITY FUNCTION DEPLOYMENT

How do lean Six Sigma organizations know what their customers want?

They ask them.

Imagine three people sitting at a table. Two manage a business, the third is trying to sell a shop scheduling software package. The salesperson spends nearly an hour talking nonstop, providing detailed information about his company's history and services. Without breaking stride or allowing time for a question, the salesperson moves on to a lecture about the features available in the software. At the end of the hour, the salesperson leaves without ever asking such key questions as: What issues are you facing? How can we help? Or what can our organization do for you? Lean Six Sigma leaders know that a disconnect between customer needs, wants, expectations, and requirements and what a company has to offer can drive customers away. **Quality function deployment (QFD)** *is a technique that seeks to bring the voice of the customer into the process of designing and developing a product or service.* Using this information, effective organizations align their processes to meet their customers' needs the first time and every time. Companies use the voice of the customer information obtained by QFD to drive changes to the way they do business. Information taken directly from the customer is used to modify processes, products, and services to better conform to the needs identified by the customer.

Developed in Japan in the 1970s by Dr. Yoji Akao, QFD was first used in the United States in the 1980s. Essentially, *QFD is a planning process for guiding the design or redesign of a product or service. The principal objective of a QFD is to enable a company to organize and analyze pertinent information associated with its product or service.* A QFD can point out areas of strengths as well as weaknesses in both existing and new products. QFDs help capture three aspects of the voice of the customer:

- Needs, wants, and expectations known and voiced by customers.
- Needs, wants, and expectations known by but not voiced by customers.
- Needs, wants, and expectations not yet identified by customers.

Utilizing a matrix, information from the customer is organized and integrated into the product or process specifications. QFD allows for preventive action rather than a reactive action to customer demands. When a company uses the QFD format when designing a product or service, it stops developing products and services based solely on its own interpretation of what the customer wants. Instead, it utilizes actual customer information in the design and development process. Two of the main benefits of QFD are the reduced number of engineering changes and fewer production problems. QFD provides key action items for improving customer satisfaction. A QFD can enable the launch of a new product or service to go more smoothly because customer issues and expectations have been dealt with in advance. Gathering and utilizing the voice of the customer is critical to the success of world-class companies.

A QFD has two principal parts. The **horizontal component** records information related to the customer. The **vertical component** records the technical information that responds to the customer inputs. Essentially, a QFD matrix clearly shows what the customer wants and how the organization is going to achieve those wants. The essential steps to a QFD are shown in Figures 4.1 and 4.2.

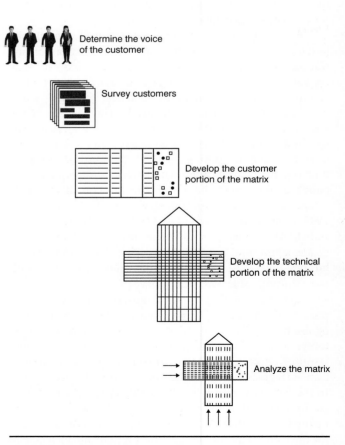

FIGURE 4.1 The QFD Process

Creating a Customer Focus

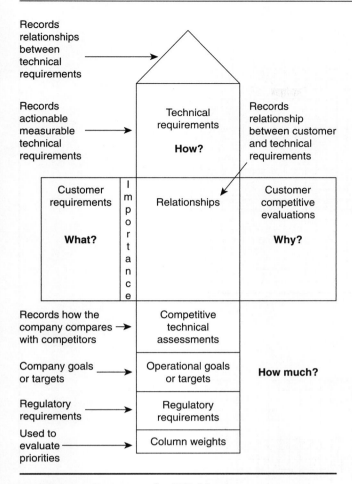

FIGURE 4.2 Summary of a QFD Matrix

QFD begins with the customer. Surveys and focus groups are used to gather information from the customers about their wants, needs, and expectations. Several key areas that should be investigated include performance, features, reliability, conformance, durability, serviceability, aesthetics, and perceived quality. To guide the surveys and focus groups, researchers can gain preliminary customer information from field reports, comment cards, complaint systems, warranty analysis, after-order follow-ups, customer hospitality days, focus groups, and undercover customers. Often, customer information, specifically, the way they say it, must be translated into actionable wording for the organization. A customer's comment that "I can never find parking" needs to be interpreted as "close, convenient parking readily available." In the first statement, the customer is expressing a need. The second statement turns that need into something the organization can act on.

Once this information is organized into a matrix, the customers are contacted to rate the importance of each of the identified wants and needs. Information is also gathered about how customers rate the company's product or service against the competition. Following this input from the customers, technical requirements are developed. These technical aspects define how the customer needs, wants, and expectations will be met. Once the matrix is constructed, the areas that need to be emphasized in the design of the product or service will be apparent. The example below shows the steps associated with building a QFD matrix.

Asking customers what they want, need, and require is a time-consuming process. As was seen in the quality function deployment exercise, translating customers' wants into an organization's hows is paramount to the success of any organization seeking to align its products, services, and the processes that provide it with what the customer wants. An organization that ignores the relationship between what a customer wants and how it is going to provide that want can never be truly effective.

LEAN SIX SIGMA TOOLS at WORK

Creating a QFD

AM Corporation sells sports drinks to the general public. It has always been very in tune to the health and nutritional needs of its customers. Recently its focus has turned to another aspect of its business: the drink containers. AM Corp. has decided to utilize a QFD when redesigning its sports drink bottles.

1. Determine the voice of the customer: What does the customer want? The first step in creating the QFD involves a survey of the customer expectations, needs, and requirements associated with sports drink bottles. AM Corp. met with several focus groups of its customers to capture the information. Following these meetings, AM team members organized and recorded the wants of the customers in the column located on the left side of the matrix (Figure 4.3).

2. Have the customer rank the relative importance of his or her wants. After the team members organized the data, they reconvened the focus groups. At that time, they asked each of the participants to rank the characteristics from one to ten, with ten being the highest value awarded to the most desired characteristic. Lowest ranking characteristics were given a value of one. Ties between characteristics were allowed. The team members recorded the customer values on the matrix next to the list of recorded wants. Following the meetings, they created the final matrix (Figure 4.4) by combining the values assigned by all the customers and taking an average. For instance, having a product that fits in a cup holder is vital to a customer choosing the product. Second on their list is that the device does not leak. The numbers in this column show in a glance what the consumer values.

(Continued)

Customer Requirements

		Requirement	#
Container	Lids	Doesn't Leak	1
		Interchangable Lids	2
		Freshness	3
		Open/Close Easily	4
		Sealed When Purchased	5
		Resealable Lid	6
	Shape	Doesn't Slip Out of Hands	7
		Fits In Cupholder	8
		Doesn't Tip Over	9
		Attractive	10
		Fits In Mini-Cooler	11
		Doesn't Spill When You Drink	12
Material	Characteristics	No Dents	13
		Doesn't Change Shape	14
		Is Not Heavy (Light)	15
		Does Not Break When Dropped	16
		Clear	17
		Reusable	18
		Recyclable	19
		Stays Cool	20
		No Sharp Edges	21
MISC	Cost	Inexpensive	22

FIGURE 4.3 Customer Requirements

Customer Requirements

		Requirement	#	Ranking
Container	Lids	Doesn't Leak	1	9
		Interchangable Lids	2	5
		Freshness	3	5
		Open/Close Easily	4	8
		Sealed When Purchased	5	5
		Resealable Lid	6	3
	Shape	Doesn't Slip Out of Hands	7	7
		Fits In Cupholder	8	10
		Doesn't Tip Over	9	3
		Attractive	10	1
		Fits In Mini-Cooler	11	1
		Doesn't Spill When You Drink	12	2
Material	Characteristics	No Dents	13	1
		Doesn't Change Shape	14	3
		Is Not Heavy (Light)	15	4
		Does Not Break When Dropped	16	5
		Clear	17	1
		Reusable	18	5
		Recyclable	19	1
		Stays Cool	20	6
		No Sharp Edges	21	1
MISC	Cost	Inexpensive	22	6

FIGURE 4.4 Rankings

3. Have the customer evaluate your company against competitors. At the same meeting, the customers also evaluated AM Corp.'s competitors. In this step, the participants ranked AM Corp. and its competitors by awarding first, second, and third in the order that they felt that the companies provided the best product or service for their recorded wants. Following the meetings, AM team members created the final matrix (Figure 4.5) by combining the values assigned by all the customers. The organization with the highest overall score (the most first place scores) was marked in each category as shown in Figure 4.5.

Creating a Customer Focus

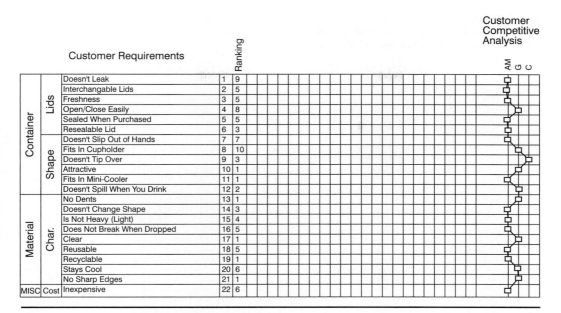

FIGURE 4.5 Customer Competitive Analysis

4. Determine how the wants will be met: How will the company provide for the wants? At this point, AM team members' efforts focused on determining how they were going to meet the customers' wants. They spent many hours in meetings discussing the technical requirements necessary for satisfying the customers' recorded wants. These were recorded at the tops of the columns in the matrix. AM team members made sure that the technical requirements or hows were phrased in terms that were measurable and actionable. Several of the wants needed two or more technical requirements to make them happen (Figure 4.6).

5. Determine the direction of improvement for the technical requirements. During the meetings discussing technical requirements, those involved also discussed the appropriate

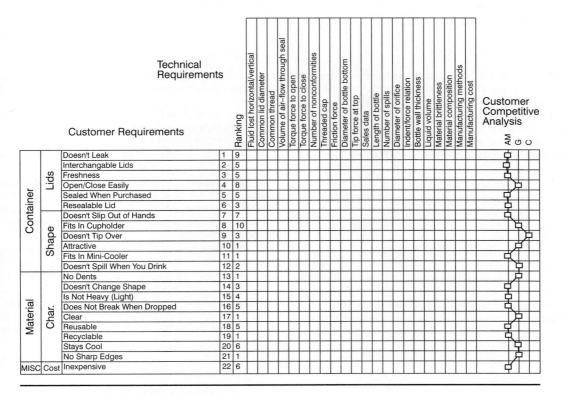

FIGURE 4.6 Technical Requirements

(Continued)

specifications for the technical requirements. They were able to identify how those technical requirements could be improved. For instance, for the comment "Fluid lost horizontal/vertical," the appropriate direction of improvement for this is "less," denoted by the downward arrow (Figure 4.7).

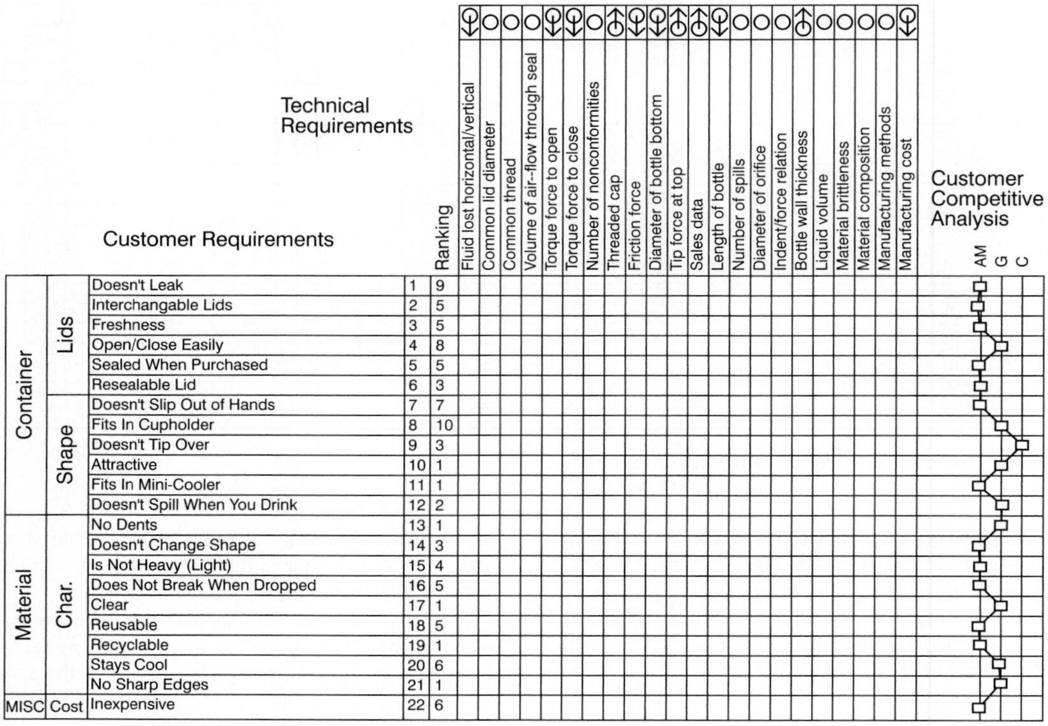

		Customer Requirements	Ranking		Customer Competitive Analysis
Container	Lids	Doesn't Leak	1	9	
		Interchangable Lids	2	5	
		Freshness	3	5	
		Open/Close Easily	4	8	
		Sealed When Purchased	5	5	
		Resealable Lid	6	3	
	Shape	Doesn't Slip Out of Hands	7	7	
		Fits In Cupholder	8	10	
		Doesn't Tip Over	9	3	
		Attractive	10	1	
		Fits In Mini-Cooler	11	1	
		Doesn't Spill When You Drink	12	2	
Material	Char.	No Dents	13	1	
		Doesn't Change Shape	14	3	
		Is Not Heavy (Light)	15	4	
		Does Not Break When Dropped	16	5	
		Clear	17	1	
		Reusable	18	5	
		Recyclable	19	1	
		Stays Cool	20	6	
		No Sharp Edges	21	1	
MISC	Cost	Inexpensive	22	6	

Technical Requirements (columns): Fluid lost horizontal/vertical, Common lid diameter, Common thread, Volume of air-flow through seal, Torque force to open, Torque force to close, Number of nonconformities, Threaded cap, Friction force, Diameter of bottle bottom, Tip force at top, Sales data, Length of bottle, Number of spills, Diameter of orifice, Indent/force relation, Bottle wall thickness, Liquid volume, Material brittleness, Material composition, Manufacturing methods, Manufacturing cost

FIGURE 4.7   Direction of Improvement

6. Determine the operational goals for the technical requirements. AM team members identified the operational goals that will enable them to meet the technical requirements (Figure 4.8).

7. Determine the relationship between each of the customer wants and the technical requirements: How does action (change) on a technical requirement affect customer satisfaction with the recorded want? The team members at AM Corp. studied the relationship between the customer wants and the technical requirements (Figure 4.9). They used the following notations:

A strong positive correlation is denoted by the value 9 or a filled-in circle.
A positive correlation is denoted by the value 3 or an open circle.
A weak correlation is denoted by the value 1 or a triangle.
If no correlation exists, then the box remains empty.
If there is a negative correlation, the box is marked with an X.

8. Determine the correlation between the technical requirements. The team members recorded the correlation between the different technical requirements in the roof of the QFD house. This triangular table shows the relationship between each of the technical requirements (Figure 4.10). They used the notations:

A positive correlation is denoted by an open circle.
A negative correlation is denoted by an x.
If no correlation exists, then the box remains empty.
If there is a negative correlation, the box is marked with an X.

9. Compare the technical performance with that of competitors. At this point, AM Corp. compared its abilities to generate the technical requirements with the abilities of its competitors. On the matrix, this information is shown in the technical competitive assessment (Figure 4.11).

10. Determine the column weights. At this point, the matrix is nearly finished. To analyze the information presented, the correlation values for the wants and hows are multiplied by the values from the customers' ranking.

For example, for the first column, a ranking of 9 for "doesn't leak" is multiplied by a value of 9 for "strong correlation," making the total 81. To this value, the ranking of 5

Creating a Customer Focus 31

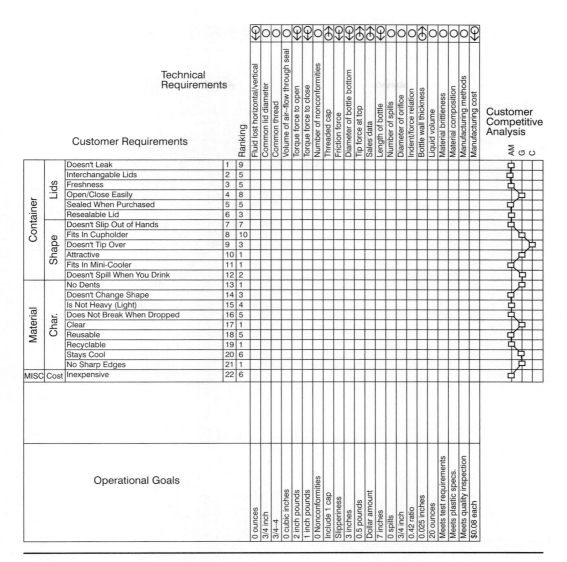

FIGURE 4.8 Operational Goals

for "sealed when purchased" is multiplied by a value of 1 for "weak correlation." The grand total for the column is 86. In the next column, the ranking of 9 for "doesn't leak" is multiplied by 9 for "strong correlation," then added to 5 multiplied by 9, then added to 8 multiplied by 3, then added to 3 multiplied by 3, then added to 5 multiplied by 3 for a total of 174. The numbers in this column show which operational goals are critical (Figure 4.12).

11. Add regulatory and/or internal requirements if necessary. Here, any rules, regulations, or requirements not set forth by the customer but by some other agency or government were identified and recorded (Figure 4.12).

12. Analyze the QFD matrix. What did the customer want? How is this supported by customer rankings and competitive comparisons? How well is the competition doing? How does our company compare? Where will our company's emphasis need to be?

(Continued)

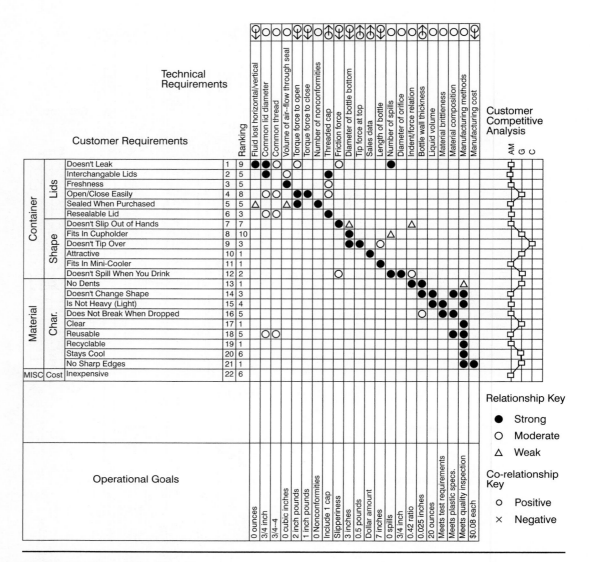

FIGURE 4.9 Correlation Matrix

AM team members studied the matrix they created and came to the following conclusions. The highest ranking columns are manufacturing methods with a score of 190, common lid diameter with a score of 174, torque to open at 144, and 3-inch diameter bottom with a score of 124. To satisfy AM Corp.'s customers and maintain a competitive advantage, it will have to focus its efforts on designing a sports drink bottle that

  Has been well manufactured with good manufacturing methods.

  Has a common lid diameter.

  Is easy to open and close, requiring no more than 2 in.-lb to open and 1 in.-lb of force to close.

  Fits into a standard cup holder in a vehicle, i.e., the base must not exceed 3 inches.

  Does not slip out of the drinker's hand easily. For this reason, the bottle diameter should be no smaller than 3 in. The type of plastic utilized must have the appropriate coefficient of friction.

Creating a Customer Focus   33

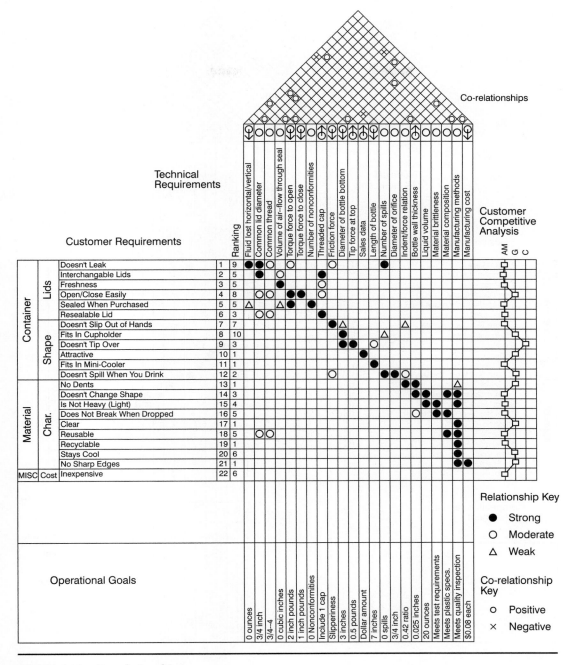

FIGURE 4.10   Co-relationships

(Continued)

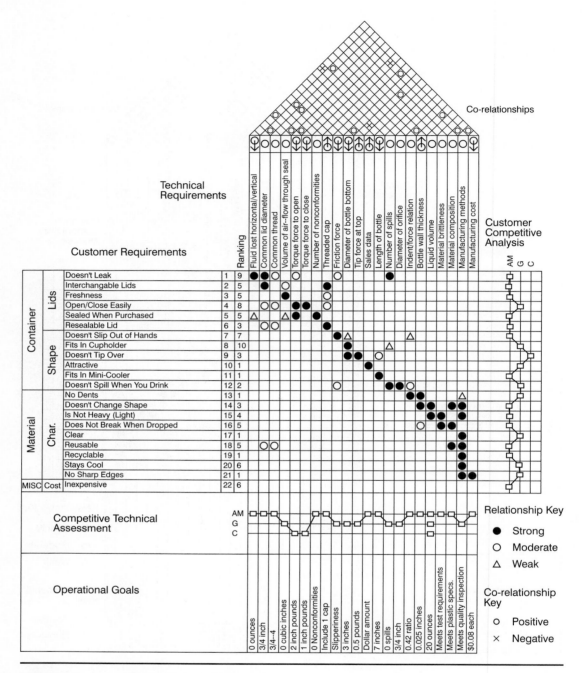

FIGURE 4.11 Competitive Technical Assessment

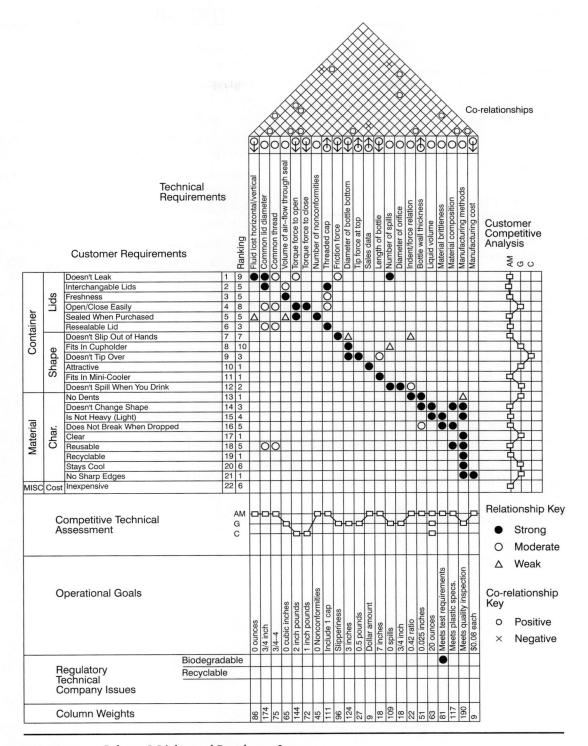

FIGURE 4.12 Column Weights and Regulatory Issues

SUMMARY

Through the use of quality function deployment lean Six Sigma organizations take the customer information they have gathered, translate it into organizational actions, and disseminate this information throughout the organization. How do lean Six Sigma organizations capture the voice of the customer?
  They ask them.
Is your organization hearing voices?

TAKE AWAY TIPS

1. Customer value perceptions are not the same as customer satisfaction perceptions.
2. Value refers to a product's or service's attributed or relative worth or usefulness.
3. Value is judged by consumers each time they trade something of worth in order to acquire the product or service.
4. Customer satisfaction centers around how they felt the last time they bought a product or service from a company. It is a comparison between customer expectations and customer experience.
5. Lean Six Sigma organizations study their processes from their customers' point of view.
6. QFD enables an organization to capture the voice of the customer.
7. QFDs translate a customer's wants into an organization's hows.

CHAPTER QUESTIONS

1. Why would an organization want to be effective at maintaining a customer focus?
2. What must an organization do to maintain a customer focus?
3. What are the benefits of maintaining a customer focus?
4. Using an example from personal experience, describe the difference between satisfaction and perceived value.
5. Describe the principal parts of a quality function deployment matrix.
6. How is each of the principal parts of a QFD matrix created? What does each part hope to provide to the users?
7. Why would a company choose to use a QFD?
8. Describe how you would begin creating a QFD.

PROCESS IMPROVEMENT TEAMS

The whole employee involvement process springs from asking all your workers the simple question, "What do you think?"

Donald Peterson, Former Chairman of Ford Motor Company

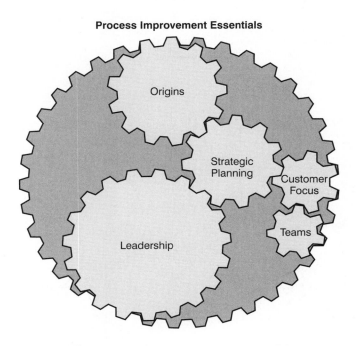

Process Improvement Essentials

 LEARNING OPPORTUNITIES

1. To discuss the importance of teams in lean Six Sigma environments
2. To discuss the importance of education and training
3. To introduce the topic of motivating employees to enhance team performance
4. To introduce the concept of change management to support process improvement
5. To discuss the role of teams in a lean Six Sigma organization

PROCESS IMPROVEMENT TEAMS

In 1962, Dr. Kaoru Ishikawa (1915–1989) became one of the first individuals to encourage the use of process improvement teams. Called **quality circles,** *these teams meet to solve quality problems related to their own work.* The quality circle concept has been adapted and modified over time to become the process improvement teams used in lean Six Sigma organizations. Membership in a process improvement team is often voluntary. Participants receive training in the key process improvement tools, determine appropriate problems to work on, develop solutions, and establish new procedures to lock in quality improvements. Dr. Ishikawa advocated that the teams use seven quality tools when tackling process improvement problems: histograms, check sheets, scatter diagrams, flowcharts, control charts, Pareto charts, and cause-and-effect (or fishbone) diagrams (Figure 5.1 and Chapter 9). Dr. Ishikawa developed the **cause-and-effect diagram** in the early 1950s. This diagram, used to find the root cause of problems, is also called the *Ishikawa diagram,* after its creator, or the *fishbone diagram,* because of its shape.

To refine an organization's approach to quality, productivity, and cost improvement, Dr. Ishikawa encouraged the use of a system of principles and major focus areas as a holistic way to achieve business performance improvement. As presented in *Quality Progress* (April 2004), his system

37

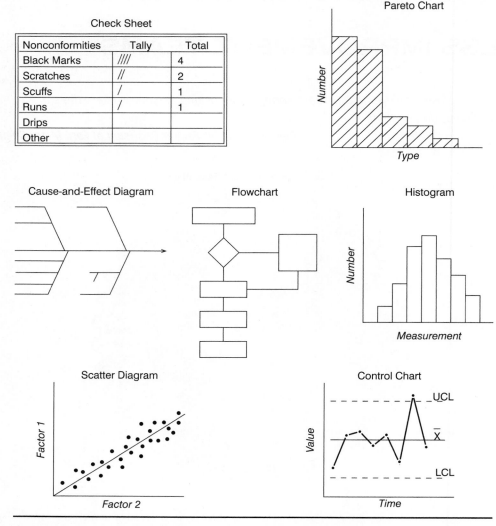

FIGURE 5.1 Seven Tools of Quality

includes six fundamentals that form the Japanese quality paradigm:

1. All employees should clearly understand the objectives and business reasons behind the introduction and promotion of company-wide quality control.
2. The features of the quality system should be clarified at all levels of the organization and communicated in such a way that the people have confidence in these features.
3. The continuous improvement cycle should be continuously applied throughout the whole company for at least three to five years to develop standardized work. Both statistical quality control and process analysis should be used, and upstream control for suppliers should be developed and effectively applied.
4. The company should define a long-term quality plan and carry it out systematically.
5. The walls between departments or functions should be broken down, and cross-functional management should be applied.
6. Everyone should act with confidence, believing his or her work will bear fruit.

The system also includes four major focus areas designed to influence process improvement through leadership:

1. Market-in quality: Leadership should encourage efforts that enable the organization to determine external customer needs, wants, requirements, and expectations. By focusing on these elements and designing processes to deliver value to the market, an organization can increase its business competitiveness.

2. Worker involvement: Quality improvement through the use of cross-functional teams enhances an organization's ability to capture improvements to the work processes. Appropriate training in problem-solving tools and techniques is a must.

3. Quality begins and ends with education: Education enhances an individual's ability to see the big picture. Education creates a deeper understanding of the activities that must take place in order for the organization to be successful.

4. Selfless personal commitment: Dr. Ishikawa lived his life as an example of selfless personal commitment. He encouraged others to do likewise, believing that improving the quality of the experience of working together helps improve the quality of life in the world.

Leaders of lean Six Sigma organizations understand and use the fundamental truths of Dr. Ishikawa's teachings. They know that the use of lean and Six Sigma tools and techniques will improve the processes that provide products and services for customers. Lean Six Sigma interdisciplinary problem-solving teams attack costs, productivity, and quality issues. Teams consisting of people who have knowledge of the process or problem under study are given the task of investigating, analyzing, and finding a solution to the problem situation within a specified time frame. Generally, this team is composed of those closest to the problem, representatives of those affected by the change, as well as a few individuals from middle management with the power to effect change. During the problem-solving process, the team can be supplemented on an as-needed basis with people who have expertise in the areas directly related to the problem solution. Upon the resolution of a project, the team will be disbanded or reorganized to deal with another problem.

Because a team is composed of a group of individuals who are united by a common goal, the best teamwork will occur when the individuals focus on the team's objectives rather than personal motives. While working together, team members must understand and agree on the goals of the team. They must establish and adhere to team ground rules for behavior and performance expectations. To ensure harmony in the team, all members must participate and the responsibilities and duties must be fairly distributed. This means that each team member must understand what his/her role is in the completion of the project. Knowledge of how internal or external constraints affect the project is also helpful. Team members must possess a variety of skills, including problem-solving skills, planning skills, facilitation and communication skills, and feedback and conflict management skills.

For teams to work, management must set clear goals that are aligned with the mission and strategic direction of the firm. Leadership must communicate the scope and boundaries of the project and how the team's progress and performance will be measured. When leadership sets the direction, the team is much more focused and tends not to get bogged down in the problem-selection process. To be successful, teams need the appropriate skills in a supportive organizational culture and the authority to do the job that they have been asked to do. Leadership can do a lot to rid the team of the barriers that inhibit its performance. These barriers include inadequate release time, territorial behavior from involved functional areas, lack of training, inadequate support systems, lack of guidance or direction, and lack of recognition. Leadership's sincere interest and support in the resolution of the problem is evidenced by their willingness to commit money and time for training in problem solving. The teams will quickly become ineffective if the solutions they propose are consistently turned down or ignored. Leadership support will be obvious in management's visibility, diagnostic support, recognition, and limited interference.

EDUCATION AND TRAINING

Lean Six Sigma organizations employ effective people. These employees are the ones who perform their jobs well and understand how their jobs fit into the overall scheme of providing products and services to customers. Their knowledge, skills, and efforts are invaluable to the firm. Lean Six Sigma organizations tap into the knowledge and skills of their employees to improve company competitiveness. They ask their employees, "What do you think?" The answers they receive are put to work to improve productivity, reduce costs, and enhance quality.

As presented in this chapter, training is one of the four key ways to sustain organizational gains. Lean Six Sigma organizations take care to ensure that employees have the appropriate education and training that enables them to perform their job requirements at a level that supports the customers' needs, wants, and expectations. Part of this education and training includes information about how quality, cost, schedule, and profit expectations for their jobs affect customer satisfaction. Through regular feedback, employees monitor the impact that their job has on each of these concepts.

Two of Dr. Deming's 14 points—*institute training on the job* and *institute a vigorous program of education and self-improvement*—focus on education and training. Training refers to job-related skill training and is usually a combination of on-the-job training and classroom-type instruction. Job training includes process improvement techniques as well as job specific concepts. These job specific concepts are evaluated

Yuzo Yasuda, *40 Years, 20 Million Ideas: The Toyota Suggestion System*

Much has been written about the "two pillars" of the Toyota production system—just-in-time and automation—designed by the late Taiichi Ohno. Although this remarkable combination played a large role in creating the efficiency and success of the Toyota Motor Company, these technological breakthroughs aren't the whole story. The fine-tuning that made the Toyota production system really work came not from upper management, nor from the engineers, but from the shop floor in the form of employee suggestions—over 20 million ideas in the last 40 years.

Yuzo Yasuda, 40 Years, 20 Million Ideas: The Toyota Suggestions System

following training in order to assess whether the training was effective. Testing is done under actual work conditions and is verified appropriately. Compared with training, education provides individuals with a broader base of knowledge. This allows individuals to look at a situation from other viewpoints. The education an individual receives may not be immediately applicable to the activities they are currently performing.

Lean Six Sigma employees are provided with the appropriate job-related skills training to give them the skills and knowledge set needed to excel in their jobs. The remaining chapters in this text provide an introduction to many of the lean and Six Sigma tools needed to make cost, quality, and productivity improvements. One key skill that lean Six Sigma organizations make sure their employees have is problem solving. Employees are trained to isolate the root causes of problems, utilize lean and Six Sigma tools and techniques to gain improvements, and lock in the gains they achieve to make the improvements permanent. Chapter 9 covers basic problem solving and introduces many root cause analysis Six Sigma tools. Follow-up or refresher training is also key to skill acquisition and retention. This type of training enables employees to maintain higher skill and performance levels by refamiliarizing employees with the best practices and eliminating poor habits.

ENHANCING TEAM PERFORMANCE

Quality, cost, and productivity improvements are about changing the mind-sets and behaviors of people to enable them to enhance the performance of business. Employees come to work to do a good job. They want a job that allows them to use their knowledge, skills, and abilities so that they can take pride in their work. Lean Six Sigma leaders recognize these traits in their employees and build on them to motivate their people. If people don't come to work to do a bad job, why do things go wrong? Why are products produced and services provided that are unsatisfactory? Usually, faulty systems are at the root of the problem. Dr. Deming warned that if employees must utilize faulty systems, they will not be able to help performing poorly. As one of Dr. Deming's 14 points says, leadership must *remove barriers that rob people of their right to pride of workmanship*. Employees often face difficulties in performing their jobs that can only be removed by leadership. Lean Six Sigma organizations recognize that it is leadership's responsibility to design and maintain non-faulty systems, thus allowing employees to work at their greatest level of productivity.

Changing conditions in the marketplace put pressure on organizations to make modifications to the way they do business. Cost reduction, productivity, and quality improvements require process changes that often involve a human element. Leadership must communicate to the employees the desired change and motivate individuals to make the change. Dr. Deming once said, "*Nothing will happen without change. Your job as a leader is to manage the change necessary.*" In the second of his 14 points—*adopt the new philosophy*—Dr. Deming encourages management to practice leadership for change. Change is a cycle that requires momentum and clear direction from leadership (Figure 5.2). Change is rarely easy. People resist change because humans are control oriented, and when their environments are disrupted, they perceive that they have lost the ability to control their lives. Some people are more resistant to change

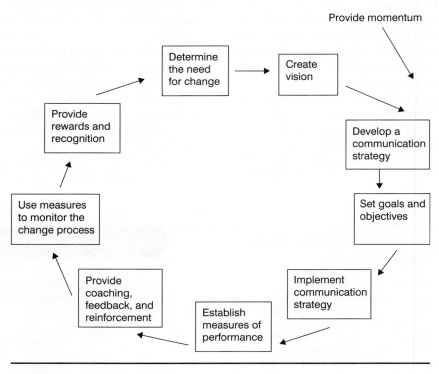

FIGURE 5.2 The Change Cycle

LEAN SIX SIGMA TOOLS at WORK

Making the Change

It is easy to tell someone else they have to change. But change isn't easy. Try to change one thing about yourself. Consider exercising more, eating healthier, or breaking an undesirable habit. Is your attitude about the change wishful thinking? To help with the change process, consider the following points and questions.

What is the desired end result? Can you picture it? To change, a person must understand what the ultimate outcome of that change will be.

What actions will you take to make the change? To change, a person takes a series of actions that produce a result that moves the individual toward the desired outcome. This result may be seen and interpreted as either positive or negative. Based on the interpretation of the result, the person modifies his or her attitude and acts differently the next time in order to produce a more appropriate result.

What is the time frame for the change? Change takes time, yet if no timetable exists, a person is likely to say, "I'll get to that tomorrow." Effective change takes place when the person works toward the change according to a schedule.

How will you stay motivated? To change, it is necessary to stay motivated. Often, it is not easy to maintain the motivation and direction required to bring about change.

How will you know you have changed? What will your indicators be? A person needs to have some sort of feedback that enables them to understand how they are progressing toward the desired outcome. Otherwise, they may not recognize when the change is complete.

In the future, use these questions or a change matrix (Figure 5.3) to guide you when making a change.

Desired Result	Supporting Actions	Timing	Motivation	Verification

FIGURE 5.3 Change Matrix

---

than others. Resistance can be based on the individual's frame of reference, his/her individual values, emotions, knowledge, and behavior. People can have a variety of attitudes about change: I wish I could change, I want to change, I will make this change, or I will make this change no matter what. It is only this last attitude, making the change happen no matter what, that actually results in change. The rest are merely wishful thinking.

Making change in any organization can be difficult if not impossible without the involvement and cooperation of the employees. It only takes moments to proclaim a new culture or a new method, but it takes a great deal longer to get people to act differently. As discussed in Chapter 2, a lean Six Sigma culture emphasizes making and sustaining gains. Employees in lean Six Sigma organizations practice continuous improvement daily as a part of their culture. Lean Six Sigma leaders recognize that employees mold their behavior according to their interpretations of the signals leadership sends them. These signals may come from policies, requests, edicts, or the day-to-day actions taken by leadership. Because actions speak louder than words, communicating through leadership actions and examples is paramount to changing behavior. To maximize the change process, lean Six Sigma leaders ensure that the worker-machine-computer interface, as well as the worker-to-worker interface, is compatible with the needs, capabilities, and limitations of the worker. Further, lean Six Sigma organizations ensure that the reward system matches the desired expectations in order to change behavior. Alignment must exist among goals and expectations, performance, reinforcement, and rewards and recognition (Figure 5.4).

Communication is a vital part of motivating employees. No leader can handle conflict or negotiate successfully without being a good communicator. ***Communication*** *occurs when information is imparted by audible, written, and/or visual (postures/actions) means.* Communication is more than just presenting information. Effective communication seeks to both exchange information and provide understanding. How people perceive information is critical. Noise in the form of personal agendas, needs, values, fears, and defenses affects communication too. Communicating in the recipients' language and from their point of view can enhance understanding. People must be able to recognize that the information affects them and that they are to act on that information. Good communication is evidenced by whether the message got through and got acted on. To ensure that everyone gets the same meaning from the message, good communicators ask questions to clarify the message. They paraphrase and restate what they think they heard or read.

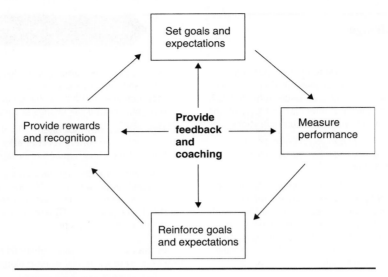

FIGURE 5.4 Supporting Change

Effective communicators take the time to practice their skills and learn to express themselves well. They also listen. Practicing active listening involves taking time to listen not only for message content but also for feelings, responses, and cues sent by the other person.

Effective leaders realize that they must listen to their employees to determine what motivates them. Effective leaders do not rely on hunches or guesses. It is impossible to know what motivates an individual without asking him or her and then truly listening to the response. While listening, a leader learns about and comes to understand an employee's attitudes toward work, pay, benefits, supervision, appraisals, and other work-related topics. This information can be used to place employees in a work situation they find motivating.

LEAN SIX SIGMA TEAMS

Many types of teams exist, including the following:

- Management teams comprised of heads of departments to do strategic planning
- Cross-functional teams with representatives from a large variety of areas for the design or development of complex systems
- Self-directed work teams made up of employees grouped by complementary skills in order to carry out production processes
- Project or problem-solving teams composed of temporary groups of individuals from the appropriate functional areas with the necessary skills to work on a specific task

---

 Clear Communication

LEAN SIX SIGMA TOOLS at WORK

Employees need to know where they stand. Annual reviews are one way of communicating expectations and providing feedback. At JF, no one received reviews of any sort. When the new owner took over, he established a system where all employees, regardless of job title, are reviewed annually based on their hire date. Reviews may occur more frequently. Employees in their first two years of employment are reviewed every 90 days. Raises and bonuses, besides the quarterly bonus based on profitability, may occur at any time. The new owner also made it clear that the reviews and any subsequent raises would focus on two issues:

How have you helped improve the productivity for JF?

How have you improved something about yourself?

The emphasis is on what the employee is doing to help JF get good product out the door on a timely basis. It is the answers to these questions that determine whether an employee receives a raise. Employees are encouraged to show productivity improvements by discussing how they trained another employee on their workstation, how they were able to use their own skills to move within the plant on an as-needed basis when staffing issues arose, how they discovered and implemented a productivity improvement, or how they worked to solve a quality, safety, customer, or productivity issue. Self-improvements include learning a new skill, learning how to operate another machine or work another job, or taking training off-site to increase their job knowledge. These questions encourage workers to look beyond being a good employee, practicing safe work methods, and being at work when scheduled. They reflect Dr. Deming's 14 points to constantly and forever improve the process of providing a product or service. They also reflect Dr. Deming's desire to encourage self-improvement. The management at JF wants employees to succeed. They provide many opportunities for training, skills assessment, mentoring, and apprenticeship.

LEAN SIX SIGMA TOOLS at WORK

Process Improvement Through Teamwork

Starbucks has more than 10,000 sites worldwide. Though improvements made at one site can be shared throughout the organization, individual locations face their own issues. Teams at each site use lean tools and techniques to identify and refine improvements to their activities. The August 4, 2009, *Wall Street Journal* article "Starbucks' New Buzzword: 'Lean' Japanese Techniques," provides several examples of process improvement. One example described a team working to improve barista performance in the blended drinks preparation area. Team members noted how employees performed a lot of reaching and moving to grasp items that were never in the same place. This resulted in long drink preparation times. The team made modifications to the work area that included placing the most commonly used items, such as syrup flavors and pitchers, closer to where the drinks are made. They also moved toppings, such as whipped cream, chocolate, and caramel drizzle closer to where the drinks are handed to the customers. These changes took eight seconds off of the 45-second process. A 18 percent improvement!

Lean Six Sigma project teams are given the task of investigating, analyzing, and finding a solution to problems. Project teams consist of people who have been given a mandate to focus on a particular process, area, or problem. Generally, this team is composed of those closest to the problem as well as a few individuals from middle management who have the power to effect change. The team may consist of people from a variety of departments, depending on the problem facing them. The team may even include an outside vendor or a representative from the customer base. Upon resolution of the problem, the team will be disbanded or reorganized to focus on another project.

Henry Ford, founder of Ford Motor Corporation, said:

Coming together is a beginning. Keeping together is progress. Working together is success.

Lean Six Sigma organizations recognize that teams are crucial to solving issues and problems facing the organization. People working together find new ways to enhance productivity and quality while reducing costs. Like Henry Ford, they also realize that teams do not always coalesce into highly functional entities without help. There are several stages of team development. Recognizing that teams experience growth throughout their existence helps leaders guide and direct team activities.

During the first stage, the team forms. This *formation* stage is usually experienced in the first few meetings. During this time, the team establishes its goals and objectives. It also determines the ground rules for team performance. For teams to work well, leadership must set clear goals that are aligned with the mission and strategic direction of the firm. When leadership sets the direction, the team is much more focused and tends not to get bogged down in the problem-selection process. The team must know the scope and boundaries that it must work within. Leadership must communicate how the team's progress and performance will be measured.

Oftentimes, following their formation, teams experience a rocky period where team members work out their individual differences. This is the time that the team gets acquainted with each other's idiosyncrasies and the demands of the project. During this *stormy* stage, the goals and scope of the team may be questioned. Because a team is composed of a group of individuals who are united by a common goal, the best teamwork will occur when the individuals focus on the team's objectives rather than personal motives. While working together, team members must understand and agree on the goals of the team. They must establish and adhere to team ground rules for behavior and performance expectations. To ensure harmony in the team, all members must participate and the responsibilities and duties must be fairly distributed. This means that each team member must understand what his/her role is in the completion of the project. Knowledge of how internal or external constraints affect the project is also helpful. Team members must possess a variety of skills, including problem-solving skills, planning skills, facilitation and communication skills, and feedback and conflict management skills.

The third stage of team development occurs when the team starts to work together smoothly. It is during this *performing* stage that things get accomplished. To be successful, teams need the appropriate skills in a supportive organizational culture and the authority to do the job that they have been asked to do. Leadership can do a lot to rid the team of the barriers that inhibit its performance. These barriers include inadequate release time, territorial behavior from involved functional areas, lack of training, inadequate support systems, lack of guidance or direction, and lack of recognition. Senior leadership's sincere interest and support in the resolution of the problem is evidenced by their willingness to commit money and time for training in problem solving and facilitation. In any case, senior leadership must monitor and encourage teams to solve problems. The teams will quickly become unmotivated if the solutions they propose are consistently turned down or ignored. Leadership support will be obvious in management's visibility, diagnostic support, recognition, and limited interference.

As the team finishes its project, the final stage occurs, *concluding*. During this phase, team members draw the project to its conclusion, verify the results, and disband the team. Several key events take place during this time. The team members, having taken action, perhaps by implementing a solution to a problem, must verify that what they planned to do got done and what they did actually worked. Teams are not finished when they have proposed a plan of action; teamwork is finished when the plans have been acted on and the results judged effective. Until then, the team cannot be disbanded.

SUMMARY

Lean Six Sigma organizations engage in a continuous cycle of learning and training. To support this cycle of continuous productivity, cost, and quality improvement, lean Six Sigma organizations design recognition and reward systems that motivate their employees to change. These rewards are aligned with customer needs, wants, and expectations as identified in the strategic plan and supported leadership actions.

TAKE AWAY TIPS

1. Lean Six Sigma organizations follow good human resource practices.
2. Lean Six Sigma organizations provide employees with the knowledge and skills they need to do their jobs well in order to provide value for the customer.
3. Change isn't easy; an individual must be motivated to change.
4. Leaders are responsible for managing change.
5. Leaders are responsible for removing the barriers that prevent their employees from performing well.
6. Lean Six Sigma organizations utilize teams to solve problems and enhance processes.
7. Participants on lean Six Sigma teams can expect their talents, skills, and personality types to be appreciated and used to their best advantage.

CHAPTER QUESTIONS

1. Who is Dr. Kaoru Ishikawa and what did he add to process improvement and the lean Six Sigma methodology?
2. Describe the difference between education and training. Why is it important to have both?
3. Describe a situation where you received (or did not receive) job skill training. Was it adequate? Why? Why not? What would you have done differently?
4. Describe a situation where you received (or did not receive) an educational experience. Was it adequate? Why? Why not? What would you have done differently?
5. Describe a situation where a leader motivated you. What did he or she do? How did you react? Why?
6. Describe a change you were required to make. How did it go? What did the change accomplish? How was it accomplished? What motivated you?
7. Describe the phases of team development.
8. For each of the phases of team development, provide an example from your own experience describing how your team got through the phase.
9. How will you guide people who work for you or with you through a change process?
10. How will you motivate the people who work for or with you?

COSTS OF QUALITY

Costs of quality are the costs associated with providing customers with a product or service that conforms to their expectations.

Philip Crosby

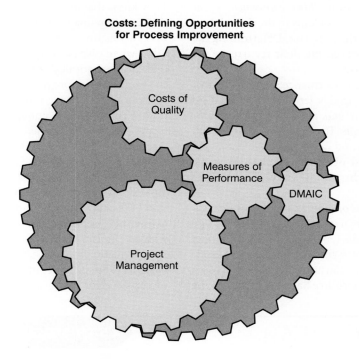

 LEARNING OPPORTUNITIES

1. To familiarize the reader with the concept of quality costs
2. To create an understanding of the interrelationships between and among the different types of quality costs
3. To gain an understanding of how quality costs can be used in decision making

COSTS OF QUALITY

What drives a company to improve? Increased customer satisfaction? Greater market share? Enhanced profitability? There are many reasons why companies seek to improve the way that they do business, and one of the most important is the cost of quality. Pressured between rising costs and consumer demand for lower prices, competing in today's global market is not easy. Quality costs, the costs that would disappear if every activity were performed without defects every time, can have a big impact on a company's financial statement.

In his book *Quality Is Free*, Philip Crosby focused management attention on the costs of quality, the cost of not doing things right the first time. Based on his understanding of the costs of quality, Crosby (1926–2001) sent a message to management that emphasizes four absolutes (Figure 6.1). The four

Quality Definition: Conformance to Requirements
Quality System: Prevention of Defects
Quality Performance Standard: Zero Defects
Quality Measurement: Costs of Quality

FIGURE 6.1 Crosby's Four Absolutes of Quality

absolutes of quality management set expectations for a continuous improvement process. The first absolute defines *quality as **conformance to requirements**.* Crosby emphasizes the importance of determining customer requirements, defining those requirements as clearly as possible, and then producing products or providing services that conform to the requirements as established by the customer. Crosby felt it necessary to define quality in order to manage quality and lower costs.

Prevention of defects, *the second absolute, is the key to the system that needs to be in place to ensure that the products or services provided by a company meet the requirements of the customer.* Prevention of quality problems in the first place is much more cost effective in the long run. Determining the root causes of defects and preventing their recurrence are integral to reducing quality costs.

According to Crosby, the performance standard against which any system must be judged is zero defects. This third absolute, ***zero defects,*** *refers to making products correctly the first time, with no imperfections.* Traditional quality control centered on final inspection and "acceptable" defect levels. Systems must be established or improved that allow the worker to do it right the first time. Preventing defects reduces the costs of quality.

His fourth absolute, ***costs of quality,*** *are the costs associated with providing customers with a product or service that conforms to their expectations.* Quality costs are found in prevention costs; detection costs; costs associated with dissatisfied customers; rework, scrap, downtime, and material costs; and costs involved anytime a resource has been wasted in the production of a quality product or the provision of a service. Once determined, costs of quality can be used to justify investments in equipment and processes that reduce the likelihood of defects. Costs of quality are the focus of this chapter.

Lean Six Sigma organizations pay careful attention to the costs associated with providing poor quality. They use quality cost information to improve the way they do business. Organizations following the lean Six Sigma methodology are well aware of quality costs. They determine these costs and use them to justify their lean Six Sigma projects. They know that by investing in the prevention of problems, they can reap financial rewards.

Two factions often find themselves at odds with each other when discussing quality. There are those who believe that no "economics" of quality exists, that it is never economical to ignore quality. There are others who feel it is uneconomical to have 100 percent perfect quality all of the time. Cries of "Good enough is not good enough" will be heard from some, whereas others will say that achieving perfect quality will bankrupt a company. Should decisions about the level of quality of a product or service be weighed against other factors such as meeting schedules and cost? To answer this question, an informed manager needs to understand the concepts surrounding the costs of quality. Investigating the costs associated with quality provides managers with an effective method to judge the economics and viability of a quality improvement system. Quality costs serve as a baseline and a benchmark for selecting improvement projects and for later evaluating their success.

Costs of Quality

LEAN SIX SIGMA TOOLS at WORK

Over an 18-year period, airlines have been trying to reduce the number of lost bags (Figure 6.2). Some airlines have even placed self-service baggage kiosks in major airports. These kiosks allow air travelers to trace their checked luggage by scanning their bar coded luggage tags. This system also allows travelers to report their missing bags in less than two minutes versus the average of 45 minutes it takes to report lost bags to an agent. In the year 2009, the average number of lost bags per 1,000 passengers was 3.99. This means that 99.6 percent of the passengers and their luggage arrived at the same airport at the same time. Why would an airline company be interested in improving operations from 99.6 percent? Why isn't this good enough? What are the costs of quality in this situation?

In the case of lost luggage, the types of quality costs are numerous. For instance, if a passenger is reunited with his or her luggage after it arrives on the next flight, the costs of quality include passenger inconvenience, loss of customer goodwill, and perhaps the cost of gesture of goodwill on the part of the airline in the form of vouchers for meals or flight ticket upgrades. As the scenario becomes more complex, a wider array of quality costs exists. For example, if an airline must reunite 100 bags with their owners per day at a hub airport such as Atlanta or New York, the costs can be enormous. The expenses associated with having airline employees track, locate, and reroute lost bags might run as high as 40 person-hours per day at $20/hour, for a cost of $800/day. Once located, the luggage must be delivered to the owners, necessitating baggage handling and delivery charges. If a delivery service were to charge $1/mile for the round trip, in larger cities, the average cost of delivering that bag to a customer might be $50. One hundred deliveries per day at $50 per delivery equals $5,000. Just these two costs total $5,800/day or $2,117,000/year! This doesn't include the cost of reimbursing customers for incidental expenses while they wait for their bags, the costs of handling extra paperwork to settle lost bag claims, loss of customer goodwill, and negative publicity. Add to this figure the costs associated with bags that are truly lost ($1,250 for domestic flights and $9.07 per pound of checked luggage for international travelers), and the total climbs even higher. And these figures are merely the estimated cost at a single airport for a single airline!

Consider the more intangible effects of being a victim of lost luggage. If customers view airline baggage handling systems as a black hole where bags get sucked in only to resurface sometime later, somewhere else, if at all, then consumers will be hesitant to check bags. If the ever-increasing amount of carry-on luggage with which people board the plane is any indication, lost luggage fears exist. Carry-on baggage presents its own problems, from dramatically slowing the process of loading and unloading of the plane, to shifting in flight and falling on an unsuspecting passenger when the overhead bins are opened during or after the flight, to blocking exits from the airplane in an emergency. These costs, such as fees for delayed departure from the gate or liability costs, can be quantified and calculated into a company's performance and profit picture.

Is it any wonder that airlines have been working throughout the past decade to reduce the number of lost bags?

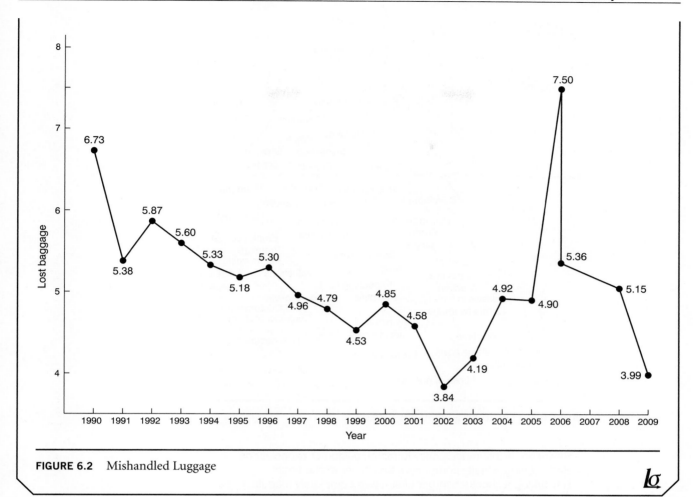

FIGURE 6.2 Mishandled Luggage

DEFINING QUALITY COSTS

Cost of quality, poor-quality cost, and *cost of poor quality* are all terms used to describe the costs associated with providing a quality product or service. A **quality cost** is considered to be any cost that a company incurs to ensure that the quality of the product or service is perfect. Quality costs are the portion of the operating costs brought about by providing a product or service that does not conform to performance standards. Quality costs are also the costs associated with the prevention of poor quality. The most commonly listed costs of quality include scrap, rework, and warranty costs. As Figure 6.3 shows, these easily identified quality costs are merely the tip of the iceberg.

Quality costs can originate from anywhere within a company. No single department has the corner of the market on making mistakes that might affect the quality of a product or service. Even departments far removed from the day-to-day operations of a firm can affect the quality of a product or service. The receptionist, often the first person the customer has contact with, can affect a customer's perceptions of the firm. The cleaning people provide an atmosphere conducive to work. Anytime a process is performed incorrectly, quality costs are incurred. Salespeople must clearly define the customer's needs as well as the capabilities of the company. Look up *re* in the dictionary and it is defined as again or anew. Doing something again is waste, additional costs that need not be incurred. Companies investigate quality costs in order to uncover these sources of waste and eliminate them. They do not want to readjust, realign, reappraise, reassemble, reassess, reassign, rebroadcast, rebuild, recheck, reconvene, recopy, redirect, reenact, reestablish, reexamine, reexplain, reformulate, reinoculate, reinsert, reinspect, reinvest, relocate, renegotiate, reopen, reorder, repack, repaint, repair, retrain, or rework anything. Figure 6.4 provides a few more examples of quality costs. Note how many *re* words, as in redo, are in this list. Every department within a company should identify, collect, and monitor quality costs within its control.

Quality, in both service and manufacturing industries, is a significant factor in allowing a company to maintain and increase its customer base. As a faulty product or service finds its way to the customer, the costs associated with the error increase. Preventing the nonconformity before it is manufactured or prepared to serve the customer is the least costly approach to providing a quality product or service. Potential problems should be identified and dealt with during the design and planning stage. An error at this phase of product development will require effort to solve, but in most cases the changes can be made before costly investments in equipment or customer service are made.

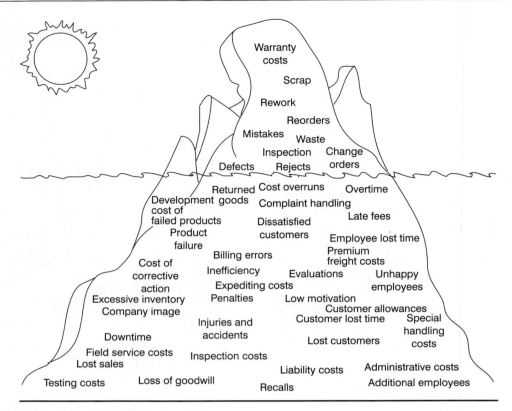

FIGURE 6.3 The Iceberg of Quality Costs

Rewriting an insurance policy to match customer expectations
Redesigning a faulty component that never worked right
Reworking a shock absorber after it was completely manufactured
Retesting a computer chip that was tested incorrectly
Rebuilding a tool that was not built to specifications
Repurchasing because of nonconforming materials
Responding to a customer's complaint
Refiguring a customer's bill when an error was found
Replacing a shirt the dry cleaner lost
Returning a meal to the kitchen because the meat was overcooked
Retrieving lost baggage
Replacing or repairing damaged or lost goods
Correcting billing errors
Expediting late shipments by purchasing a more expensive means of transportation
Providing on-site assistance to customers experiencing problems in the field
Providing credits and allowances

FIGURE 6.4 Examples of Quality Costs

If the nonconformity with the product or service is located during the manufacturing cycle, or behind the scenes at a service industry, then the nonconformity can be corrected internally. The cost is greater here than at the design phase because the product or service is in some stage of completion. Product may need to be scrapped or, at a minimum, reworked to meet the customer's quality expectations.

If the nonconforming product or service reaches the customer, the company providing it incurs the greatest costs. Customers who locate this nonconformity will experience a variety of feelings, the least of which will be dissatisfaction.

Even if the customers are later satisfied by providing them with a new product, repairing their existing product, or repeating the service, the damage has already been done to the company's reputation. Disgruntled customers rarely fail to tell others about their experience. Mistakes that reach the customer result in the loss of goodwill between the present customer, future customers, and the provider. Failure cost increases as a function of the failure's detection point in the process (Figure 6.5).

In this situation, an unsuitable product reached a consumer. Quality service from the store could not compensate

The Cost of Poor Quality

LEAN SIX SIGMA TOOLS at WORK

Jim and Sue Homeowners recently purchased a home on a large lot. Using their small push mower to cut the lawn proved to be a tedious, time-consuming job, so over the winter, they saved to purchase a riding lawn mower. Early in spring, they checked a variety of lawn mower sales and finally settled on Brand X. Unknown to them, this particular mower had a design flaw that had gone unnoticed by the manufacturer. The mower deck, which covers the blades, is held onto the mower by bolts. Because there is too little clearance between the mower-deck fastener bolts and the blade, under certain circumstances the blades come in contact with the bolts, shearing them off and releasing the mower deck. When this happens, the damage to the blades is irreparable.

This design error was not apparent in the store lot where the Homeowners tried the mower. The lot was smooth, and the problem only becomes obvious when the mower is used on uneven ground. The Homeowners went home to try out their new lawn mower and within minutes the mower was moving along uneven terrain. A loud shearing sound was heard, and the mower deck fell off, mangling the blades in the process. The Homeowners immediately called the salesperson, who, understanding the importance of happy customers, quickly sent a technician to investigate.

The technician identified the problem, returned to the store with the mower, and replaced the first mower with a second.

Even though the technician knew what had caused the failure, he had no way of informing the manufacturer. Previous attempts to contact the manufacturer to give them information about other problems had not been successful. The manufacturer thus lost a valuable bit of information that could have improved the design.

Although they had used most of a Saturday getting the problem resolved, the Homeowners felt satisfied with the service they had received . . . until they tried the new mower. The new mower experienced the same problem on a different patch of grass. This time, the store was unable to replace the mower until the next weekend. The Homeowners resorted to their push mower.

The next Saturday was not a good day. Another mower was promptly delivered and tried out in the presence of the salesperson and the technician. Same problem. The Homeowners returned the mower and resolved to purchase no other yard-care products—no weed whips, wagons, push mowers, or edging tools—from this company. At a barbecue that weekend, they told their neighbors about their troubles with Brand X products. Over the next few weeks, the Homeowners investigated the types of lawn mowers used by yard-service firms and golf courses. Even though it was a bit more expensive, they bought Brand Y because of its proven track record and quality reputation.

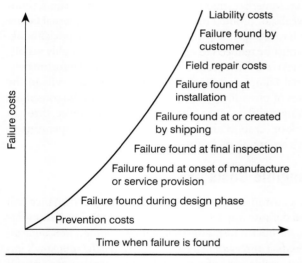

FIGURE 6.5 Increases in Quality Costs as Faulty Product Reaches the Customer

| Lawn mower manufacturer notices design flaw during design review. | Lawn mower manufacturer finds design flaw at start of production. | Lawn mower manufacturer finds design flaw during assembly or testing. | Customer finds design flaw. |

FIGURE 6.6 Increase in Failure Costs

for the original design flaw. As Figure 6.6 shows, failure to spot the design flaw during the design review, during production and assembly, or during testing dramatically increased costs. The quality costs associated with providing a poorly designed lawn mower included not only the three damaged lawn mowers but also sales of related equipment to the original customers and future sales to their friends.

TYPES OF QUALITY COSTS

The costs associated with preventing nonconformities, with appraising products or services as they are being produced or provided, and with failure can be defined. With an understanding of these costs, a manager can make decisions concerning the implementation of improvement projects. Using quality costs, a manager can determine the usefulness of investing in a process, changing a standard operating procedure, or revising a product or service design.

Prevention Costs

Prevention costs are those costs that occur when a company is performing activities designed to prevent poor quality in products or services. Prevention costs are often seen as front-end costs designed to ensure that the product or service is created to meet the customer requirements. Examples of such costs are design reviews, education and training, supplier selection and capability reviews, process capability reviews, and process improvement projects. Prevention activities must be reviewed to determine whether they truly bring about improvement in the most cost-effective manner.

Unfortunately, many organizations focus on taking corrective actions, fixing problems, and reporting savings. An

> **Lean Six Sigma Tools at Work**
>
> ### Prevention Costs in Drug Development
>
> The Food and Drug Administration (FDA) and insurance providers made headlines in April 2008 when they announced the launch of a combined effort to track the relationship between medical claims and prescription drug use. According to the April 15, 2008, article in the *Wall Street Journal*, this information will be used for tracking drug safety. The collaborative effort will use charts to look for hidden patterns or spikes in medical problems that could potentially signal that there is an interaction between the patient's condition, the medication, or the combination of drugs that they are using.
>
> Drug safety is a key issue in today's society. Consumers expect new drugs to combat a wide variety of illnesses. The FDA wants to make sure that the drugs released on the market do not inadvertently harm the consumers using them. In the past, monitoring drug reactions and problems has not been easy. When problems occurred with the Merck & Co. painkiller Vioxx, it took five years to amass enough data to provide a clear link between the drug and heart attacks. This new system has the potential to prevent similar situations by using the information of nearly 35 million insured individuals to follow drug use. The quality prevention costs associated with this effort are high. Information sharing of computerized databases does not happen easily. The data must be sorted, graphs drawn, and interactions determined. The benefits for investing in these prevention costs make it a worthwhile investment. The system hopefully will be able to detect problems with drugs much more quickly and reliably than ever before. These drug safety surveillance capabilities will enhance patient safety.

emphasis on the prevention of defects is much more cost effective, as no money is lost making mistakes in the first place. Prevention efforts try to determine the root causes of problems and eliminate them at the source so recurrences do not happen. Preventing poor quality stops companies from incurring the cost of doing it over again. Essentially, if they had done it right the first time, they would not have to repeat their efforts. In his book *Quality Is Free* (McGraw-Hill, 1979), Philip Crosby emphasizes the need to invest in preventing problems in order to reap the benefits of reduced appraisal and failure costs. The initial investment in improving processes is more than compensated by the resulting cost savings, so, in essence, quality is free.

Appraisal Costs

Appraisal costs are the costs associated with measuring, evaluating, or auditing products or services to make sure that they conform to specifications or requirements. Appraisal costs are the costs of evaluating the product or service during the production of the product or the providing of the service to determine whether, in its unfinished or finished state, it is capable of meeting the requirements set by the customer. Appraisal activities are necessary in an environment where product, process, or service problems are found. Appraisal costs can be associated with raw materials inspection, work-in-process (activities-in-process for the service industries) evaluation, or finished product reviews. Examples of appraisal costs include incoming inspection, work-in-process inspection, final inspection or testing, material reviews, and calibration of measuring or testing equipment. When the quality of the product or service reaches high levels, then appraisal costs can be reduced.

Failure Costs

Failure costs occur when the completed product or service does not conform to customer requirements. Two types exist: internal and external. *Internal failure costs are those costs associated with product nonconformities or service failures found before the product is shipped or the service is provided to the customer.* Internal failure costs are the costs of correcting the situation. This failure cost may take the form of scrap, rework, remaking, reinspection, or retesting. *External failure costs are the costs that occur when a nonconforming product or service reaches the customer.* External failure costs include the costs associated with customer returns and complaints, warranty claims, product recalls, or product liability claims. Because external failure costs have the greatest impact on the corporate pocketbook, they must be reduced to zero. Because they are highly visible, external costs often receive the most attention. Unfortunately, internal failure costs may be seen as necessary evils in the process of providing good-quality products to the consumer. Nothing could be more false. Doing the work twice, through rework or scrap, is not a successful strategy for operating in today's economic environment.

Intangible Costs

How a customer views a company and its performance will have a definite impact on long-term profitability. *Intangible costs, the hidden costs associated with providing a nonconforming product or service to a customer, involve the company's image.* Intangible costs of poor quality, because they are difficult to identify and quantify, are often left out of quality-cost determinations. They must not be overlooked or disregarded. Is it possible to quantify the cost of missing an important deadline? What will be the impact of quality problems or schedule delays on the company's image? Intangible costs of quality can be three or four times as great as the tangible costs of quality. Even if these costs can only be named, and no quantifiable value can be placed on them, it is important for decision makers to be aware of their existence.

The four types of quality costs are interrelated. In summary, **total quality costs** *are considered to be the sum of prevention costs, appraisal costs, failure costs, and intangible costs.* Figure 6.7 shows some of the quality costs from Figure 6.3 in their respective categories. Investments made to prevent poor quality will reduce internal and external failure costs. Consistently high quality reduces the need for many appraisal activities. Suppliers

Prevention Costs

    Quality planning
    Quality program administration
    Supplier-rating program administration
    Customer requirements/expectations market research
    Product design/development reviews
    Quality education programs
    Equipment and preventive maintenance

Appraisal Costs

    In-process inspection
    Incoming inspection
    Testing/inspection equipment
    Audits
    Product evaluations

Failure Costs

Internal:
    Rework
    Scrap
    Repair
    Material-failure reviews
    Design changes to meet customer expectations
    Corrective actions
    Making up lost time
    Rewriting a proposal
    Stocking extra parts
    Engineering change notices

External:
    Returned goods
    Corrective actions
    Warranty costs
    Customer complaints
    Liability costs
    Penalties
    Replacement parts
    Investigating complaints

Intangible Costs

    Customer dissatisfaction
    Company image
    Lost sales
    Loss of customer goodwill
    Customer time loss
    Offsetting customer dissatisfaction

FIGURE 6.7 Categories of Quality Costs

with strong quality systems in place can reduce incoming inspection costs. High appraisal costs combined with high internal failure costs signal that poor-quality products or services are being produced. Efforts made to reduce external failure costs will involve changes to efforts being made to prevent poor quality. Internal failure costs are a portion of the total production costs, just as external failure costs reduce overall profitability. A trade-off to be aware of when dealing with quality costs is the need to ensure that appraisal costs are well spent. Companies with a strong appraisal system need to balance two points of view: Is the company spending too much on appraisal for its given level of quality performance, or is the company risking excessive failure costs by underfunding an appraisal program? In all three areas—prevention, appraisal, and failure costs—the activities undertaken must be evaluated to ensure that the efforts are gaining further improvement in a cost-effective manner. Figure 6.8 reveals that as quality costs are reduced or—in the case of prevention, quality costs—invested wisely, overall company profits will increase. The savings associated with doing it right the first time show up in the company's bottom line.

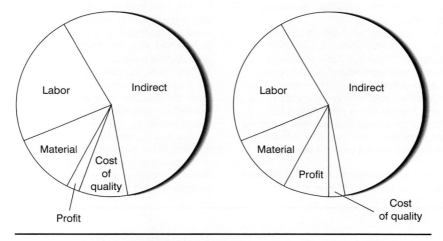

FIGURE 6.8 Cost of Quality Versus Profit

QUALITY-COST MEASUREMENT SYSTEM

In the struggle to meet the three conflicting goals of quality, cost, and delivery, identifying and quantifying quality costs help ensure that quality does not suffer. When quality costs are discussed as a vague entity, their importance in relation to cost and delivery is not understood. By quantifying quality costs, all individuals producing a product or providing a service understand what it will cost if quality suffers. True cost reduction occurs when the root causes of nonconformities are recognized and eliminated.

A quality-cost measurement system should be designed to keep track of the different types of quality costs. Being able to define and quantify quality costs enables quality to be managed more effectively. Once quantified, these costs can be used to determine which projects will allow for the greatest return on investment and which projects have been most effective at improving processes, lowering failure rates, and reducing appraisal costs. A quality-cost measurement system should use quality costs as a tool to help justify improvement actions.

Quantifying quality costs will help zero in on which problems, if solved, will provide the greatest return on investment. A quality-cost measurement system should try to capture and reduce significant quality costs. Effective cost reduction occurs when the processes providing products or services and their related support processes are managed correctly. Measures (Chapter 7) are key to ensuring that the processes are performing to the best of their capability. After all, what is measured can be managed. Once cost of quality data has been measured and tabulated, this data can be used to select quality improvement projects as well as identify the most costly aspects of a specific problem. Failure costs are prevalent in all aspects of providing goods and services, from design and development to production and distribution. Support processes, such as marketing and human resources, can also be the source of quality costs. When justifying improvement projects, look for rework, waste, returns, scrap, complaints, repairs, expediting, adjustments, refunds, penalties, wait times, and excess inventory. Cost-improvement efforts should be focused on locating the largest cost contributors to the improvement. Try not to be tempted to spend time chasing insignificant, incremental measures of quality costs. Use a Pareto chart (Chapter 9) to identify the projects with the greatest return on investment; those should be tackled first.

Those establishing a quality-cost measurement system need to keep in mind that the success of such a system is based on using quality-cost information to guide improvement. Each cost has a root cause that can be identified and prevented in the future. Remember that preventing problems in the first place is always cheaper. Those implementing a quality-cost measurement system need to establish a basic method of identifying correctable problems, correcting the problems, and achieving a new level of performance.

Uncovering Quality Costs at JF

LEAN SIX SIGMA TOOLS at WORK

JF works with a number of suppliers. Recently, one of them, a sheet metal stamping firm, bid $75,000 as the price of providing an order of aluminum trays. After stamping a large number of these trays, the company investigated its costs and determined that if production continued, the cost of providing the order would be $130,000. This was nearly double what it had bid. The company wanted JF to cover this unexpected cost. Naturally, JF balked at the higher price. It recommended that the company investigate its costs of quality.

The company set out to figure out what was causing such a dramatic increase in its costs. It investigated its product design and found it to be satisfactory. Prevention costs associated with product design and development were negligible. The customer had not returned any of the completed trays. There were no customer complaints and therefore no corrective-action costs. The investigation showed that external failure costs were zero. Appraisal costs were normal for a product of this type. Incoming material and in-process inspection costs were similar to the amount projected. The product had not required any testing upon completion.

It was only when the company investigated the internal failure costs that the problem became apparent. At first, it felt that internal failure costs were also negligible because there was almost no scrap. It wasn't until a bright individual observed that the operators were feeding the trays through the stamping operation twice that the company was able to locate the problem. When questioned, operators told the investigators that they knew that due to inappropriate die force, the product was not stamped to the correct depth on the first pass. Because the operators knew that the trays would not pass inspection, they decided among themselves to rework the product immediately, before sending it to inspection. The near doubling of the manufacturing costs turned out to be the internal failure costs associated with reworking the product.

Essentially, the company incurred a cost of quality equal to the standard labor and processing costs of the operation. To meet specifications, a significant number of trays were being handled and processed twice. Only by preventing the need for double processing can the costs of quality be reduced.

This type of internal failure cost actually began as a prevention cost. If the manufacturer had investigated the capability of the stamping operation more closely, it would have determined that maintenance work was necessary to improve the machine's performance and therefore improve product quality.

LEAN SIX SIGMA TOOLS at WORK

Costs of Quality Reduced by Problem-Solving Process

Farmer Friendly Inc. manufactures and sells farm and lawn equipment. It follows the Six Sigma methodology, so management understands the importance of tracking the costs of quality. As part of its quality-cost measurement system, Farmer Friendly monitors the costs associated with the prevention of defects, product appraisal, internal and external failure costs, and, to a certain extent, the intangible costs. By monitoring cost of quality values on a regular basis, management is able to spot issues before they become a serious problem. Recent charts have shown that despite Farmer Friendly's best prevention and appraisal efforts, external failure costs, as reflected by the number and costs of warranty claims, have risen slightly on its mid-sized farm tractor (Figure 6.9). External failure costs like these are expensive when the cost of complaint investigation, complaint handling, repair, and corrective action efforts is taken into account. Because the customer has found the problem, there are also intangible costs—lost sales, loss of customer goodwill, and customer dissatisfaction—to be considered.

Warranty claims show that customers are complaining that the steering wheel pulls to the right while the tractor is in motion. No in-depth investigation occurred at the dealership; most repair personnel opted to merely realign the front wheels. This would satisfy the customer briefly. However, most problems recurred. Several customers have brought their tractor in for alignment issues as many as three times. One tractor, deemed irreparable, was bought back by the company for the original purchase price. Current external failure costs associated with this problem are shown in Figure 6.10.

Recognizing the potential liability costs associated with steering problems, though the number of warranty claims is still low, Farmer Friendly immediately assigned a process improvement team to look into the situation.

Following a problem-solving approach similar to that outlined in Chapter 9, the team began by studying the warranty information provided from the field. To aid them in their search for a root cause, they created a cause-and-effect diagram of the reasons why *steering pull* may happen (Figure 6.11). The discussions surrounding the cause-and-effect diagram, combined with additional information on key product dimensions, pointed to suspension control arm bushing angularity discrepancies (Figure 6.12). The parts saved from the repaired tractors were studied, and this bushing angularity discrepancy was verified.

Having found the problem and in order to determine how this problem could occur, the team created a WHY-WHY diagram (Figure 6.13) and mapped the process (Figure 6.14). These documents pointed to a loose work-holding device, which worked intermittently. Without the support of the work-holding device, the control arm bushing could not be correctly mounted, thus affecting the angularity of the part. The team reported its findings by completing the Warranty Investigation Report shown in Figure 6.15.

Countermeasures were implemented to repair the work-holding device. The operator was also alerted to the important role the device plays in making the part to specification. To drive home its importance, information concerning the costs of quality was also shared. Careful tracking of cost of quality data alerted Farmer Friendly to a problem that had the potential to become quite costly.

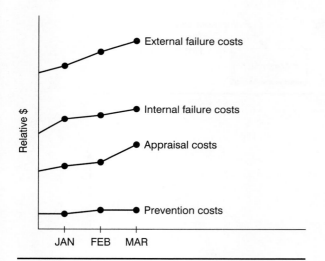

FIGURE 6.9 Warranty Claims

Returned goods (Buy back)	$31,000
Warranty costs	$600/tractor + $60/hr × 4 hr labor
Customer complaint processing	$200/complaint
Replacement parts/alignment	$500
Corrective actions (team)	$8,000 (5 people × 80 hrs × salary)
Complaint investigation	$1,200 to date

FIGURE 6.10 External Failure Costs

(Continued)

CHAPTER SIX

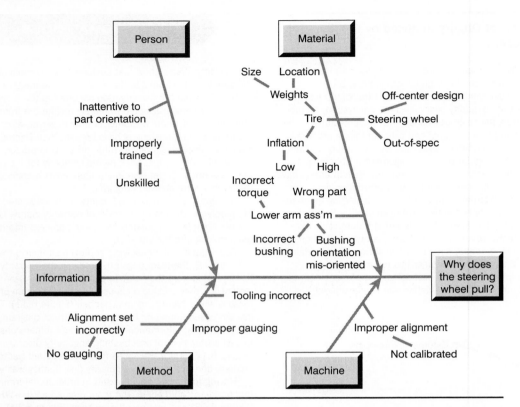

FIGURE 6.11 Cause-and-Effect Diagram

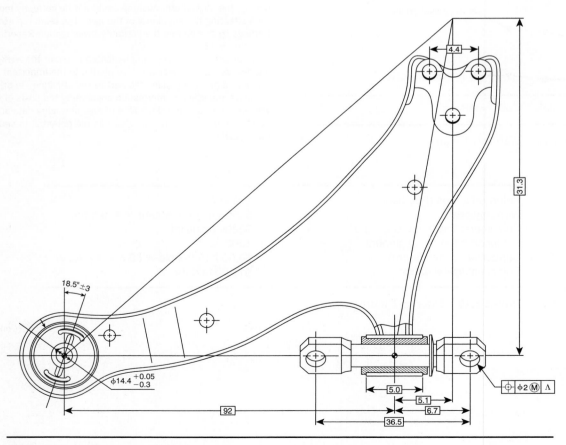

FIGURE 6.12 Product Dimensions

Costs of Quality

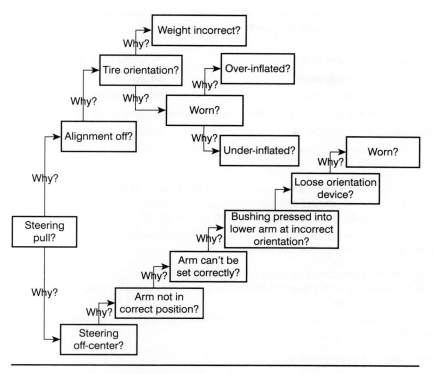

FIGURE 6.13 WHY-WHY Diagram

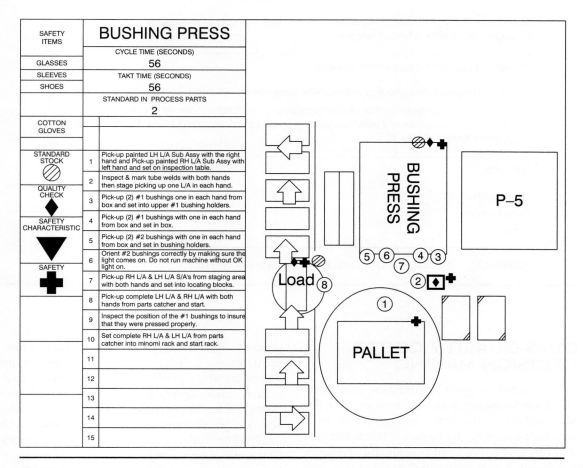

FIGURE 6.14 Process Map

(Continued)

Warranty Investigation Report

Subject: **Steering wheel pull**   Responsibility: Team 12
Model Name: Mid-size Tractor
Products: Steering Assembly   Part Numbers: 1, 2, 3, 5, 8, 11, 19
Problem: Customers requesting tractor service for steering wheel pull to right.
Problem Identified: 10/21/05   Solution Due Date: 12/1/05

Observation in field:

Mechanics report that when driven, the tractors pull slightly to the right, causing the tractor to steer off-center. No abnormal sounds noticed. No vibrations either. Visual inspection of steering components revealed no immediately visible cause.

Temporary countermeasure/containment:

Steering column/component replacement in affected tractors. Check tractor production records to determine if this is an isolated problem or one affecting a series of tractors produced at the same time.

Investigation:

See attached charts and report

Cause:

Suspension control arm bushing angularity discrepancy due to working holding device slippage. Improperly oriented raw part.

Permanent countermeasure/containment:

Process changes to ensure that the work-holding device orients part correctly.

Dealer alert sent requesting that dealers conduct a bushing angularity check if tractor steering pull is not corrected after completing other troubleshooting steps. Report findings to Farmer Friendly, Team 12.

Sign-offs:

General Manger:

Team Leader:

Date:

FIGURE 6.15   Warranty Investigation Report for Tractor Steering Wheel Pull

UTILIZING QUALITY COSTS FOR DECISION MAKING

All initiatives undertaken by an organization must contribute to the financial success of the organization. Lean Six Sigma leaders understand that better communication with management happens when quality data is translated into business and financial terms. Quality goals and objectives should be included in an organization's business plan, but they must be stated in financial terms. Quality costs can be used as a justification for actions taken to improve the product or service. Typically, investments in new equipment, materials, or facilities require the project sponsor to determine which projects will provide the greatest return on investment. These calculations traditionally include information on labor savings, production time savings, and ability to produce a greater variety of products with better quality. The "better quality" aspect of these calculations can be quantified by investigating the costs of quality, particularly the failure costs. It is important to determine the costs of in-process and incoming material inspection, sorting, repair, and scrap as well as the intangible costs associated with having a nonconforming product or service reach the customer. Making a decision with more complete quality information, such as product appraisal costs, can help determine the true profitability of a product or service.

Once quality costs are identified, those reviewing the project can determine whether the money for the project is being well spent to prolong the growth of the company. Identifying and quantifying quality costs has a twofold benefit. Cost savings are identified and quality is improved. By improving the quality performance of a company, the company also improves its quality costs.

SUMMARY

Developing and deploying an effective quality-cost management system is critical to success in today's global work environment. Determining and then dealing with the costs associated with external and internal failures can significantly contribute to an organization's overall financial health. Systematically enhancing the systems that prevent errors leads to great rewards. An increase in prevention costs leads to much more significant decreases in failure costs. The savings gained by preventing quality problems in the first place can be used for additional growth and profitability. Organizations focused on prevention can take advantage of these improvements to enhance customer satisfaction and thus their position in a global marketplace.

TAKE AWAY TIPS

1. Collecting quality costs provides a method to evaluate system effectiveness.
2. Quality costs can serve as a basis or benchmark for measuring the success of internal improvements.
3. Prevention costs are the costs associated with preventing the creation of errors or nonconformities in the first place.
4. Appraisal costs are the costs incurred when a product or service is studied to determine whether it conforms to specifications.
5. Internal failure costs are the costs associated with errors found before the product or services reach the customer.
6. External failure costs are the costs associated with errors found by the customer.
7. Intangible failure costs are the costs associated with the loss of customer goodwill and respect.
8. Once quantified, quality costs enhance decision making if they are used to determine which projects will allow for the greatest return on investment and which projects are most effective at lowering failure and appraisal quality costs.
9. Quality-cost information should be used to guide improvement.

CHAPTER PROBLEMS

1. Which of the following statements about quality costs is true? Why or why not?
   a. Quality-cost reduction is the responsibility of the quality control department.
   b. Quality costs provide a method of determining problem areas and action priorities.
   c. The efficiency of any business is measured in terms of the number of defective products.
2. There are four specific types of quality costs. Clearly define two of them. Give an example of each kind of quality cost you have defined.
3. What is a prevention cost? How can it be recognized? Describe where prevention costs can be found.
4. Describe the two types of failure costs. Where do they come from? How will a person recognize either type of failure cost?
5. How do the four types of quality costs vary in relation to each other?
6. Describe the relationship among prevention costs, appraisal costs, and failure costs. Where should a company's efforts be focused? Why?
7. A lamp manufacturing company has incurred the following quality costs. Analyze how this company is doing on the basis of these costs. Creating a graph of the values may help you see trends. Values given are in percentage of total cost of lamp.

	Costs			
Year	Prevention	Appraisal	Internal Failure	External Failure
1	1.2	3.6	4.7	5.7
2	1.6	3.5	4.3	4.6
3	2.2	3.8	4.0	3.8
4	2.5	2.7	3.5	2.2
5	3.0	2.0	2.8	0.9

8. How can quality costs be used for decision making?
9. Where should dollars spent on quality issues be invested in order to provide the greatest return on investment? Why?
10. Choose two industries and list specific examples of their costs of quality.
11. Why is the following statement true? The further along the process that a failure is discovered, the more expensive it is to correct.
12. What should a quality-cost program emphasize?
13. What are the benefits of having, finding, or determining quality costs?
14. On what premise is the strategy for using quality costs based?

15. For the Farmer Friendly case, discuss how quality costs were used for decision making. Be sure to establish a relationship between the identified quality costs and the decision they affected.

16. Companies that choose to ignore the costs of quality end up spending 20 to 30 percent of their total costs on reworking and redoing. These companies are also ones that may face product liability suits. Locate a company and review its costs of quality. Answer the following questions:
    a. Identify as many different types of quality costs as you can. Categorize your costs of quality according to the five categories presented in the text.
    b. Discuss how quality costs can be used for decision making in this company. Be sure to establish a relationship between the identified quality costs and the type of decision they play a role in for this company.
    c. In your review of the company, were you able to find any information about the costs this company has or will incur related to product recalls, warranty claims, product liability suits, or any other costs of quality? If so, provide a summary of those costs and their sources.

17. For each of the following activities, identify whether the costs incurred should be allocated to the prevention, appraisal, failure (internal, external), or intangible costs category:
    a. Customer inquires about incorrect billing statement.
    b. Maintenance technician is unable to gain access to customer's premises during a service call because the service is not being performed when scheduled and no one notified the customer.
    c. Workers check/test their work before allowing it to leave their work area.
    d. Management carries out a quality systems audit.
    e. The accounts department is having difficulty obtaining monthly figures from the production department.
    f. A customer's purchase breaks because of inadequate training in the use of the item and must be replaced.
    g. Management must deal with written complaints from a customer about continuous poor service.
    h. A technician keeps an appointment after driving across town to pick up equipment from an outside supplier upon finding it out of stock in his own warehouse.
    i. An instruction manual and set of procedures are developed for maintaining a new piece of equipment.
    j. The company conducts a survey of customer requirements.

PROCESS PERFORMANCE MEASURES

We cannot manage what we do not measure.

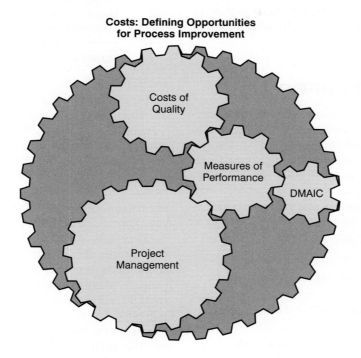

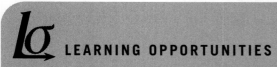

LEARNING OPPORTUNITIES

1. To show the link between achieving lean Six Sigma performance levels and the use of measures
2. To familiarize the reader with the design of measures

DEFINING PROCESS MEASURES

Very simply, what gets measured gets done. For that reason, measures play a significant role in lean Six Sigma organizations. Lean Six Sigma organizations recognize that if they cannot measure, they cannot manage. They understand that measures tell them how they are progressing toward their strategic goals and objectives. They know that if they do not have sufficient information about a process, product, or service, they cannot control it. And if a process cannot be controlled, the organization is at the mercy of chance. Lean Six Sigma organizations manage their employees, processes, scheduling, production cycle times, supplier partnerships, shipping, and service contracts far better than their competition. Dr. Deming's favorite question was, "How do you know?" Measures of performance answer this question. Effective performance measurement systems are used for understanding, aligning, and improving performance at all levels and in all parts of the organization.

Measures are indicators of performance. Properly designed measures, by comparing past results with current performance, enable leaders to answer the questions, "*How are we doing?*" and "*How do we know how well we are doing?*" Effective leaders understand that without measures, they don't know what they don't know. They can't act on what they don't know. They won't search for answers unless they question, and they won't question what they don't measure. Leaders receive a wide variety of information each day. Having enough information usually isn't the problem. Having *useful* information is. Humorist Mark Twain once said, "Data is like garbage. You had better know what you are going to do with it before you collect it." As discussed in Chapter 3, effective strategic plans contain performance measures chosen for their ability to quantify information about the critical success factors. Information concerning these critical success factors enables leaders to make better, more informed decisions. Measures of performance enable effective organizations to define the meaning of success numerically. As Lord Kelvin, the British physicist, stated in 1891, "When you can measure what you are speaking about, and express it in numbers, you know something about it; but when you cannot measure it, when you cannot express it in numbers, your knowledge is of a meager and unsatisfactory kind: it may be the beginning of knowledge, but you have scarcely, in your thoughts, advanced to the stage of science."

MEASURES OF PERFORMANCE

Employees at lean Six Sigma organizations recognize the importance of working on activities valued by leadership. Leaders use measures of performance linked to strategic plans to communicate what activities are important. Good measures of performance are designed based on what is valued by the organization and its customers. Well-designed measures encompass the priorities and values of both. Essentially, performance measures enable an organization to answer the following questions:

- How well is something performing its intended purpose?
- Is the organization able to measure the impact of the changes being made?
- How does the organization know that it has allocated its assets correctly?

Well-constructed measures are aligned with the strategic goals of an organization as well as with its customers' priorities. Willingness to use performance measures increases when the measures are relevant to the organization operationally and, where applicable, to the individual personally. Usability is also a function of understandability. Measures that are clearly written and focused are more powerful than ones that are oblique or lengthy. Return to Figure 3.5 in Chapter 3 and see how measures are linked to goals and objectives.

Customer needs, wants, requirements, and expectations must translate into measures of performance. Effective organizations ask themselves the question, "Have we identified, selected, and measured factors that reflect what the customers want?" Critical to the success of any organization is the ability to determine what its customers want and to find the best way to get it to them. Well-designed measures align strategic objectives with the priorities of the customer. These indicators measure what is of value to the customer. Measures should also provide information about undesired customer outcomes that must be avoided, eliminated, or at least minimized. Measures should also define the product, service, or process characteristics valued by the organization's customers.

Measures of performance can be found in two categories: processes and results. Processes exist to get work done. They are the activities that must take place to produce a product or provide a service. Because processes are how organizations do their work, process measures monitor operational activities, or how the work is done (Figure 7.1).

Results relate to both organizations and their customers. To an organization, results are the objectives the organization wants to achieve. From a customer point of view, results represent what the customer hopes to obtain by doing business with the organization, whether it be a product or a service. Performance measures related to organizational results focus on strategic intent (Figure 7.2). Performance measures related to customer results concentrate on the attributes of products and services (Figure 7.3). Products and services are created by the organization and purchased by customers. Products can be tangible manufactured items or information products such as reports, invoices, designs, or courses. Services received by customers are even more varied. Organizations may provide anything from information to dental work to entertainment products such as movies, games, or amusement park rides.

Schedule/delivery performance
Throughput
Quality
Downtime
Idle time
Expediting costs
Inventory levels
Work-in-process levels
Safety, environment, cleanliness, order
Use of space
Frequency of material movement

FIGURE 7.1 Process Measures

Market share
Repeat/retained customers
Product line growth
Name recognition
Customers/employees ratio
Pretax profit

FIGURE 7.2 Organizational Results Performance Measures Focused on Strategic Intent

---

Scrap
Rework
Downtime
Repair costs
Warranty claims
Complaints
Liability costs

FIGURE 7.3 Customer Results Performance Measures Related to Product or Service Attributes

Traditional Measures	Overall Organization Measures
Revenues	Employee satisfaction, growth, and development
Profits	Customer survey results
Growth	Number of completed improvement projects
Earnings	Cost of poor quality (COPQ) reduction
Return on investment (ROI)	Vendor quality rating
Sales revenue	Return on process improvement investment
Total expenses	Safety, environment, cleanliness, order
Number of customers	Condition and maintenance of equipment/tools
Number of repeat buyers	
Payroll as a percent of sales	
Number of customers per employee	
Number of customer complaints	
Complaint resolution rating	
Schedule achievement	

FIGURE 7.4 Traditional and Broader Performance Measures

Measures need to be integrated and utilized throughout the entire organization. Traditionally, organizations have focused their attention on measures related to the financial aspects of doing business, such as revenues, profits, and earnings. Lean Six Sigma organizations realize that the outcomes, processes, products, and services of their business must be measured. Figure 7.4 provides some traditional as well as broader measures.

For both lean and Six Sigma, the three key performance measures are cycle time, value-creation time, and lead time. *Cycle time represents how often a product is completed by a process or the time it takes an operator to complete all the steps in the work cycle before repeating them. Value-creation time is the part of the cycle time during which the work activities actually transform the product in a way the customer is willing to pay for. Lead time is the time it takes for one piece to move all the way through a process from start to finish.*

Leaders of lean Six Sigma projects use key process input variables (KPIV) and key process output variables (KPOV) as measures of performance to ensure alignment among the organization's mission, objectives, and strategic plan. KPIVs and KPOVs are the measures of performance that define a lean Six Sigma project's success numerically by linking it to the strategic plan as well as the organization's bottom line.

 Justifying a Project

LEAN SIX SIGMA TOOLS at WORK

Queensville Manufacturing Corporation creates specialty packaging for automotive industry suppliers. The project team has been working to improve a particularly tough packaging problem involving transporting finished transmissions to the original equipment manufacturer (OEM). Company management has told the team that several key projects, including theirs, are competing for funding. To improve the chances of their project being selected, team members developed strong metrics to show how investment in their project will result in significant cost savings and improved customer satisfaction through increased quality. After brainstorming about this project, the team members developed the following list of objectives and metrics for their project.

By retrofitting Packaging Machine A with a computer guidance system, the following will improve:

Capacity:

KPOV: Downtime reduction from 23 percent to 9 percent daily
(downtime cannot be completely eliminated due to product changeovers)

KPIV: Resource consumption reduction
Twenty percent less usage of raw materials such as cardboard and shrink-wrap due to an improved packaging arrangement allowing five transmissions per package instead of four previously achieved

Customer satisfaction:

KPOV: Improved delivery performance
Improved packaging arrangement integrates better with customer production lines, saving 35 percent of customer production time

KPIV: Reduced space required for in-process inventory
Improved packaging arrangement allowing five transmissions per package instead of four previously achieved, saving 10 percent of original factory floor space usage

KPOV: Reduced defect levels due to damage from shipping
Improved packaging arrangement provides better protection during shipment, saving 80 percent of damage costs

Revenue:

KPIV: Reduced costs
$600,000 by the third quarter of the following fiscal year given a project completion date of the fourth quarter of this fiscal year

KPOV: Reduced lost opportunity cost
Increased customer satisfaction will result in an increased number of future orders

This project was selected as a focus for lean Six Sigma improvement because it could be justified by the impact it would have on overall organization performance.

Key process input variables are measures of things coming into a process. They try to predict the result of using a thing. Key process output variables indicate how well things were done. They are measures of things going out of the process. KPOVs are after-the-fact measures and help answer the question, "Were customer needs being met?" Key business metrics include revenue dollars, labor rates, fixed and variable unit costs, gross margin rates, operating margin rates, inventory costs, general and administrative expenses, cash flow, warranty costs, product liability costs, and cost avoidance.

Any organization recognizes the importance of working on activities that enhance bottom line performance. Leaders use measures of performance to select appropriate lean Six Sigma projects.

The Balanced Scorecard method, introduced by Robert Kaplan and David Norton, goes beyond financial measures and integrates measures from four areas. These measures focus on key business processes and are aligned into a few manageable indicators of performance so that management is able to quickly access the short-term and long-term health of the organization. The Balanced Scorecard combines and categorizes process and results measures into four areas: Customer Focus, Internal Processes, Learning and Growth, and Financial Analysis.

When designing measures related to Customer Focus, the emphasis is on connecting with the customer, determining what the customer is interested in achieving, and using that information to translate the mission and strategy statements into specific market-based and customer-based objectives. In other words, what must the organization deliver to its customers to achieve high degrees of satisfaction, retention, acquisition, and eventually, market share? What is the organization going to do for its customers that will make them want to buy from the organization time and time again? These measures identify and monitor the level of value propositions the organization delivers to targeted customers and market segments. *Value propositions are the attributes that the product or service has that meets customer needs, wants, and expectations.* They are indicators that the customers are satisfied. These measures are in three categories:

1. Product or service attributes related to functionality, price, and quality
2. Customer relationship attributes such as delivery, response time, convenient access, responsiveness, and long-term commitment
3. Image and reputation attributes, which are more intangible factors that attract a customer

Internal Processes measures study the effectiveness and efficiency of the processes performed by the organization in order to fulfill customer requirements. When designing these measures, effective organizations identify the internal processes that are most critical for achieving customer and shareholder objectives. Once these key processes have been identified, measures are developed that concentrate on monitoring the improvement efforts in the areas of quality, response time, and cost. This information is used to determine whether the current processes serve the customer effectively, whether these are the best processes, and whether they are operating at their best.

Learning and Growth measures monitor the individual and group innovation and learning occurring within an organization. These measures focus on the ability of the organization to enhance the capability of its people, systems, and organizational processes. Recognizing that this is a long-term effort, the organization places emphasis on the development of employee capabilities, the development of information system capabilities, and employee motivation, retention, productivity, and satisfaction. These measures are used to judge whether the organization has employees who have sufficient information and the right equipment to do their jobs well. The measures also serve those interested in finding out whether employees are involved in the decisions affecting them, and how much recognition and support employees receive. They can also be used to evaluate employee skills and competencies and compare this information with what employees will need in the future. These measures can also be used to monitor the morale and climate within the organization.

Financial Analysis measures are perhaps the measures with which people are the most familiar. These measures are focused on monitoring financial performance. Examples of financial measures include revenue levels, cost levels, productivity, asset utilization, and investment risk. Examples of each of the four types of measures are provided in Figure 7.5.

Measures are not without their pitfalls. The number of measures used by an organization must be reasonable in order to be effective. It is critical to determine what needs to be measured and why it needs to be measured before designing a measure of performance. Measures for measurement's sake must be avoided. Too many measures, or unfocused measures, lead to hesitation in the users. Care must be taken to establish meaningful and actionable measures. For instance, if the measure is stated as "Improve contact with customers," there is little room for defining what is actually being measured or the effectiveness of activities designed to enhance contact with the customers. Measures need to be specific and quantitative to be effective. For example, "Visit five customer accounts per month" or "Decrease scrap on Line 2 by 5 percent in six months" is significantly more specific and actionable as well as more measurable when compared with the generic statements "improve contact" or "reduce waste." It is also critical to ensure that the right things are being measured. Sometimes the choice is made to measure the easy actions rather than the important actions. This leads to measures that are merely counts of actions rather than indicators of business opportunities. In other cases, measures may be too abstract. Though the measure may sound good on paper, it cannot be achieved because few understand what it means. One key question to ask is, "If only one item, parameter, or type of result could be measured to assess the performance of the organization at a given level, what would it be?"

When developing measures, consider the following:

What does the organization need to know?

What is currently being measured?

Process Performance Measures

Financial Measures	Internal Measures
Sales revenues	Payroll as a percent of sales
Total expenses	Number of customers per employee
Pretax profit	Cost of poor quality
Return on investment	Employee survey results
Customer Measures	Throughput yield
	Quality level
Number of customers	Product/service cost
Number of repeat buyers	Product/service cost
Customer survey results	Productivity
Number of customer complaints	Employee morale
Complaint resolution rating	**Learning and Growth Measures**
Name recognition	
Price differentials	Number of teams
On-time delivery	Number of completed projects
Response time	Number and percentage of employees involved
	Number and percentage of employees involved in educational opportunities

FIGURE 7.5 Measures of Performance

Indicators or Measures of Performance

LEAN SIX SIGMA TOOLS at WORK

Leaders of PM Printing and Design created their strategic plan in order to communicate to their employees the importance of creating and maintaining a customer-focused process orientation as they improve the way they do business. Their indicators or measures of performance reflect this desire.

Customer Measures
Results measures: Overall customer satisfaction (KPOV)
Market share (KPOV)
- Number of customers
- Number of repeat customers
- Number of new customers

Process measures: Time to complete customer order (KPIV)

Financial Measures
Results measures: Cost per printed unit (impression) (KPOV)
Profitability (KPOV)
Return on investment (KPOV)

Process measures: Cost avoidance (performing work in-house versus contracting work) (KPIV)

Internal Measures
Results measures: Improvements in hours paid vs. hours billed (KPIV)
Process measures: Improvements in job turnaround time (KPIV) (cycle time reduction/removal of non-value-added activities)
Improvements in billing lag time (KPIV) (cycle time reduction/removal of non-value-added activities)
Improvements in first time through quality (KPIV) (reduction in rework/scrap)

Learning and Growth Measures
Results measures: Improvements in employee retention (KPIV)
Process measures: Progress toward cross-training goals for critical processes as identified by customers (KPIV)

How does what the organization needs to know compare with what is currently being measured?
How is this information being captured?
Is the information currently being captured useful?
Is the information currently being captured being used?
What old measures can be deleted?
What new measures are appropriate?
Do the identified, selected, and measured factors reflect what the customers need, require, and expect?
Are these measurements being captured over time?
Are these selected factors actionable within the organization?
Can the impact of the changes made be measured?
Have the organization's assets been allocated correctly?

Measures must be defined in objective terms. The best measures are easy to express as a numerical value. Valuable measures show progress toward the milestones of strategic goals and objectives. Measures are useful and effective when

> ## Measures of Patient Care
>
> **LEAN SIX SIGMA TOOLS at WORK**
>
> In response to customer complaints about substandard treatment, hospitals are taking steps to measure patient satisfaction and make improvements based on this information. To create more patient-centered care, hospitals need to understand patient experiences by looking at processes from the patients' point of view. Measuring patient-centered care is more complex than the traditional measures of performance that hospitals have used in the past, requiring a new set of performance measures. The new measures of performance focus on more subjective patient issues, such as how patients feel about the way they are treated, and how the kindness and courtesy of the staff affected their experience. Previously, most of the measures have focused on safety and quality standards, such as how many infections were contracted during patient hospital stays or the number of heart-bypass surgeries or C-sections a hospital performs in a year. Though there is a tendency to view the "soft stuff" as unimportant, more and more hospitals are developing measures of performance that focus on the patient. These patient-care focused measures of performance may include how well-informed of their condition and treatment patients felt they were, the patient's physical comfort and pain relief, the emotional support they received, the respect they received for their preferences and expressed needs, and how well hospitals prepared them for caring for themselves at home. Hospitals use this information to improve the level of care they offer their patients. Some examples of measures of performance are given in Figure 7.6.
>
> | Financial Measures | Internal Measures |
> |---|---|
> | Total expenses | Medicine-dispensing error rate |
> | Return on investment (ROI) | Number of patients per employee |
> | **Customer Measures** | Length of stay |
> | | Quality level |
> | Number of patients treated | Safety rating |
> | Patient survey results | Number of surgeries |
> | Number of patient/customer complaints | Infection rates |
> | Complaint resolution rating | **Learning and Growth Measures** |
> | Kindness rating | |
> | Staff courtesy rating | Number of teams |
> | Patient information availability | Number of completed projects |
> | Patient comfort rating | Number and percentage of employees involved |
> | Patient pain relief | Number and percentage of employees involved in educational opportunities |
> | Patient readmittance rate | |
>
> **FIGURE 7.6** Measures of Performance for a Hospital

those using them can identify how the measure enables them to make better decisions in relation to their day-to-day activities.

MEASURES OF PERFORMANCE AND GAP ANALYSIS

One of the chief goals of a measurement system is to provide leaders with a multidimensional and qualitative view of their organization. A measurement system is a critical element in the strategic planning process because it allows an organization to measure progress toward goals and objectives. Performance measures are decision tools that enable leaders to link their strategy with day-to-day operations. Effective organizations measure the performance of areas that the organization values the most. Measurement systems allow effective organizations to:

- Determine that a gap exists between desired and actual performance
- Determine the root cause of the gap
- Determine the necessary corrective action to eliminate the root cause of the gap
- Determine whether the corrective actions eliminated the root cause and closed the gap between the actual and desired performance

Gap analysis—the study of the difference between actual and planned progress—allows effective organizations to learn where their efforts should be focused concerning their strategic plans. Gap analysis is a critical component of a measurement system because it drives organizational change. For instance, suppose an organization's strategic plan focused on five activities: health and safety, quality, cost reduction, throughput, and environmental issues. Then in order to know whether progress was being made toward these goals, measures must be in place. Effective organizations utilize measures to change the way they do business. Measures of performance that reinforce the strategic plan enable these organizations to guide the direction their business is taking.

Information is instrumental to guiding an effective organization. Leaders of effective organizations analyze

	→ Goal A	→ Objective A1, A2, ...	→ Measures A1, A2, ...	→ Action (Project) A1, A2, ...
Critical Success Factors as Identified by Customers → Vision → Mission	→ Goal B	→ Objective B1, B2, ...	→ Measures B1, B2, ...	→ Action (Project) B1, B2, ...
	→ Goal C	→ Objective C1, C2, ...	→ Measures C1, C2, ...	→ Action (Project) C1, C2, ...
	→ Goal D	→ Objective D1, D2, ...	→ Measures D1, D2, ...	→ Action (Project) D1, D2, ...

FIGURE 7.7 Strategic Planning Diagram

information provided by performance measures to assess and understand their organization's overall performance. This information enables leaders to make appropriate decisions about the actions they must take to ensure organizational success. Performance measures can be used to align day-to-day activities with the strategic plan and improve organizational performance at all levels and in all parts of the organization (Figure 7.7).

The quality management system and lean Six Sigma programs used by effective organizations emphasize good management of information and knowledge. Through the use of performance measures, leaders use the systems and programs to select projects, ensuring that all decisions are tied to business results. The measures used by these systems and programs seek to optimize overall business results by balancing cost, quality, features, and availability of products and their production. The measurement phase of any problem-solving process should require a complete evaluation of key process variables through the use of measures. This enables those working on projects to understand how the measures work together within a complex system to produce good products in a timely manner, at the best cost, and in a way that meets the needs of the customer and the company.

Measuring Emergency Room Performance

LEAN SIX SIGMA TOOLS at WORK

When a medical emergency occurs, the last thing a person wants to do is wait in an emergency room to be seen. On average it takes 49 minutes for a doctor to see a patient in a U.S. emergency room, according to studies by VHA Inc., an alliance of community hospitals that tracks such information. Hospitals are responding to patients' desire to be seen quickly by listening to the customer, acting on the information they gather, and making significant changes to the way they do business. Because of these efforts, short waits in the emergency room may become the norm rather than the exception. As reported in the *Wall Street Journal*, July 3, 2002, one hospital, Oakwood Hospital and Medical Center in Dearborn, Michigan, went so far as to promise that anybody taken to the emergency department would be seen by a doctor within 30 minutes—or he/she would get a written apology and two free movie passes.

To reduce emergency room wait times, hospitals may seek to improve patient check-in procedures, billing, record keeping, and laboratory processes; they may also upgrade their technical staff. Measures of performance guide improvement efforts by enabling hospitals to track indicators such as patient wait time before being seen; causes for delay, such as incomplete paperwork or doctor unavailability; and the amount of time a physician spends with patients. These measures also serve as a guideline for hospitals, pointing out where improvements need to be made and how effective their changes have been. These measures will also tell a hospital whether it can actually offer the guarantee. Some hospitals are redesigning their emergency departments along a dual-track system to speed up their handling of lesser problems. Acute-care beds are set aside for the real emergencies, and an adjacent "fast track" section is designed to treat lesser emergencies and to get patients treated and released quickly. This type of arrangement has been shown to reduce the length of stay in the emergency unit by as much as half. Since Oakwood implemented the 30-minute guarantee, at all four of Oakwood Healthcare System's hospitals, patient satisfaction levels have soared. The performance measures revealed that only 0.9 percent of the 191,000 emergency room patients (about 1,700 people) were eligible for free tickets. By studying its performance measures, Oakwood was able to determine that the average wait is now 17 minutes between patient arrival and a physician's examination. Who knows—with further process improvements, maybe they will start offering a 15-minute guarantee.

SUMMARY

If you don't drive your business, you will be driven out of business.

B. C. Forbes

Lean Six Sigma organizations, through the use of performance measures, select projects that are tied to customer requirements and business results. These measures help optimize overall business results by balancing cost, quality, features, and availability considerations for products and their production.

Measuring Performance

JF leaders develop measures of performance to support the goals and objectives that they laid out in their strategic plan. Their measures relate to all aspects of their business: customers, employees, and finances. Figure 7.8 provides a brief summary of several of their goals and measures. Using these measures, they are able to respond to such questions as:

How do customers view JF?
What must JF excel at as an organization?
How can JF's people continue to improve and enhance customer value, both internally and externally?
How does JF appear to our investors?

Improvements and changes cost money. JF's investors have raised concerns over the money being spent, but management response is to show how the cost of not running parts, not making quality parts, and not meeting customer expectations far outweighs wise improvements to the business. As effective leaders, they watch their cash flow. Positive cash flow means that the organization has money to grow on. They won't need to borrow for growth, and they have enough to meet their debt obligations. Several of their measures support this, including cash flow and inventory. Cash flow comes from what an organization sells. It represents money coming in. Once the product is complete, they ensure that customers are paying on a timely basis. A high value in the accounts receivable column isn't always good. Effective organizations can collect on these accounts. They don't have long-past-due accounts.

Because inventory that sits around unused ties up money, effective leaders find improvements to increase inventory turnover. This means increasing the flow of product through the plant. It also means that material and component purchasing is timed carefully with customer orders. In service-related activities, careful management of human resources and the supplies or information they need is important. Having who you need and what you need in order to service customers means that customers are served more quickly. Serving more customers quickly means that more customers can be served.

Their measures of performance support investment in the organization, too. Sales have risen from $4 million to $8 million a year. The number of new, long-term customers has risen from 30 to 36. The number of orders handled weekly by the plant has risen from 30 to 60 per week. Lead time has fallen from four to six weeks to nearly two weeks. First-pass quality has increased from 80 percent to 87 percent. Profit has risen from 2 percent to 8 percent, a figure that includes money spent on improvements. The number of employees has risen from 26 to 38, primarily due to adding a second shift to handle the increase in customer orders. Even with the additional employees, the ratio between the JF's sales and the number of people they employ is strong, $4 million/26 to $8 million/38. By making improvements to their billing and account receivable processes, despite the significant increase in sales, the dollar amount of outstanding receivables has not increased.

Health and Safety

Goal: Reduce reportable and lost-time incident rate by 33%
  Measure: Lost-time incident rate
Goal: 100% employee compliance in audit of ear and eye protection
  Measure: Count of employees without ear and eye protection during random monthly inspection

Quality

Goal: Reduce PPM by 50% from last year's level
  Measure: PPM defect rate
Goal: Complete PDSA problem-solving training for all members of engineering department in four weeks
  Measure: Number of people who have completed training per week

Cost

Goal: Reduce waste from last year's level by 33%
  Measure: Scrap rate
  Measure: Rework rate

Throughput Objectives

Goal: Up-time objective for Line 1—improve 10% from actual performance level of 70% to 80% by year-end
  Measure: Up-time performance level for Line 1

Environment

Goal: Reduce PPM exhaust rate by 33% by year-end
  Measure: PPM exhaust rate

FIGURE 7.8   JF Measures of Performance

TAKE AWAY TIPS

1. The measurement phase of any problem-solving process should require a complete evaluation of key process variables.
2. Measurement systems allow lean Six Sigma organizations to perform a gap analysis, studying the difference between actual to planned progress by:
   - Determining that a gap exists between desired and actual performance
   - Determining the root cause of the gap
   - Determining the necessary corrective action to eliminate the root cause of the gap
   - Determining whether the corrective actions eliminated the root cause and closed the gap between the actual and desired performance
3. Quality costs can be used as justification for actions taken to improve products and services.
4. Identifying quality costs will allow management to judge improvement investments and profit contributions.
5. When designing a measure, ask: If only one item, parameter, or type of result could be measured to assess the performance of the organization at a given level, what would it be?

CHAPTER QUESTIONS

1. Now that you have read the chapter, what does B. C. Forbes mean by, "If you don't drive your business, you will be driven out of business"?
2. How do effective organizations use performance measures?
3. What is the difference between process and results measures?
4. Why is an effective performance measurement system necessary?
5. What performance measures does the organization you work for use?
6. How does your organization use performance measures?
7. Create a set of measures based on the Balanced Scorecard and the Lean Six Sigma Tools at Work feature Indicators or Measures of Performance for a fast-food restaurant.
8. Create a set of measures based on the Balanced Scorecard and the Lean Six Sigma Tools at Work feature Indicators or Measures of Performance for a movie theater.

CHAPTER EIGHT

PROJECT MANAGEMENT

How do you eat an elephant?
One bite at a time.
Unknown

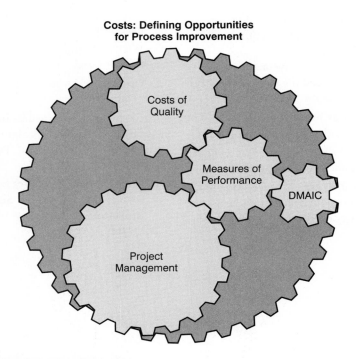

LEARNING OPPORTUNITIES

1. To introduce the topic of project management
2. To familiarize the reader with project proposals
3. To familiarize the reader with project plans
4. To familiarize the reader with project schedules
5. To introduce the concept of Gantt charts
6. To introduce the concept of PERT charts
7. To introduce the concept of critical path method

LEAN SIX SIGMA PROJECTS

Lean projects focus on reducing waste and decreasing lead times. Six Sigma projects focus on reducing sources of process or product variation. Depending on the situation, a project may focus on applying lean tools and techniques, take a Six Sigma analysis approach, or try a combination of both. Lean projects are chosen because of their ability to improve process performance by eliminating waste and non-value-added activities. Lean projects attack such waste as overproduction, excess inventory, unnecessary transportation or movement, defects, and process deficiencies. Lean projects seek to improve lead times by eliminating waste and non-value-added activities. Six Sigma projects investigate process variation with an aim to reducing variation and making processes and products more consistent. Lean Six Sigma projects are chosen based on their ability to contribute to the bottom line on a company's income statement by being connected to the strategic plans, objectives, and goals of the corporation. Projects are essential for a lean Six Sigma organization to set its strategic plans in motion. By identifying projects that support meeting the goals and objectives established in the plan, an organization can more toward its ultimate vision of world-class performance.

PROJECT CHARACTERISTICS

Projects come in various shapes and sizes. A project may be large in scale, requiring the full involvement of either full- or part-time project personnel. Projects may be short or long in duration. Kaizen projects (Chapter 13) are often rapid, week-long projects requiring the full attention of those working on the project. Other project work is spread out over time and requires that a person balance his/her everyday work with project work. Regardless of the size or scope of a project, lean Six Sigma projects are selected for their ability to add value to the organization.

Lean Six Sigma projects have three basic characteristics: performance, cost, and time. Performance refers to what the project seeks to accomplish. Unlike day-to-day activities, projects are unique, one-time occurrences created to fulfill specific goals for the organization. Cost refers to the resources needed to complete a project. Most projects must be completed with a limited set of resources. These resources may be related to people, skills, time, money, equipment, facilities, or knowledge. Projects are complex and typically involve a variety of people from a number of areas within an organization. Projects normally have a specific time frame for completion with beginnings and endings clearly defined. Tasks within a project are sequenced, with one phase or activity being completed before another begun.

PROJECT SELECTION

Lean Six Sigma projects are easy to identify. Lean Six Sigma methodologies seek to reduce the variability present in processes, eliminate sources of waste, and focus on customer and environmental issues. Projects that do not directly tie to customer issues or financial results are often difficult to sell to management. Before selecting a project, lean Six Sigma leaders ask a lot of questions:

- Does the project fulfill a key business need?
- Will the process targeted for improvement directly affect the organization's internal or external customers?
- Will the project add value to the organization?
- Will the project solve a significant issue for the organization and its customers?
- Does the project solve a problem or issue that affects quality, productivity, or cost?
- Does the project make financial success—does it provide a realistic return on investment?

Questions like these help leaders select projects that will provide the greatest amount of value to their organization and their customers. When choosing a lean Six Sigma project, care should be taken to avoid poorly defined objectives or metrics. Key business metrics include revenue dollars, labor rates, fixed and variable unit costs, gross margin rates, operating margin rates, inventory costs, general and administrative expenses, cash flow, warranty costs, product liability costs, and cost avoidance.

Lean Six Sigma organizations select projects based on the project's ability to contribute to one or all three of the following: customer perceived value and satisfaction, the organization's financial strength, or operational necessities (Figure 8.1). For customer-focused organizations, many projects will be selected based on their ability to increase customer value, satisfaction, and retention. Projects may be chosen to enable the organization to maintain its competitive edge. These projects may involve the development of a new product or service or an extension to a product line or the development of a product- or service-enhancing feature.

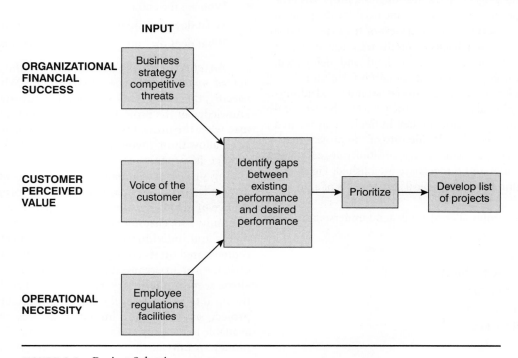

FIGURE 8.1 Project Selection

Computer Technology Transition and Upgrade Project Selection

LEAN SIX SIGMA TOOLS at WORK

Max's Munchies maintains manufacturing facilities in 4 cities and sales offices in 12. The majority of the computers, 1,600 in all, are located at the manufacturing facilities. The sales offices have 400 more. Leaders believe significant benefits can be achieved by enhancing communication and information sharing between their employees worldwide. The company currently uses two computer platforms: desktops and laptops. Though these are limited to a narrow range of models, one of the goals of the Computer Technology Transition and Upgrade project is to create more uniformity company wide by selecting a single desktop model and a single laptop model. The company has selected the desired operating system and software applications that these computers must run.

Max's Munchies recently invited vendors to submit proposals for a technological upgrade of all of its desktop and laptop computers for all of its nationwide locations. The project has been titled Computer Technology Transition and Upgrade.

These projects will ultimately enhance an organization's financial success. Some projects are operational necessities, such as meeting government regulations or repairing or replacing aging equipment. Regardless of the reasons behind a project being selected, effective organizations recognize that a project must be financially sound and provide a payback for their investment.

PROJECT PROPOSALS

*A **project proposal** is a document that provides clear information concerning the goals and objectives that a particular project hopes to achieve.* Along with this information, the project proposal discusses how the project supports the overall mission, goals, and objectives of the organization. An effective lean Six Sigma project proposal provides readers with insight into what needs to be accomplished and how it will get accomplished. Through the use of clearly stated mission, deliverables, goals, and objectives associated with the project, proposals sell the project. As an introduction, proposals provide background information about the need for the project. They contain a description or overview of the expectations of the project, including details about the technical aspects of the project. Essential tasks are outlined and delineated. A thorough project proposal will contain information concerning financial requirements, time constraints, and administrative and logistical support for the project. The proposal contains information about the key individuals associated with the project, including the identity of the project manager. Basic areas of performance responsibility are assigned. Tentative schedules and budgets are established. Figure 8.2 gives a brief summary of the typical components of a project proposal.

The project proposal creates a general understanding of the following:

- What is needed
- What is going to be done
- Why it is going to be done
- Who is going to do it
- When it will be done
- Where it will be done
- How it will be done

PROJECT GOALS AND OBJECTIVES

Effective project proposals state clear project goals. Project goals are established for three reasons:

- To state what must be accomplished to complete the mission
- To create commitment and agreement about the project goals among participants
- To create clarity of focus for the project

Lean Six Sigma project managers recognize the importance of establishing clear goals and objectives for a project. They are careful to make sure that anyone working on the project, regardless of the level of their involvement, understands and supports these goals and objectives. Lean Six Sigma project goals are stated in terms of the users' needs. Key questions to ask to ensure that a project proposal has a customer focus and clear project goals include:

Who are the end users?
What does the end user want from the project?
What does the end user say the project should do?

An effective goal statement is specific, measurable, agreed upon, realistic, and time-framed. When goals are specific, they are so well-defined that anyone with a basic knowledge of the project can understand them and recognize what the project is trying to accomplish. Measurable goals allow those involved in the project to judge how the project is progressing toward its mission. Agreement among project participants about the overall goals of a project is critical because without it, the project has little chance of achieving success. Participants include, but are not limited to, customers, leadership, and affected departments and individuals. Given that a project is to be accepted based on its ability to meet the mission, goals, and objectives of an organization, insisting that a project be realistic seems unnecessary. Here the word *realistic* refers to the need to be aware of the time frame established for the project, as well as the human resources and financing available to the project. Limited resources, whether time or money, or unrealistic expectations given the time, money, and talent available, are detrimental to a project. When a

---

The Project Proposal

General Project Description

Provide a detailed description of the project that includes a statement of the project goals. Describe the major project subsystems or deliverables. Include a preliminary layout design if applicable. Address any special client requirements, including how they will be met.

- What is the purpose of the project?
- What is the scope of the project?
- Why should the project be selected?
- What is the life cycle (time from beginning to end) of the project?
- What is the complexity level of the project?
- Is there anything that makes the project unique?
- What measures will be used to judge the project's performance?

Implementation Plan

The implementation plan section contains a brief description listing the major components of the project, with time estimates for each component. This section also includes preliminary cost estimates and a preliminary schedule for the major project components.

- What are the project deliverables?
- How will these deliverables be met?
- What level of risk is associated with the project?
- What are the costs associated with the project?
- What are the time estimates for the components of the project?

Logistic Support and Administration

This section describes the facilities, equipment, and skills that are needed for the entire project.

- What difficulties may be encountered during the construction or implementation phase of the project?
- What contingency plans exist?
- Who will be the project manager?
- Who else will be necessary to help with this project?

FIGURE 8.2 Project Proposal Guidelines

project is realistic, it is achievable given the resources, knowledge, and time available. Projects are selected to support the overall objectives of an organization. If a project is not completed within established time periods, chances are the organization has missed an opportunity for success. For this reason, it is critical that the project goals have clearly stated time frames for accomplishment of the project.

Project objectives are the specific tasks required to accomplish the project goals. Project objectives clearly align with and support the project goals. In some instances, several project objectives will be necessary to ensure that the organization accomplishes a specific goal. They define who is responsible for accomplishing the goal, what resources are necessary, and what inputs will be needed. As with project goals, project objectives must also be specific, measurable, agreed upon, realistic, and time-framed.

Establishing clear goals with supporting objectives helps lean Six Sigma project managers keep projects on track. Project managers use these goals and objectives as a way to reinforce the commitment of individuals to the project and the team. Well-written goals and objectives enhance communication, keeping everyone associated with a project aware of their role and what they need to do in order to keep the project on track. Goals and objectives also make it easy to see how far the project has progressed and what still needs to be done. Project proposals are submitted to organizational leadership who will judge each project based on its ability to help the organization meet its mission of achieving world-class performance.

> **Computer Technology Transition and Upgrade Project Goals and Objectives**
>
> **LEAN SIX SIGMA TOOLS at WORK**
>
> Max's Munchies has been working with vendors to clearly define the mission, goals, and objectives of the Computer Technology Transition and Upgrade project. The request for proposals sent to vendors states that the mission is:
>
> > to upgrade all desktop and laptop computers to new models so the company can process more advanced applications in accounting, computer-aided design, and new business systems.
>
> This mission is supported by the following goals and objectives.
>
> **Goal: Upgrade Existing Computers**
> Supporting objectives:
> - Specify a single desktop and laptop model that runs the appropriate software
> - Install appropriate software on the new computers
>
> **Goal: Remove Existing Computers**
> Supporting objectives:
> - Delete software on the old computers
> - Dispose of existing computers in an environmentally friendly manner
>
> **Goal: Exchange All Computers in 15 Weeks**
> Supporting objective:
> - Handle the logistics of the exchange with the users

PROJECT PLANS

Once a proposal has been accepted, it becomes the framework or foundation of a project plan. Project plans are significantly more detailed than project proposals. Projects have three interrelated objectives: meeting the budget, finishing on schedule, and meeting the performance specifications. A good project plan enables an organization to accomplish all three. Though a project plan may be modified several times during a project, effective project plans remain key to organizational success. The framework of a lean Six Sigma project often follows Six Sigma's Define, Measure, Analyze, Improve, Control (DMAIC) cycle or Dr. Deming's Plan-Do-Study-Act (PDSA) process (Chapter 9). Project plans provide information about:

- Mission and deliverables
- Specific goals and objectives supporting the mission and its deliverables
- Tasks required to meet the goals and objectives
- Technicalities of who, what, where, when, why, and how
- Schedules—the time needed to support each aspect of the plan
- Resources—what is needed to support each aspect of the plan
- Cost analysis
- Value analysis
- Personnel—who is needed to support each aspect of the plan
- Personnel—responsibilities and assignments
- Evaluation measures for keeping the project on track
- Risk analysis—what could go wrong and how it will be dealt with
- Project change management process

Figure 8.3 gives a brief summary of the typical components of a project plan.

The Project Plan

The Project Plan

In the project plan, the mission and deliverables are clarified. The plan also identifies the who, what, where, when, why, and how aspects of the project. The plan details how the project will be accomplished.

Project Plan Elements

- Overview (the mission and the deliverables; what will the final outcome will be?)
- Objectives (specific objectives supporting the mission)
- General Approach (technicalities of who, what, where, when, why, how)
- Contractual Aspects (specifics of who is required to do what)
- Schedules (what time is needed to support each aspect of the plan)
- Resources (what is needed to support each aspect of the plan)
- Personnel (who is needed to support each aspect of the plan)
- Evaluation Measures (performance, effectiveness, cost; how will the project be kept on track?)
- Potential Problems (what could go wrong? how will it be dealt with?)

FIGURE 8.3 Project Plan Guidelines

Computer Technology Transition and Upgrade Project Plan

LEAN SIX SIGMA TOOLS at WORK

The contractor whose proposal was accepted for the Computer Technology Transition and Upgrade project submitted the following project plan.

PROJECT MISSION

Contractor is responsible for developing and deploying a process for the replacement of all 2,000 computers—laptop and desktop—in a manner that minimizes user and productivity disruption. This includes retrieving the old equipment from the user, setting up the new equipment for the user by loading saved files and new software, and disposing of the old equipment. This service must be accomplished in 15 weeks from the start date.

PROJECT GOALS AND SUPPORTING OBJECTIVES

The specific duties of the contractor are as follows:

Goal: Preparation Services

Objectives:

1. Assist Max's Munchies with the process of receiving and warehousing new computers. Max's Munchies will provide computer hardware and software and network cable connections.
2. Assist Max's Munchies in receiving old computers from users.
3. Perform the work on-site. This includes setup and connection of server cabling and tabletop equipment placement. Space where the contractor can perform the upgrade and exchange must be provided at each Max's Munchies location. Contractor's equipment must be utilized as much as possible during the changeover procedure.

Goal: Deployment Services

Objectives:

1. Burn a CD of the user's data from the old PC for backup. Provide to user.
2. Transfer data from user's old computer to new computer.
3. Load hardware and software onto new computers.
4. Maintain records of the asset exchange with the user.
5. Provide "morning after" assistance to users following setup.
6. Provide a Help Desk while the new computers are being deployed.

Goal: Remediation Services

Objectives:

1. Remove all data/computer programs from old computers before disposal.
2. Dispose of all old computers in an environmentally friendly manner.

Goal: Project Control

Information about project control is provided in a later example.

Goal: Schedule

The mission is to be accomplished in 15 weeks. The schedule is provided in a later Lean Six Sigma Tools at Work feature.

Measures of Performance

- Price based on per computer cost
- Number of user difficulties experienced during changeover
- Amount of user downtime hours
- Time to effect changeover
- Changeovers per shift

Contingency Plans

Five extra computers have been planned for during rollout in case the count of computers exceeds 2,000.

Extra staffing will be available in case the rates of changeover are not high enough.

PROJECT SCHEDULES

Lean Six Sigma project planners need to know how much time is available to complete the project. They also need to know whether any flexibility exists with this deadline. Schedules convert a project plan into an operating timetable. This timetable is used to monitor and control project activity by showing the relationships among dates, times, activities/tasks, and people. For some people, combining longer term project expectations with pressing day-to-day activities is often detrimental to the completion of a project. Some people have a tendency to put the project off just a little longer because of the perception that more time will be available in the future. Project schedules help remind people of the importance of working on the project on a regular basis.

To create a schedule, project planners must know the following: the tasks, the order in which they must be completed, when they must be completed, and the rate at which they can be completed. To schedule a project, project planners should lay out the tasks and activities associated with a project plan according to the time they will take to complete. Once they have created a list of activities, they can derive time estimates for each activity. Starting with the project due date, the activities are stepped backwards in time, eventually reaching a start date as all the activities are accounted for. Realistic time estimates need to be created in order to determine how much total time it will take to complete the activities and tasks associated with a project. As the project progresses, it is not unusual for technical difficulties to arise. These often take longer to solve than originally planned. In other cases, materials and human resources are unavailable or late resulting in changes to schedules and task sequencing that throw off time estimates.

Monitoring a project schedule is often done through the use of checkpoints and milestones. Milestones represent long-term or major events that have been or need to be completed for a project. To reach a milestone, a series of smaller activities or tasks will have taken place. Projects unfold as a logical sequence of activities takes place or tasks are completed; therefore, the relationship between these activities is critical to any project. When taken out of order, these activities or tasks waste time and effort. Checkpoints are smaller points throughout a project that are used to judge how far the project is toward completion.

Computer Technology Transition and Upgrade Project Schedule

LEAN SIX SIGMA TOOLS at WORK

The contractor whose project proposal was accepted for the Computer Technology Transition and Upgrade project submitted the following schedule with the project plan.

SCHEDULE

Major milestones and deployment rates are as follows. This schedule outlines the time frame and pace of activity.

Phase One: Project Preparation and Pilot Rollout

Week 1 Contractor's project manager begins.
Week 2 Pilot rollout begins: 10 PCs per day exchanged. Goal: 50 for the week.
Week 3 Pilot rollout continues: 12 PCs per day exchanged. Goal: 60 for the week.
Week 4 Pause for review of rollout. Some rollout will occur as necessary for priority needs (estimate 20 for the week).
Phase one checkpoint: 130
Phase one milestone: 130

Phase Two: Rollout

Week 5 Main rollout begins. 20/day, 100 for the week.
Week 6 Main rollout: 25/day, 125 for the week.
Week 7 Main rollout: 30/day, 150 for the week.
Week 8 Main rollout: 30/day, 150 for the week.
Phase two checkpoint: 525
Phase two milestone: 655

Phase Three: Rollout

Week 9 Main rollout: 40/day, 200 for the week.
Week 10 Main rollout: 40/day, 200 for the week.
Week 11 Main rollout: 40/day, 200 for the week.
Week 12 Main rollout: 40/day, 200 for the week.
Phase three checkpoint: 800
Phase three milestone: 1,455

Phase Four: Wind-Down

Week 13 Main rollout: 40/day, 200 for the week.
Week 14 Main rollout: 40/day, 200 for the week.
Week 15 Finishing activities: 30/day, 150 for the week.
Phase four total: 550
Phase four milestone: 2,005

Gantt charts, the program evaluation and review technique (PERT), and the critical path method (CPM) are excellent tools for monitoring the complex links of activities associated with a project. Available on Microsoft Project, these charts are invaluable when scheduling a project. *A Gantt chart, shown in Figure 8.4 for the Computer Technology Transition and Upgrade Project, enables the user to keep track of the flow and completion of various tasks associated with a project.* The chart promotes the identification and assignment of clear-cut tasks while enabling users to visualize the passing of time. Divisions on the chart represent both an amount of time and a task to

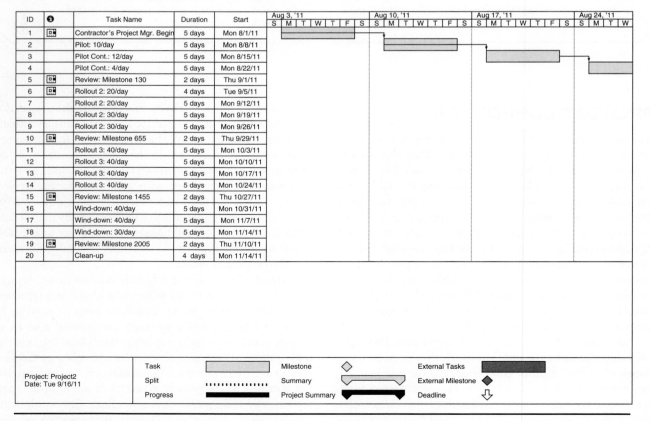

FIGURE 8.4 Gantt Chart

Project Management 75

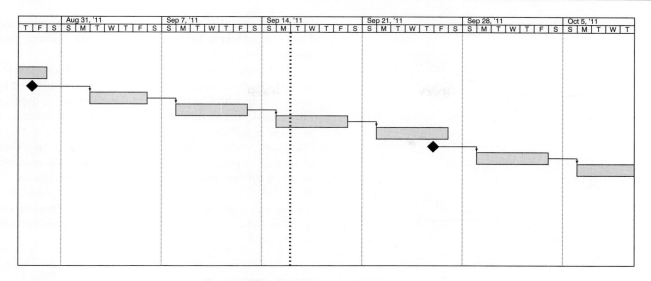

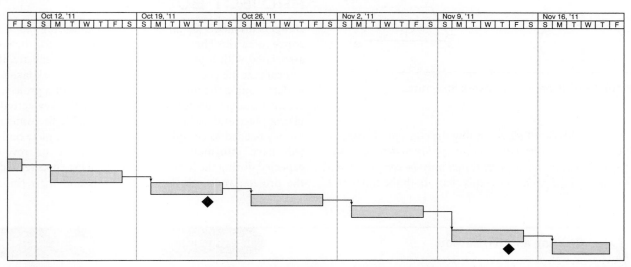

FIGURE 8.4 Continued

be done. A line drawn horizontally through a space shows the amount of work actually done compared to the amount of work scheduled to be done.

A PERT chart improves upon a Gantt chart by showing the relationships between tasks (Figure 8.5). Unlike the Gantt chart, which is a list of tasks, *the PERT chart enables the project to be viewed as an integrated whole.* Because it coordinates and synchronizes many tasks, it is well designed to handle complex projects. To create a PERT network:

1. Compile a list of events/tasks/activities.
2. Determine the relationships between the activities (predecessors, successors).
3. Begin constructing the diagram from the end, working back to the beginning. The key events/tasks/activities identified in step 1 are placed on the diagram between the nodes. Related nodes, those with predecessors and successors, are linked as shown in Figure 8.6.

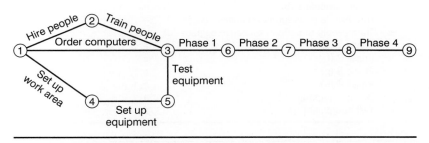

FIGURE 8.5 PERT Network for Computer Installation

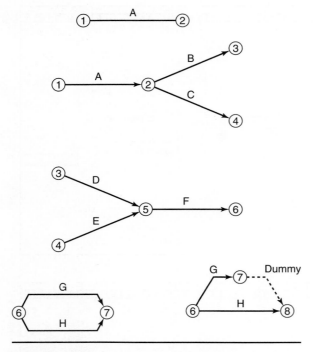

FIGURE 8.6 Common Network Structures

CPM builds on PERT by adding the concept of cost per unit time. *CPM gets its name from its ability to determine the longest series of interrelated events that must be completed in a project, the critical path.* To use this chart, both the times and the costs associated with the activities must be estimated. The Lean Six Sigma Tools at Work feature on the next page describes the creation of a CPM.

If the speed of the project needs to be increased for any reason, then the path can be "crashed" or shortened by determining which activities can be done more quickly. The critical path method also reveals the cost of doing each of these activities more quickly. For instance, if Max's Munchies wishes to complete the project in less than 28 weeks, the project managers must decide whether or not to speed up the ordering of the computers or one of the four phases. Each week of increased speed in order processing will cost them $5,000 compared to increasing the speed of the actual transition at a cost of between $20,000 and $40,000.

PROJECT BUDGETS

Budgets are plans for allocating and monitoring the use of scarce resources. They set the overall estimates for the costs associated with a project. For budgets to be realistic, it is critical that the people actually closest to the work take part in determining the time and money needed for a project. To create a budget, project managers study the key activities taking place and estimate as best as possible the time and money needed to complete the project. A complete budget will have information about any income or revenue expected during the life of the project, all expenses related to the project, and cash flow projections and their timing.

LEAN SIX SIGMA TOOLS at WORK

Creating a PERT Network

The successful contractor for Max's Munchies' Computer Technology Transition and Upgrade project submitted the following PERT network with the project plan. To create the network, the contractor completed the following steps:

1. Compile a list of events/tasks/activities.
   - Set up work area
   - Order computers
   - Hire people to perform exchange/upgrade
   - Train people to perform exchange/upgrade
   - Phase 1
   - Phase 2
   - Phase 3
   - Phase 4
   - Set up equipment
   - Test equipment

2. Determine the relationships between the activities (predecessors, successors).

3. Begin constructing the diagram from the end, working back to the beginning. The key events/tasks/activities identified in step 1 are placed on the diagram between the nodes as shown in Figure 8.5.

Task	Predecessor
Set up work area (a)	—
Order computers (b)	—
Hire people to perform exchange/upgrade (c)	—
Train people to perform exchange/upgrade (d)	c
Phase 1 (e)	a, b, c, d, i
Phase 2 (f)	e
Phase 3 (g)	e, f
Phase 4 (h)	e, f, g
Set up equipment (i)	a
Test equipment (j)	a, i

Creating a CPM

LEAN SIX SIGMA TOOLS at WORK

The steps to create a critical path begin much the same as creating a PERT network. Building on the previous example, to finish the CPM, follow these steps:

1. Compile a list of tasks/activities.
2. Determine the relationships between the activities (predecessors, successors; see Figure 8.7).
3. Determine the costs and times associated with the activities (Figure 8.7).
4. Begin constructing the diagram from the end, working back to the beginning (Figure 8.8).
5. Add up each of the individual paths to determine the critical path. The critical path for this example is 1-3-6-7-8-9 and equals 28 weeks. The other paths are 1-4-5-6-7-8-9 (24 weeks) and 1-2-3-6-7-8-9 (26 weeks).

Task	Predecessor	Weeks Normal	Weeks Crash	Cost Normal	Cost Crash
Set up work area (a)	—	2	2	$25	0
Order computers (b)	—	13	7	$20	$5/wk
Hire people to perform exchange/upgrade (c)	—	6	3	$12	$6/wk
Train people to perform exchange/upgrade (d)	c	5	4	$25	$10/wk
Phase 1 (e)	a, b, c, d, i	4	3	$100	$20/wk
Phase 2 (f)	e	4	3	$100	$20/wk
Phase 3 (g)	e, f	4	3	$100	$30/wk
Phase 4 (h)	e, f, g	3	2	$100	$40/wk
Set up equipment (i)	a	5	4	$13	$2/wk
Test equipment (j)	a, i	2	1	$10	$3/wk

FIGURE 8.7 Precedents, Timing, and Costs (1000)

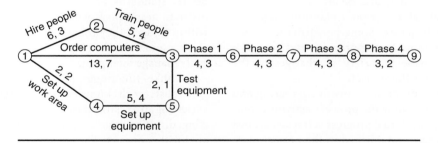

FIGURE 8.8 CPM for Computer Installation

Budgets are not arbitrary; effort should be taken to ensure that realistic cost and timing information has been found. Budgets should be monitored throughout the life of a project to keep project costs on track. Variances should be reported as soon as they are found. These variances can point to problems and recommended actions necessary to keep the project on target. Budgets can enable those working on a project to be aware of how their actions add to or take away from the end result of the project.

PROJECT CONTINGENCY PLANS AND CHANGE CONTROL SYSTEMS

Projects, due to their very nature, are complex. For every project, there is a risk of failure. Organizing a project into manageable sections can reduce the fear of failure as well as the potential for failure. Effective project managers don't forget to ask: "What if?" They work with their project team members to identify potential activities or events that may derail a project. Contingency plans are created to ensure that the project team is ready to handle potential problems. Although all problems cannot be foreseen, a project plan that includes contingency plans keeps the team members flexible and aware that they may be asked to make adjustments to their project plan some time during its lifetime.

Effective project managers recognize that clients make changes to the project as it progresses. All project proposals and plans should contain a description of how requests for changes in the project's plan, budget, schedule, or performance deliverables will be handled. An effective change control system will have steps in place that review the requested changes for both content and procedure and identify how the change will affect the project. This change must be reflected in adjustments to the project's performance objectives, the schedule, and the budget. Once accepted, change orders become part of the overall project plan. Part of the job of a change control system is to clearly communicate any changes to any person or part of the project affected by the change. The best way to ensure that this critical communication occurs is to have all changes approved in writing by all appropriate representatives of the impacted areas. Ultimately, the change should only be made if its benefits outweigh the costs of implementing the change.

PROJECT CONTROL

Throughout a project, effective project managers monitor the progress a project is making toward completion. Performance, cost, and time—the three aspects of a project—all need to be monitored and controlled to ensure project success. Performance of the project refers to the end result and the steps taken to get to it. The performance of a project may be affected by unexpected technical problems, quality or reliability problems, or insurmountable technical difficulties. For instance, a building may have its completion delayed because the foundation must be redone due to technical problems with the concrete being used. Performance of a project may also be affected by insufficient resources brought about by poor planning, logistical problems, or underestimating. Some performance problems may be brought about by the project's end user. The end user may have made a variety of changes to the original specifications for the project.

Project control and monitoring involve gathering and appraising information on how the project's activities compare with the project plan. Actual progress is tracked against the performance measures established in the project plan. These performance measures help a project manager assess how time, money, and other resources have been used to produce the expected outcomes. Many reasons exist why costs need to be monitored during a project. Technical difficulties may require more resources than originally planned. Client-related changes in specifications may have changed the scope of the work significantly, thus affecting the total costs of the project. Or the budget may have been inadequate in the first place due to inadequate estimates, poor projections for inflation, or additional costs due to client-related changes. Costs can also get out of control when the project costs are not watched closely and when corrective cost control is not exercised in time.

By closely monitoring the performance measures associated with the project, an alert project manager can be prepared to respond quickly to deviations in order to keep the project on track and under control. Though very few projects have not had their goals and objectives modified in some way or another from their beginning to their end, careful project control enables a project manager to minimize the effects of these changes on the overall project.

PROJECT MANAGERS

An effective lean Six Sigma project manager achieves the desired results within budget and on time and according to the desired standards. Effective project managers realize that in order to accomplish what needs to be done on time and within budget, they must take time to plan their projects. Once a good plan has been created, effective project managers manage their plan.

Unlike functional managers, project managers are generalists with knowledge and experience in a wider variety of areas. Project managers are responsible for organizing, directing, planning, and controlling the events associated with a project. They deal with budgets and schedules.

Computer Technology Transition and Upgrade Project Control

LEAN SIX SIGMA TOOLS at WORK

The project plan submitted by the contractor contained this section detailing project control goals and objectives.

GOAL: PROJECT CONTROL

Objectives:

1. The contractor will provide a project manager who is responsible for the entire scope of the project and the direction of the contractor's personnel involved in the project. This individual will be the single point of contact.
2. The contractor's project manager will develop the schedule and logistical planning for the Computer Technology Transition and Upgrade project. The schedule will include milestones based on the number of computers to be upgraded. The milestones will become the basis for the forecast on each phase of the project. The schedule will include resource planning indicating the level of staffing necessary to achieve project milestones.
3. The contractor's project manager will schedule and oversee all phases on the Computer Technology Transition and Upgrade project.
4. The contractor's project manager will supervise on-site personnel directly or have assigned supervisory person(s) during periods of critical processes.
5. The contractor's project manager will participate in review and problem-solving meetings scheduled over the course of the project.
6. The contractor will create processes and procedures that meet the objectives of schedule, user satisfaction in the equipment exchange, and inventory control.

Responsibility for the project rests on their shoulders, and they must understand what needs to be done, when it must be done, and where the resources will come from. Throughout a project, the manager will be the one who must clarify misunderstandings; calm upset clients, leaders, and team members; and meet the client's demands while keeping the project on time and within budget. Project managers are responsible for finding the necessary resources, motivating personnel, dealing with problems as they arise, and making project goal trade-offs. In essence, effective project managers are individuals who do whatever is necessary to keep the project on schedule, within budget, and able to meet performance expectations. Project managers must be prepared to make adjustments to schedules, budgets, and resources in order to deal with the unexpected. For this reason, they must be good at recognizing the early signs of problems and be able to cope with stressful situations. As discussed under the heading of project plans, effective project managers utilize the checkpoints, activities, and time estimates established in the project plan to guide those working on the project. Following a clearly laid out project schedule with clearly delineated responsibilities enables effective project managers to keep their projects on track in terms of time, performance, and cost. Clear project plans enable effective project managers to direct people both individually and as a team.

Project managers manage people as well as projects. To do this, effective project managers schedule frequent progress reports. These meetings allow the project managers to react quickly when they recognize that a difficulty has arisen. Effective meetings are essential when working on a project.

Due to the very nature of a project, the people associated with a project are temporary, and the project work they do is often in addition to their regular jobs. If this is the case, how does a project manager maintain the commitment and involvement of these individuals in order to get the project done? A project manager must motivate these individuals. This can be done in a variety of ways, including increasing the person's visibility in the organization. In other words, make sure people working on the project are recognized for their work and what they have accomplished. Project managers also have it within their power to create interesting and challenging possibilities for their team members. As discussed in the chapter on teams, people are much more motivated to perform when their assigned tasks enable them to use and stretch the talents, skills, and knowledge they already possess. Another powerful tool a project manager possesses is praise. People like being recognized publicly and privately for a job well done.

SUMMARY

Projects, when chosen because they support the overall goals and objectives of an organization, fulfill the strategic plan and move the organization toward its vision and mission. Without projects, organizations lack a long-term vision and instead focus solely on doing day-to-day activities. Conversely, poorly managed projects are costly for an organization. A project that does not achieve its goals and objectives, or one that is not aligned with the goals and objectives of the organization, wastes time, money, and other resources.

TAKE AWAY TIPS

1. Lean Six Sigma organizations recognize that projects managed by people with good project management skills are crucial to the success of an organization.
2. Lean Six Sigma organizations select projects that support the organization's mission and goals.
3. Projects have three objectives: performance, cost, and schedule.
4. Project proposals are the documents that sell a project. They concentrate on the goals and objectives that a project hopes to achieve.
5. Project goals create clarity of focus for a project.
6. Project plans provide details related to how a project will be accomplished, including:
    - Mission and deliverables
    - Goals and objectives
    - Tasks
    - Technicalities
    - Schedules
    - Resources
    - Cost analysis
    - Value analysis
    - Personnel
    - Evaluation measures
    - Risk analysis
    - Change management process
7. Project plans should include contingencies.
8. Gantt charts show the flow and completion of tasks within a project.
9. PERT charts show the relationships between tasks in a complex project.
10. CPM builds on PERT by adding the concept of cost per unit time.

CHAPTER QUESTIONS

1. Why are project management skills important to apply in order to be effective?
2. Describe the three basic project characteristics in terms of a project you have worked on.
3. Why is completing a project like eating an elephant?
4. How is a project selected to be worked on?
5. What are the components of an effective project proposal?
6. Consider a project you are working on either at work or at school. Using the guidelines presented in this chapter,

write a project proposal. Why should your project proposal be accepted? Are the reasons for selecting your project apparent in your proposal?

7. What are the components of an effective project plan?

8. Consider a project you are working on either at work or school. Using the guidelines presented in this chapter, write a project plan. Are the project's goals and objectives clearly stated? How do you know?

9. Create a Gantt chart for a project you are working on either at work or at school.

10. What differentiates a Gantt chart from a PERT chart from a CPM?

11. A not-for-profit organization is interested in buying caramels and selling them to raise money. Create a PERT chart for the following information.

		Weeks		Cost	
Task	Predecessor	Normal	Crash	Normal	Crash
Design ads (a)	—	2	2	$250	0
Order stock (b)	—	12	8	$200	$35/wk
Organize salespeople (c)	—	6	3	$120	$60/wk
Place ads (d)	a	3	2	$ 25	$10/wk
Select distribution sites (e)	c	4	3	$100	$20/wk
Assign distribution sites (f)	c, e	4	3	$100	$10/wk
Distribute stock to salespeople (g)	c, e, f	2	1	$100	$25/wk
Sell caramels (h)	e, f, g	5	3	$100	$40/wk

12. Complete a CPM for the information in Question 11.

13. The leaders of the project in Question 12 want to speed up their project by three weeks. What would be the most cost effective way of accomplishing that?

14. What does it mean to keep a project under control? How is a project controlled?

15. What are contingency plans? Why is it important to have contingency plans?

16. What is a change control system? How is it structured? What is it used for?

17. What does it take to be an effective project manager?

PROBLEM SOLVING USING DEFINE, MEASURE, ANALYZE, IMPROVE, CONTROL

Have you ever been lost? What is the first thing you need to do before proceeding? If you answered "find out where you are," then you realize the importance of knowing your current status in order to get to your desired destination. Lean Six Sigma is about knowing where you are and where you are going.

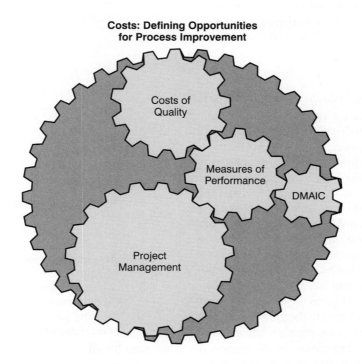

LEARNING OPPORTUNITIES

1. To understand and utilize the define, measure, analyze, improve, control (DMAIC) problem-solving process in order to learn to ask the right questions, present information clearly, and make judgments based on the information
2. To understand and utilize a variety of techniques for effective problem diagnosis and problem solving
3. To learn to diagnose and analyze problems that cause variation in the manufacturing, process, and service industries

PROBLEM SOLVING

Problem solving, the isolation and analysis of a problem and the development of a permanent solution, is an integral part of the lean and Six Sigma methodologies. To achieve the desired Six Sigma levels of performance, people need to be trained in correct problem-solving procedures. Lean Six Sigma practitioners recognize that to locate and eliminate the root cause of a problem, problem solving should follow a logical, systematic method. Other, less structured attempts at problem solving run the risk of attempting to eliminate the symptoms associated with the problem rather than eliminating the problem at its source. Lean Six Sigma problem-solving efforts focus on finding root causes. Proposed solutions prevent a recurrence of the problem. Controls are put in place to monitor the solution. Teamwork, coordinated and directed problem solving, problem-solving techniques, and statistical training are all part of ensuring that problems are isolated, analyzed, and corrected. The tools and techniques presented in this chapter and used in lean Six Sigma problem solving come from many sources. Dr. Deming encouraged the use of Dr. Shewhart's Plan-Do-Study-Act (PDSA) cycle for isolating and removing the root causes of process variation (Figure 9.1). Lean Six Sigma organizations use a modified version of the PDSA cycle: design, measure, analyze, improve, control (DMAIC) (Figure 9.2). Dr. Ishikawa felt that all individuals employed by a company should become involved in quality problem solving. He advocated the use of seven quality tools: histograms, check sheets, scatter diagrams, flowcharts, control charts, Pareto charts, and cause-and-effect (or fishbone) diagrams (Figure 9.3). These tools and others are presented in this chapter and in Chapters 10, 17, and 18.

DEFINE

In problem solving, Six Sigma's DMAIC steps place a strong emphasis on studying the current conditions and planning how to approach a problem. At this stage, problem solvers seek to identify the problem or project, define the requirements, and establish the goals to be achieved. Information concerning the problem(s) may have come from a number of different sources, departments, employees, or customers. Management involvement and commitment is crucial to the success of any major problem-solving process. Managers should participate in the recognition and identification of problems, because they are ultimately responsible for selecting lean Six Sigma projects.

Once a problem situation has been recognized and before the problem is attacked, the lean Six Sigma improvement team is created. This team will be given the task of investigating, analyzing, and finding a solution to the problem situation within a specified time frame. This problem-solving team consists of people who have knowledge of the process or problem under study. Generally, this team is composed of those closest to the problem as well as a few individuals from middle management with enough power to effect change.

MEASURE

Once a project has been identified, team members gather information about the existing process. During this phase of problem solving, critical information is obtained about the key metrics including key process input and output variables. This information is tracked over time and is used to identify potential root causes of the problem. Several tools can aid in this process including check sheets, Pareto charts, process flow diagrams (Chapter 10), cause-and-effect diagrams, WHY-WHY diagrams, and force field diagrams.

Technique: Check Sheets

Several techniques exist to help team members determine the true nature of their problem. The most basic of these is the check sheet. *A **check sheet** is a data-recording device, essentially a list of categories.* As events occur

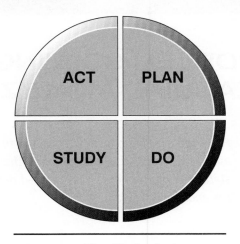

FIGURE 9.1   Plan-Do-Study-Act

Define (Plan)
1. Identify the problem/project.
2. Define the requirements.
3. Establish goals.

Measure (Plan)
4. Gather information about the current process.
5. Define and measure key process steps and inputs.
    a. Select appropriate metrics: key process output variables (KPOV).
    b. Determine how these metrics will be tracked over time.
    c. Determine current baseline performance of project/process.
    d. Determine the key process input variables (KPIV) that drive the key process output variables (KPOV).

Analyze (Plan)
6. Identify potential root causes of the problem.
7. Validate the cause-and-effect relationship.
    a. Determine what changes need to be made to the key process input variables in order to positively affect the key process output variables.

Improve (Do)
8. Implement the solutions to address the root causes of the problem.
9. Test solutions.
    a. Determine whether the changes have positively affected the KPOVs.
10. Measure results.

Control (Study, Act)
11. Evaluate and monitor improvements.
    a. If the changes made result in performance improvements, establish control of the KPIVs at the new levels; if the changes have not resulted in performance improvement, return to step 5 and make the appropriate changes.
12. Establish standard operating procedures.

FIGURE 9.2   Six Sigma Problem-Solving Steps

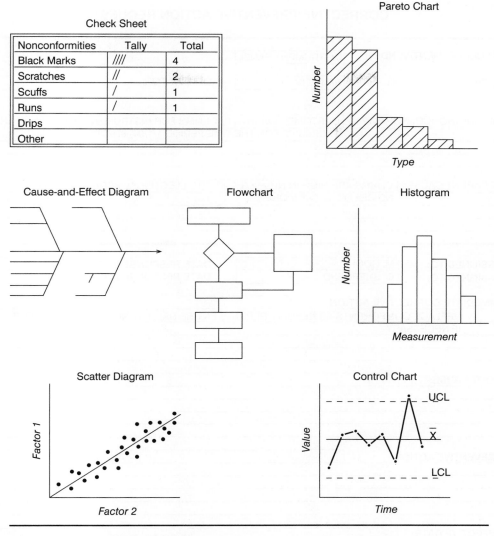

FIGURE 9.3 Seven Tools of Quality

in these categories, a check or mark is placed on the check sheet in the appropriate category. Given a list of items or events, the user of a check sheet marks down the number of times a particular event or item occurs. In essence, the user checks off occurrences. Check sheets are often used in conjunction with other quality assurance techniques. Be careful not to confuse a check sheet with a checklist. The latter lists all of the important steps or actions that need to take place, or things that need to be remembered.

Technique: Pareto Analysis

One very useful lean Six Sigma tool is Pareto analysis. Dr. Joseph Juran (1904–2008), a quality contemporary of Dr. W. Edwards Deming, popularized the Pareto principle for

---

lσ Plastics and Dashes: Identifying the Problem   **LEAN SIX SIGMA TOOLS at WORK**

Plastics and Dashes Inc. (P&D) supplies instrument panels and other plastic components for automobile manufacturers. Recently their largest customer informed them that there have been an excessive number of customer complaints and warranty claims concerning the P&D instrument panel. The warranty claims have amounted to over $200,000, including the cost of parts and labor. In response to this problem, Plastics and Dashes' management has initiated a corrective action request (Figure 9.4) and formed an improvement team to investigate. The steps the team will take to solve this problem are detailed in the Lean Six Sigma Tools at Work features throughout this chapter.

(Continued)

CORRECTIVE/PREVENTIVE ACTION REQUEST

TO DEPARTMENT/VENDOR: INSTRUMENT PANEL

DATE: 8/31/2004 **ORIGINATOR:** R. SMITH

FINDING/NONCONFORMITY: CUSTOMER WARRANTY CLAIMS FOR INSTRUMENT PANEL 360ID ARE EXCESSIVE FOR THE TIME PERIOD 1/1/04–8/1/04

APPARENT CAUSE: CLAIMS ARE HIGH IN NUMBER AND CITE ELECTRICAL PROBLEMS AND NOISE/LOOSE COMPONENTS

ASSIGNED TO: M. COOK **DATE RESPONSE DUE:** 10/1/04
ASSIGNED TO: Q. SHEPHERD **DATE RESPONSE DUE:** 10/1/04

IMMEDIATE CORRECTIVE ACTION: REPLACE COMPONENTS AS NEEDED; REPLACE ENTIRE DASH AT NO COST TO CUSTOMER IF NECESSARY

ROOT CAUSE:

PREVENTIVE ACTION:

EFFECTIVE DATE:

ASSIGNEE	DATE	ASSIGNEE	DATE
QUALITY ASSURANCE	DATE	R. SMITH ORIGINATOR	8/3/2004 DATE

COMMENTS/AUDIT/REVIEW: SATISFACTORY ☐ UNSATISFACTORY ☐

NAME **DATE**

FIGURE 9.4 Plastics and Dashes: Corrective Action Request Form

Plastics and Dashes: Using Check Sheets

LEAN SIX SIGMA TOOLS at WORK

The problem-solving team at Plastics and Dashes has a great deal of instrument panel warranty information to sort through. To gain a better understanding of the situation, team members have decided to investigate the key process output variable of warranty claims from the preceding six months. A check sheet has been chosen to record the types of claims and to determine how many of each type exists.

To create a check sheet, they first brainstormed a list of potential warranty problems. The categories they came up with include loose instrument panel components, noisy instrument panel components, electrical problems, improper installation of the instrument panel or its components, inoperative instrument panel components, and warped instrument panel. The investigators will use the check sheet created from this list to record the types of warranty problems. As the investigators make their determination, a mark is made in the appropriate category of the check sheet (Figure 9.5). Once all the warranty information has been reviewed, these sheets will be collected and turned over to the team to be tallied. The information from these sheets will help the team members focus their problem-solving efforts.

Loose instrument panel components	///// ///// ///// ///// ///// ///// //
Noisy instrument panel components	///// ///// /////
Electrical problems	///// ///
Improper installation of the instrument panel or its components	///// ///// /////
Inoperative instrument panel components	///// ///// //
Warped instrument panel	/////
Other	

FIGURE 9.5 Plastics and Dashes: Warranty Panel Information Partially Completed Data Recording Check Sheet

use in problem identification. Dr. Juran's approach to process improvement involves creating awareness of the need to improve, making quality improvement an integral part of each job, providing training in quality methods, establishing team problem solving, and recognizing results. Dr. Juran (Figure 9.6) recognized how important it is to focus on solving the problems that will provide the greatest gains for a company. Because the **Pareto chart** is *a graphical tool for ranking causes of problems from the most significant to the least significant*, it serves of valuable purpose ranking problems by their importance.

First identified by Vilfredo Pareto, a 19th-century engineer and economist, the Pareto principle, also known as the 80–20 rule, originally pointed out that the greatest portion of wealth in Italy was concentrated in a few families. Dr. Juran, stating that 80 percent of problems come from 20 percent of causes, used Pareto's work to encourage management to focus its improvement efforts on the 20 percent "vital few." Pareto charts are a graphical display of the 80–20 rule. These charts are applicable to any problem that can be separated into categories of occurrences.

Although the split is not always 80–20, the Pareto chart is a visual method of identifying which problems are most significant. Pareto charts allow users to separate the vital few problems from the trivial many. The use of Pareto charts also limits the tendency of people to focus on the most recent problems rather than on the most important problems.

FIGURE 9.6 Dr. Joseph Juran
(**Source:** Juran Institute, Inc.)

Plastics and Dashes: Constructing a Pareto Chart

LEAN SIX SIGMA TOOLS at WORK

At Plastics and Dashes, the team members working on the instrument panel warranty issue have decided to begin their investigation by creating a Pareto chart.

Step 1. *Select the subject for the chart.* The subject of the chart is the key process output variable: instrument panel warranty claims.

Step 2. *Determine what data need to be gathered.* The data to be used to create the chart are the different reasons customers have brought their cars in for instrument panel warranty work. Cost information on instrument panel warranty work is also available.

Step 3. *Gather the data related to the quality problem.* The team has determined that it is appropriate to use the warranty information for the preceding six months. Copies of warranty information have been distributed to the team.

Step 4. *Make a check sheet of the gathered data and record the total numbers in each category.* Based on the warranty information, the team has chosen the following categories for the x axis of the chart: loose instrument panel components, noisy instrument panel components, electrical problems, improper installation of the instrument panel or its components, inoperative instrument panel components, and warped instrument panels (refer to Figure 9.5).

Step 5. *Determine the total number of nonconformities and calculate the percentage of the total in each category.* From the six months of warranty information, the team also has the number of occurrences for each category:

1. Loose instrument panel components   355   41.5%
2. Noisy instrument panel components   200   23.4%
3. Electrical problems   110   12.9%
4. Improper installation of the instrument panel or its components   80   9.4%
5. Inoperative instrument panel components   65   7.6%
6. Warped instrument panel   45   5.2%

Warranty claims for instrument panels total 855.

Step 6. *Determine the costs associated with the nonconformities or defects.* The warranty claims also provided cost information associated with each category.

1. Loose instrument panel components $115,000
2. Noisy instrument panel components $25,000
3. Electrical problems $55,000
4. Improper installation of the instrument panel or its components $10,000
5. Inoperative instrument panel components $5,000
6. Warped instrument panel $1,000

Step 7. *Select the scales for the chart.* The team members have decided to create two Pareto charts, one for number of occurrences and the other for costs. On each chart, the x axis will display the warranty claim categories. The y axis will be scaled appropriately to show all the data.

Step 8. *Draw a Pareto chart by organizing the data from the largest category to the smallest.* The Pareto charts are shown in Figures 9.7 and 9.8. A Pareto chart for percentages could also be created.

Step 9. *Analyze the charts.* An analysis of the charts shows that the most prevalent warranty claim is loose instrument panel components. It makes sense that loose components might also be noisy, and the Pareto chart (Figure 9.7) reflects this, noisy instrument panel components being the second most frequently occurring warranty claim. The second chart, in Figure 9.8, tells a slightly different story. The category of loose instrument panel components has the highest costs; however, the electrical problems category has the second-highest costs.

At this point, although all the warranty claims are important, the Pareto chart has shown that efforts should be concentrated on investigating the causes of loose instrument panel components. Solving this warranty claim would significantly affect warranty numbers and costs.

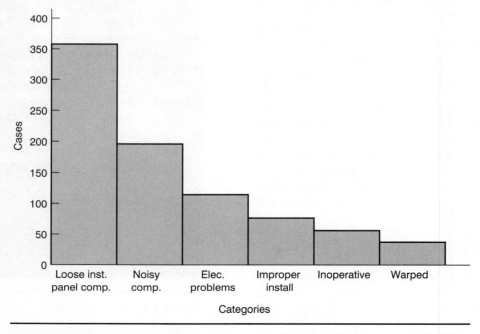

FIGURE 9.7 Plastics and Dashes: Instrument Panel Problems by Warranty Claim Type

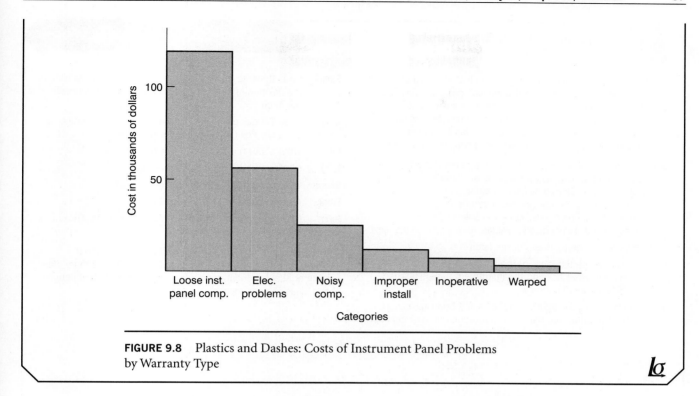

FIGURE 9.8 Plastics and Dashes: Costs of Instrument Panel Problems by Warranty Type

A Pareto chart is constructed using the following steps:

1. Select the subject for the chart. This can be a particular product line exhibiting problems, a department, or a process.
2. Determine what data need to be gathered. Determine whether numbers, percentages, or costs are going to be tracked. Determine which nonconformities or defects will be tracked.
3. Gather data related to the quality problem. Be sure that the time period during which data will be gathered is established.
4. Use a check sheet to gather data. Record the total numbers in each category. Categories will be the types of defects or nonconformities.
5. Determine the total number of nonconformities and calculate the percentage of the total in each category.
6. Determine the costs associated with the nonconformities or defects.
7. Select the scales for the chart. The y axis scales are typically the number of occurrences, number of defects, dollar loss per category, or percent. The x axis usually displays the categories of nonconformities, defects, or items of interest.
8. Draw a Pareto chart by organizing the data from the tallest category to the shortest. Include all pertinent information on the chart.
9. Analyze the chart or charts. The tallest bars represent the vital few problems. If solved, getting rid of these problems will provide the greatest gains for the organization. If there does not appear to be one or two major problems, recheck the categories to determine whether another analysis is necessary.

ANALYZE

During the analyze phase of problem solving, lean Six Sigma teams focus on identifying potential root causes of the problem they are working on. They also seek to validate a cause-and-effect relationship between the key process input variables and the key process output variables. This analysis enables the team to identify the vital few root causes of the problem.

Technique: Brainstorming

To identify potential root causes of problems, lean Six Sigma improvement teams often turn to brainstorming. *The purpose of **brainstorming** is to generate a list of problems, opportunities, or ideas from a group of people.* Everyone present at the session should participate. The discussion leader must ensure that everyone is given an opportunity to comment and add ideas. Critical to brainstorming is that no arguing, no criticism, no negativism, and no evaluation of the ideas, problems, or opportunities take place during the session. It is a session devoted purely to the generation of ideas.

The length of time allotted to brainstorming varies; sessions may last from 10 to 45 minutes. Some team leaders deliberately keep the meetings short to limit opportunities to begin problem solving. A session ends when no more items are brought up. The result of the session will be a list of ideas, problems, or opportunities to be tackled. After being listed, the items are sorted and ranked by category, importance, priority, benefit, cost, impact, time, or other considerations.

Plastics and Dashes: Brainstorming

LEAN SIX SIGMA TOOLS at WORK

The team at Plastics and Dashes Inc. conducted a further study of the causes of loose instrument panel components. The investigation revealed that the glove box in the instrument panel was the main problem area (Figure 9.9). To better understand why the glove box might be loose, the team assembled to brainstorm the key process input variables associated with the glove box.

Jerry: I think you all know why we are here today. Did you all get the opportunity to review the glove box information? Good. Well, let's get started by concentrating on the relationship between the glove box and the instrument panel. I'll list the ideas on the board here, while you folks call them out. Remember, we are not here to evaluate ideas. We'll do that next.

Sam: How about the tightness of the latch?

Frank: Of course the tightness of the latch will affect the fit between the glove box and the instrument panel! Tell us something we don't know.

Jerry: Frank, have you forgotten the rules of a brainstorming session? No criticizing. Sam, can you expand on your concept?

Sam: I was thinking that the positioning of the latch as well as the positioning of the hinge would affect the tightness of the latch.

Jerry: Okay. (Writes on board.) Tightness of Latch, Positioning of Latch, Positioning of Hinge. Any other ideas?

Sue: What about the strength of the hinge?

Jerry: (Writes on board.) Strength of Hinge.

Sharon: What about the glove box handle strength?

Frank: And the glove box handle positioning?

Jerry: (Writes on board.) Glove Box Handle Strength. Glove Box Handle Positioning.

The session continues until a variety of key process input variables have been generated (Figure 9.10). After no more ideas surface or at subsequent meetings, discussion and clarification of the ideas can commence.

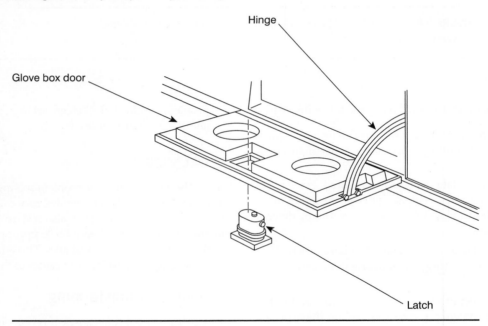

FIGURE 9.9 Plastics and Dashes: Glove Box

Positioning of the Glove Box
Strength of the Glove Box
Tightness of the Latch
Positioning of the Latch
Strength of the Latch
Positioning of the Hinge
Strength of the Hinge
Glove Box Handle Strength
Glove Box Handle Positioning
Glove Box Construction Materials

FIGURE 9.10 Key Process Input Variables Associated with the Glove Box

Technique: Cause-and-Effect Diagrams

Another excellent method of identifying potential root causes is the cause-and-effect diagram, which is also called the Ishikawa diagram after Kaoru Ishikawa (who developed it in the 1950s) or the fishbone diagram because the completed diagram resembles a fish skeleton. The ***cause-and-effect diagram*** *will help identify causes for nonconforming or defective products or services.* This chart is useful in a brainstorming session because it organizes the ideas that are presented. Problem solvers benefit from using the chart by being able to separate a large problem into manageable parts. The problem or effect is clearly identified on the right-hand side of the chart, and the potential causes of the problem are organized on the left-hand side. The cause-and-effect diagram also allows the session leader to logically organize the possible causes of the problem and to focus on one area at a time. The chart not only displays the causes of the problem, but also shows subcategories related to those causes.

The steps to construct a cause-and-effect diagram are as follows:

1. Clearly identify the effect or the problem. The succinctly stated effect or problem statement is placed in a box at the end of a line.

LEAN SIX SIGMA TOOLS at WORK

Plastics and Dashes: Constructing a Cause-and-Effect Diagram

As the Plastics and Dashes Inc. instrument panel warranty team continued its investigation, it was determined that defective latches were causing most of the warranty claims associated with the categories of loose instrument panel components and noise. The lean Six Sigma improvement team would like to identify potential root causes of defective latches.

Step 1. *Identify the effect or problem.* The team identified the problem as defective latches.

Step 2. *Identify the causes.* Rather than use the traditional methods, materials, machines, people, environment, and information, the team members felt that the potential areas to search for causes related directly to the latch. For that reason, they chose these potential causes: broken, misadjusted, binds, inoperative, loose.

Step 3. *Build the diagram.* The team brainstormed root causes for each category (Figure 9.11).

Step 4. *Analyze the diagram.* The team members discussed and analyzed the diagram. After much discussion, they came to the following conclusions. Latches that were broken, misadjusted, or inoperable or those that bind have two root causes in common: improper alignment and improper positioning. Latches that were loose or broken had a root cause of low material strength (those materials supporting the latch were low in strength). From their findings, the team determined that there were three root causes associated with defective latches: improper alignment, improper positioning, and low material strength.

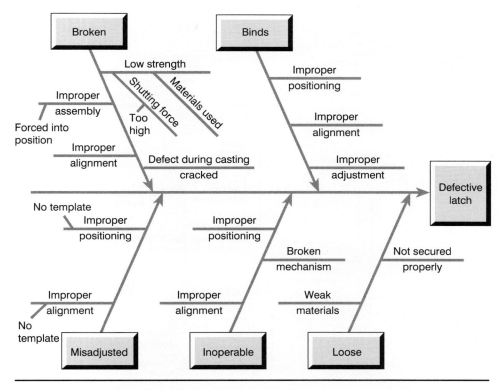

FIGURE 9.11 Plastics and Dashes: Cause-and-Effect Diagram

2. Identify the causes. Discussion ensues concerning the potential causes of the problem. To guide the discussion, attack just one possible cause at a time. General topic areas are usually methods, materials, machines, people, environment, and information, although other areas can be added as needed. Under each major area, subcauses related to the major cause should be identified. Brainstorming is the usual method for identifying these causes.

3. Build the diagram. Organize the causes and subcauses in diagram format.

4. Analyze the diagram. At this point, solutions will need to be identified. Decisions will also need to be made concerning the cost-effectiveness of the solution as well as its feasibility.

Technique: WHY-WHY Diagram

An excellent technique for identifying the root cause(s) of a problem is to ask "Why" five times. **WHY-WHY diagrams** *organize the thinking of a problem-solving group and illustrate a chain of symptoms leading to the true cause of a problem.* By asking "why" five times, the problem solvers are stripping away the symptoms surrounding the problem and getting to the true cause of the problem. This technique eliminates superficial solutions. At the end of a session it should be possible to make a positively worded, straightforward statement defining the true problem to be investigated.

Developed by group consensus, the WHY-WHY diagram flows from left to right. The diagram starts on the left with a statement of the problem to be resolved. Then the group is asked why this problem might exist. The responses will be statements of causes that the group believes contribute to the problem under discussion. There may be only one cause or there may be several. Causes can be separate or interrelated. Regardless of the number of causes or their relationships, the causes should be written on the diagram in a single, clear statement. WHY statements should be supported by facts as much as possible and not by hearsay or unfounded opinions. Ask why this outcome occurs. Figure 9.12 shows a WHY-WHY diagram the Plastics and Dashes problem-solving team completed for instrument panel warranty costs.

This investigation is continued through as many levels as needed until a root cause is found for each of the problem statements, original or developed during the discussions. Frequently, five levels of "why" are needed to determine the root cause. In the end, this process leads to a network of reasons the original problems occurred. The ending points indicate areas that need to be addressed to resolve the original problem. These become the actions the company must take to address the situation. WHY-WHY diagrams can be expanded to include notations concerning who will be responsible for action items and when the actions will be completed.

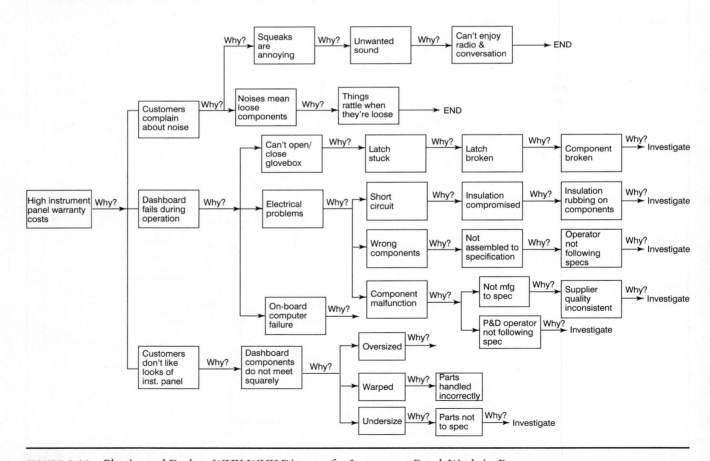

FIGURE 9.12 Plastics and Dashes: WHY-WHY Diagram for Instrument Panel, Work-in-Progress

Technique: Scatter Diagrams

*The **scatter diagram** is a graphical technique that is used to analyze the relationship between two different variables.* Two sets of data are plotted on a graph. The independent variable—i.e., the variable that can be manipulated—is recorded on the *x* axis. The dependent variable, the one being predicted, is displayed on the *y* axis. From this diagram, the user can determine whether a connection or relationship exists between the two variables being compared. If a relationship exists, then steps can be taken to identify process changes that affect the relationship. Figure 9.13 shows different interpretations of scatter diagrams.

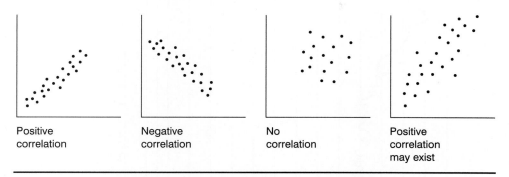

FIGURE 9.13 Scatter Diagram Interpretations

Creating a Scatter Diagram

LEAN SIX SIGMA TOOLS at WORK

Shirley is the setup operator in the shrink-wrap area. In this area, 5-ft-by-5-ft cartons of parts are sealed with several layers of plastic wrap before being loaded on the trucks for shipment. Shirley's job is to load the plastic used in the shrink-wrapping operation onto the shrink-wrap machine, set the speed at which the unit will rotate, and set the tension level on the shrink-wrap feeder. To understand the relationship between the tension level on the feeder and the speed of the rotating mechanism, Shirley has created a scatter diagram.

The rotator speed is most easily controlled, so she has placed the most typically used speed settings (in rpm) on the *x* axis. On the *y* axis, she places the number of tears (Figure 9.14).

The diagram reveals a positive correlation: As the speed increases, the number of tears increases. From this information, Shirley now knows that the tension has to be reduced in order to prevent the wrap from tearing. Using the diagram, Shirley is able to determine the optimal speed for the rotor and the best tension setting.

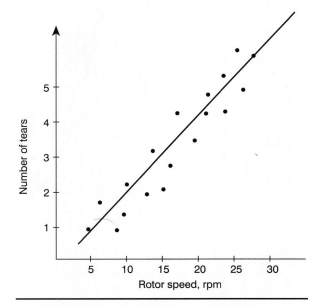

FIGURE 9.14 Scatter Diagram of Tears versus Rotor Speed

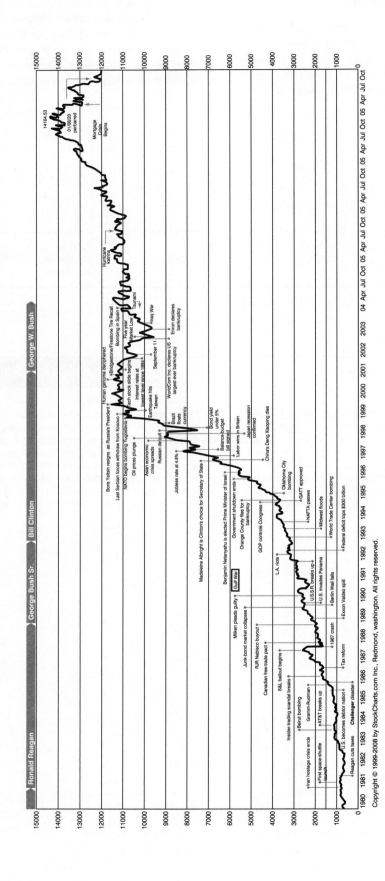

FIGURE 9.15 Run Chart of Financial Performance

(**Source:** Copyright © 1999–2008 by StockCharts.com Inc., Redmond, Washington. All rights reserved.)

To construct a scatter diagram, use these steps:

1. Select the characteristic, the independent variable, you wish to study.
2. Select the characteristic, the dependent variable, that you suspect affects the independent variable.
3. Gather the data about the two characteristics.
4. Draw, scale, and label the horizontal and vertical axes.
5. Plot the points.
6. Interpret the scatter diagram to see whether there is a relationship between the two characteristics.

Technique: Run Charts

Run charts, also called time series plots, can be used to monitor process changes associated with a particular characteristic over time. Run charts are versatile and can be constructed with data consisting of either variables or attributes. These data can be gathered in many forms, including individual measurements, counts, or subgroup averages. Time is displayed on the x axis of the chart; the value of the variable or attribute being investigated is recorded on the y axis. The median value of the data can be used as the centerline of the chart. Cycles, trends, runs, and other patterns are easily spotted on a run chart. As Figure 9.15 shows, financial performance over time is often displayed in the form of a run chart.

LEAN SIX SIGMA TOOLS at WORK

Using Run Charts in Decision Making

A U.S. state highway route through a small town has had an increase in traffic accidents in recent years. The city manager has decided to do a traffic study to support the reduction of the speed limit, in hopes of reducing the number of traffic accidents. The study will cost approximately $2,000. Upon completion, the study will be turned over to the state's Department of Transportation. Factors that weigh into studying the speed limit include accident history, traffic volume, average speed of motorists, access points, and the width of the shoulder and roadway.

Part of the study includes creating and studying run charts of the accident history:

Step 1. *Determine the time increments necessary to study the process properly.* Under their current system, the city combines and reports accident information in January, May, and December. Information is available for the past four years. Time increments based on these months were marked on the x axis of the chart.

Step 2. *Scale the y axis to reflect the values that the measurements or attributes data will take.* The y axis is scaled to reflect the number of accidents occurring each month.

Step 3. *Collect the data.* The data were collected (Figure 9.16).

Step 4. *Record the data on the chart as they occur.* The data were recorded on the chart (Figure 9.17).

January 2000	10
May 2000	15
December 2000	10
January 2001	23
May 2001	26
December 2001	22
January 2002	30
May 2002	25
December 2002	30
January 2003	35
May 2003	38
December 2003	36

FIGURE 9.16 Accident Data

Step 5. *Interpret the chart.* The chart, when interpreted, showed that accidents have definitely been increasing over the past four years. This will be very useful information when the state tries to determine whether to lower the speed limit.

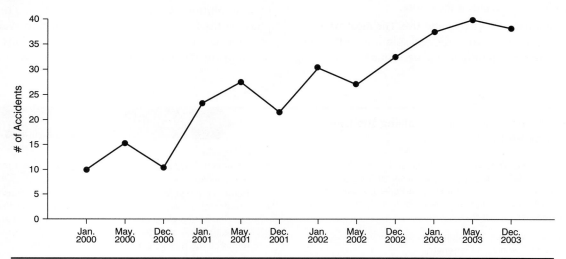

FIGURE 9.17 Run Chart of Accident Data

A run chart is created in five steps:

1. Determine the time increments necessary to properly study the process. These may be based on the rate at which the product is produced, on how often an event occurs, or according to any other time frame associated with the process under study. Mark these time increments on the x axis of the chart.
2. Scale the y axis to reflect the values that the measurements or attributes data will take.
3. Collect the data.
4. Record the data on the chart as they occur.
5. Interpret the chart. Because there are no control limits, the interpretation of a run chart is limited to looking for patterns in the data. A point that appears to be high or low cannot be judged as a special cause without control limits. It may be the worst data point in a series of points, but that does not equate with being out of control.

IMPROVE

Once the root cause of the problem has been identified, it is time to propose and implement potential solutions. The chosen solution should address the root causes of the problem and should be tested to ensure that it does truly solve the problem. By analyzing the data gathered, the project team can choose the best solution for the problem. The solution should be judged against four general criteria:

1. The solution should be chosen on the basis of its potential to prevent a recurrence of the problem. A quick or short-term fix to a problem will only mean that time will be wasted in solving this problem again when it recurs in the future.
2. The solution should address the root cause of the problem. A quick or short-term fix that focuses on correcting the symptoms of a problem will waste time because the problem will recur in the future.
3. The solution should be cost-effective. The most expensive solution is not necessarily the best solution for the company's interests. Solutions may necessitate determining the company's future plans for a particular process or product. Major changes to the process, system, or equipment may not be an appropriate solution for a process or product that will be discontinued in the near future. Technological advances will need to be investigated to determine whether they are the most cost-effective solutions.
4. The solution should be capable of being implemented within a reasonable amount of time. A timely solution to the problem is necessary to relieve the company of the burden of monitoring the current problem and its associated quick fixes.

Implementing the solution is often done by members of the problem-solving team. Critical to ensuring the success of the solution implementation is assigning responsibilities to specific individuals and holding them accountable for accomplishing the task. Knowing who will be doing what and when will help ensure that the project stays on track.

Sometimes implementing solutions is complicated by a variety of factors, including conflicting departmental needs, costs, or priorities. Using a force field analysis can help identify the issues affecting the implementation of a solution.

Technique: Force Field Analysis

A *force field analysis* is a chart that helps teams separate the driving forces and the restraining forces associated with a complex situation. These easy-to-develop diagrams help a team determine the positive or driving forces that are encouraging improvement of the process as well as the forces that restrain improvement. Teams may also choose to use force field analysis as a source of discussion issues surrounding a particular problem or opportunity. Once the driving and restraining forces have been identified, the team can discuss how to enhance the driving forces and remove the restraining forces.

CONTROL

Once a solution has been implemented, during the control phase of the DMAIC cycle, steps are taken to evaluate and monitor the improvements. At this time, decisions are made to adopt the change, abandon it, make adjustments, or if

---

lσ Plastics and Dashes: Solving the Glove Box Problem

LEAN SIX SIGMA TOOLS at WORK

Through the use of WHY-WHY and cause-and-effect diagrams, the team was able to identify the root causes of the glove box latch problem as improper alignment, improper positioning, and low material strength. The team members decided to make the following changes part of their solution:

1. Redesign the glove box latch. This solution was chosen to counteract low material strength.
2. Reposition the glove box door, striker, and hinge. This solution was chosen to counteract improper positioning and alignment. They also hoped this change would eliminate potential squeaks and rattles.
3. Reinforce the glove box latch. This solution was chosen to counteract breakage. By increasing the material at the latch position on the glove box door, they hoped to eliminate breakage. They also decided to use a stronger adhesive to reinforce the rivets securing the latch to the door.

lσ

Plastics and Dashes: Using Force Field Analysis

LEAN SIX SIGMA TOOLS at WORK

Though the lean Six Sigma team isolated the root causes of the glove box latch problems, there is some resistance from management to implement the necessary changes. The team has decided to use a force field analysis to counter this resistance.

The team members began by brainstorming the driving forces behind the change. Then they changed direction and brainstormed all the reasons why management might not want to make the change. Having completed these two sides of the diagram, the team identified ways to enhance the driving forces and minimize the effects of the restraining forces. Then the team got busy and completed the steps in the action plan. Figure 9.18 presents the completed force field diagram. Note that the numbers in the table represent the link between the action plans and the driving or restraining force.

After listening to the results of action plan steps 1 through 4, management has a clearer understanding of the issues involved in making the improvements to the glove box line. They have decided to go ahead with the improvements.

Glove Box Latch Line Improvements

Driving Forces

Less rework (1)
Lower costs (1)
Less scrap (1)
Quicker throughput (1, 2)
Improved customer satisfaction (4)
Less inspection (1, 2, 4)
Fewer non-value-added activities (2)
Improved reputation (4)
Increased business (4)
Loss of customers (3)

Restraining Forces

Time to retool line (1)
Cost to retool line (1)
Time to retrain employees (1)
Poor understanding of need for improvement (5)

Action Plan

1. Perform cost–benefit analysis on rework, scrap, and customer dissatisfaction costs versus costs required to solve the problem.
2. Perform analysis to determine cost savings associated with removal of non-value-added activities.
3. Contact customers to assess risk of losing business.
4. Use measures of performance to determine whether the implementation is effective.
5. Present information from action plan steps 1 through 4 to management to improve its understanding of the need for improvement.

FIGURE 9.18 Plastics and Dashes: Force Field Analysis

---

necessary, repeat the problem-solving cycle. If the solution proves to be viable, it is made part of the standard operating procedures to ensure that the problem does not resurface.

This phase of the DMAIC process exists to ensure that the new controls and procedures stay in place. It is easy to believe that the "new and better" method should be utilized without fail; however, in any situation where a change has taken place, there is a tendency to return to old methods, controls, and procedures. Widespread active involvement in improvement projects helps ensure successful implementation of new methods. Extensive training and short follow-up training sessions are also helpful in ingraining the new method. Follow-up checks must be put in place to prevent problem recurrences from lapses to old routines and methods.

Plastics and Dashes: Evaluating the Solution

LEAN SIX SIGMA TOOLS at WORK

The instrument panel warranty team implemented its solutions and used the measures of performance developed earlier to study the solutions to determine whether the changes were working. The team's original measures of performance were warranty costs and number and type of warranty claims. The Pareto chart in Figure 9.19 provides information about warranty claims made following the changes. When this figure is compared to Figure 9.7 showing warranty claims before the problem-solving team went into action, the improvement is obvious. Warranty costs declined in proportion to the decreased number of claims, to just under $25,000. A process measure also tracked the length of time to implement the changes, a very speedy five days.

(Continued)

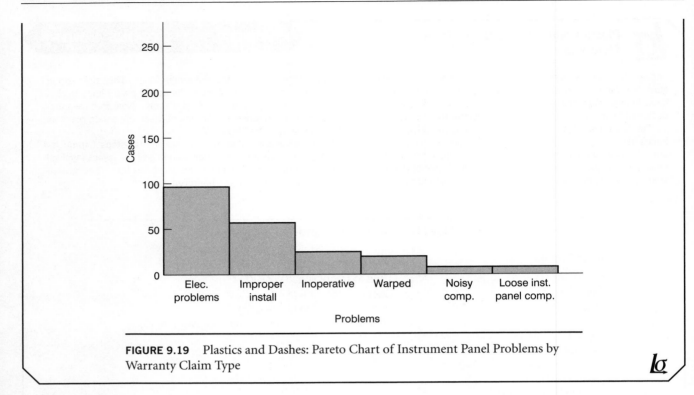

FIGURE 9.19 Plastics and Dashes: Pareto Chart of Instrument Panel Problems by Warranty Claim Type

SUMMARY

Improvement projects are easy to identify. A review of operations will reveal many opportunities for improvement. Any sources of waste, such as warranty claims, overtime, scrap, or rework, as well as production backlogs or areas in need of more capacity, are potential projects. Even small improvements can lead to a significant impact on the organization's financial statement. Having completed one project, others wait for the same problem-solving process.

TAKE AWAY TIPS

1. Problem solving is the isolation and analysis of a problem and the development of a permanent solution. Problem solving should be logical and systematic.
2. The following are the lean Six Sigma problem-solving steps:
   Step 1. Identify the problem/project.
   Step 2. Define the requirements.
   Step 3. Establish goals.
   Step 4. Gather information about the current process.
   Step 5. Define and measure key process steps and inputs.
      a. Select appropriate metrics: key process output variables (KPOV).
      b. Determine how these metrics will be tracked over time.
      c. Determine current baseline performance of project/process.
      d. Determine the key process input variables (KPIV) that drive the key process output variables (KPOV).
   Step 6. Identify potential root causes of the problem.
   Step 7. Validate the cause-and-effect relationship.
      a. Determine what changes need to be made to the key process input variables in order to positively affect the key process output variables.
   Step 8. Implement the solution to address the root causes of the problem.
   Step 9. Test solutions.
      a. Determine whether the changes have positively affected the KPOVs.
   Step 10. Measure results.
   Step 11. Evaluate and monitor improvements.
      a. If the changes made result in performance improvements, establish control of the KPIVs at the new levels. If the changes have not resulted in performance improvement, return to step 5 and make the appropriate changes.
   Step 12. Establish standard operating procedures.
3. The following are techniques used in problem solving: brainstorming, Pareto analysis, WHY-WHY diagrams, flowcharts, force field analysis, cause-and-effect diagrams, check sheets, and scatter diagrams.
4. Problem solvers are tempted to propose solutions before identifying the root cause of the problem and performing an in-depth study of the situation. Adhering to a problem-solving method avoids this tendency.
5. Brainstorming is designed for idea generation. Ideas should not be discussed or criticized during a brainstorming session.
6. Flowcharts are powerful tools that allow problem solvers to gain in-depth knowledge of the process.

7. Cause-and-effect diagrams enable problem solvers to identify the root causes of succinctly stated problems.
8. Steps must be taken to ensure that the new methods or changes to the process are permanent.
9. Don't be afraid to apply the techniques you know.

CHAPTER QUESTIONS

1. An orange juice producer has found that the fill weights (weight of product per container) of several of its orange juice products do not meet specifications. If the problem continues, unhappy customers will stop buying its product. Outline the steps that it should take to solve this problem. Provide as much detail as you can.
2. Bicycles are being stolen at a local campus. Campus security is considering changes in bike rack design, bike parking restrictions, and bike registrations to try to reduce thefts. Thieves have been using hacksaws and bolt cutters to remove locks from the bikes. Create a problem statement for this situation. How will an improvement team use the problem statement?
3. A pizza company with stores located citywide uses one order call-in phone number for the entire city. Callers, regardless of their address, can phone XXX-1111 to place an order. Based on the caller's phone number, an automated switching service directs the call to the appropriate store.
    On a recent Friday evening, the pizza company lost as many as 15,000 orders when a malfunctioning mechanical device made calls to its phone number impossible. From about 5:30 to 8 P.M., when callers hoping to place an order phoned, they were met with either silence or a busy signal. This is not the first time this malfunction has occurred. Just two weeks earlier, the same problem surfaced. The pizza company and the company that installed the system are working to ensure that the situation doesn't repeat itself. They believe a defective call switch is to blame.
    Based on what you have learned in this chapter, why is a structured problem-solving process critical to the success of finding and eliminating the problem? What steps do you recommend they follow?
4. During the past month, a customer-satisfaction survey was given to 200 customers at a local fast-food restaurant. The following complaints were lodged:

Complaint	Number of Complaints
Cold Food	105
Flimsy Utensils	20
Food Tastes Bad	10
Salad Not Fresh	94
Poor Service	15
Food Greasy	9
Lack of Courtesy	5
Lack of Cleanliness	25

Create a Pareto chart with this information.

5. A local bank is keeping track of the different reasons people phone the bank. Those answering the phones place a mark on their check sheet in the rows most representative of the customers' questions. Given the following check sheet, make a Pareto diagram:

Credit Card Payment Questions	245
Transfer Call to Another Department	145
Balance Questions	377
Payment Receipt Questions	57
Finance Charge Questions	30
Other	341

Comment on what you would do about the high number of calls in the "Other" column.

6. Once a Pareto chart has been created, what steps would you take to deal with the situation given in Question 5 in your quality-improvement team?
7. Review Question 3. What are two measures of performance that can be used to determine whether the changes they make are effective?
8. Review Question 5. What are two measures of performance that can be used to determine whether the changes they make are effective?
9. Brainstorm 10 reasons why a computer might malfunction.
10. Brainstorm 10 reasons why a customer may not feel the service was adequate at a department store.
11. Create a WHY-WHY diagram for how you ended up taking this particular class.
12. A mail order company has a goal of reducing the amount of time a customer has to wait in order to place an order. Create a WHY-WHY diagram about waiting on the telephone. Once you have created the diagram, how would you use it?
13. A customer placed a call to a mail order catalog firm. Several times, the customer dialed the phone and received a busy signal. Finally, the phone was answered electronically, and the customer was told to wait for the next available operator. Although it was a 1-800 number, he found it annoying to wait on the phone until his ear hurt. Yet he did not want to hang up for fear he would not be able to get through to the firm again. Using the problem statement "What makes a customer wait?" as your base, brainstorm to create a cause-and-effect diagram. Once you have created the diagram, how would you use it?
14. Create cause-and-effect diagrams for (a) a car that won't start, (b) an upset stomach, and (c) a long line at the supermarket.
15. Create a force field diagram for Question 2 concerning bike thefts.
16. Create a force field diagram for a restaurant where customers are waiting more than 10 minutes for their food.

CHAPTER TEN

VALUE-ADDED PROCESS MAPPING

But, that's the way we have always done it.

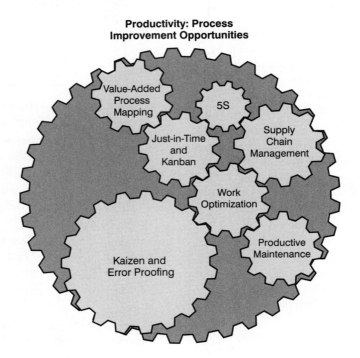

 LEARNING OPPORTUNITIES

1. To learn how to identify key processes
2. To understand how well-managed processes create value and generate customer satisfaction
3. To discover the importance of process improvement
4. To learn how to use a process map to identify all of the key activities in a process
5. To learn to separate the non-value-added activities from the value-added activities
6. To understand the importance of using the knowledge gained about the process to create the best process

DEFINING PROCESSES

A "that's the way we've always done it" attitude works only if customers' expectations are low and the competition is not hungry. Chances are in today's world, this is not true. Lean Six Sigma companies continuously improve their process performance. A *process takes inputs and performs value-added activities on those inputs to create an output* (Figure 10.1). Every business, whether in manufacturing or the service industries, has key processes that it must absolutely perform well to attract and retain customers to whom to sell their products or services.

People perform processes on a day-to-day basis without realizing it. Even something as simple as going to the movies requires a process. The input required for this process includes information about what the show times and places are, who is going, and what criteria will be used for choosing a movie. The value-added activities are driving to the movie theater, buying a ticket, and watching the movie. The output is the result, the entertainment value of the movie. Organizations have innumerable processes that enable them to provide products or services for customers. Think about the number of processes necessary to provide a shirt by mail order over the Internet. The company must have a catalog website preparation process, a website distribution process, a process for obtaining the goods it plans to sell, an ordering process, a credit-check process, a packaging process, a mailing process, and a billing process, to name a few. Other

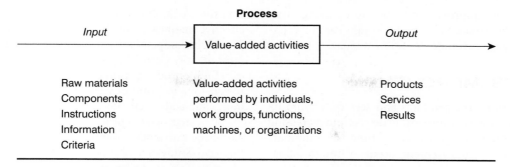

FIGURE 10.1 A Process

processes typically found in any organization include financial management; customer service; equipment installation and maintenance; production and inventory control; employee hiring, training, reviewing, firing, and payroll; software development; and product or service design, creation, inspection, packaging, delivery, and improvement. The key or critical business processes work together within an organization to carry out the organization's mission and strategic objectives. If an organization's processes do not work together, or if they work ineffectively, then the organization's performance will be less than optimal. Lean Six Sigma organizations recognize that in order to supply what the customer needs, wants, and expects, they must focus on maintaining and improving the processes that enable them to meet these needs, wants, and expectations.

IDENTIFYING KEY PROCESSES

Key processes are the business processes that have the greatest impact on customers' value perceptions about the product or service and the greatest impact on customer retention. Lean Six Sigma organizations concentrate system and process improvement efforts on business processes that will increase their competitiveness. The challenge for today's organizations is to implement systems that reduce the frequency of human errors and to devise ways of limiting the consequences of the errors that do occur. Lean Six Sigma organizations design systems that investigate and analyze process performance so that the root causes of problems can be found and corrective actions taken. By managing their business processes effectively, organizations realize significant improvements in overall organizational performance that make an impact on the bottom line of the income statement.

CREATING VALUE AND GENERATING CUSTOMER SATISFACTION

Time waste differs from material waste in that there can be no salvage.

Henry Ford

Anytime that waste occurs in a process, an organization and its customers lose. This wasteful, hidden factory consumes resources that might otherwise have been used to create valuable products or services. These costs create nothing of value yet are often hidden under the term overhead. The costs of poor quality, long process lead times, and variability in products or services significantly affect the profitability of an organization. Process management enables organizations to eliminate wasted time, effort, material, money, and human resources. Lean Six Sigma organizations concentrate

---

 Process Improvement at Hospitals　　　　**LEAN SIX SIGMA TOOLS at WORK**

When a medical emergency strikes, patients want the newest health care options available. Unfortunately, the high costs of these high-tech offerings force hospitals to continually generate funds to provide the most up-to-date technology and services to their patients. Recognizing that mistakes and waste are costly, hospitals have been changing the way they operate. Many have done so by studying the quality assurance philosophies of Dr. W. Edwards Deming and Dr. Joseph Juran, as well as investigating the activities of such Six Sigma companies as Motorola Inc. and Toyota Motor Corporation. Echoing Dr. Deming's sentiment that system failures, not employees, are at the heart of problems, many hospital administrators feel that process management is the only way to enhance hospital performance.

Dispensing medications incorrectly has the potential to create a life-threatening situation. To improve the process of medication dispensing, many hospitals are taking action to identify where errors might occur, and they are finding ways to error-proof medication dispensing processes. One technique, reported in the Wall Street Journal article "ICU Checklist System Cuts Patients' Stay in Half" (August 6, 2003), is to provide a checklist to accompany each patient's medication. This checklist is used to verify that the correct patient is receiving the correct medication in the correct dosage at the correct time. Other hospitals are using checklists developed during process improvement efforts to verify that it is appropriate to move a patient from an intensive care unit to a regular hospital room. This checklist provides a systematic method of ensuring that all the necessary actions are taken before the patient is moved. In each of these examples, the checklist serves to error-proof the process by prompting caregivers to remember vital patient needs.

on value-driven improvements by recognizing that processes should be measured and the results carefully analyzed to identify opportunities for improvement.

PROCESS IMPROVEMENT

Processes providing the products and services should be improved with the aim of preventing defects and increasing productivity by reducing process cycle times and eliminating waste. Process improvement occurs through value-added process mapping, problem isolation, root cause analysis, and problem resolution. Many processes develop over time, with little concern for whether they are the most effective manner in which to provide a product or service. To remain competitive in the world marketplace, companies must identify wasteful processes and improve them. The key to refining processes is to concentrate on the process from the customer's point of view and to identify and eliminate non-value-added activities.

Two step-by-step process improvement methodologies have been introduced in this text: Shewhart's and Deming's Plan-Do-Study-Act (PDSA) cycle and Six Sigma's Define, Measure, Analyze, Improve, Control (DMAIC) cycle. A typical process for improving processes is shown in Figure 10.2.

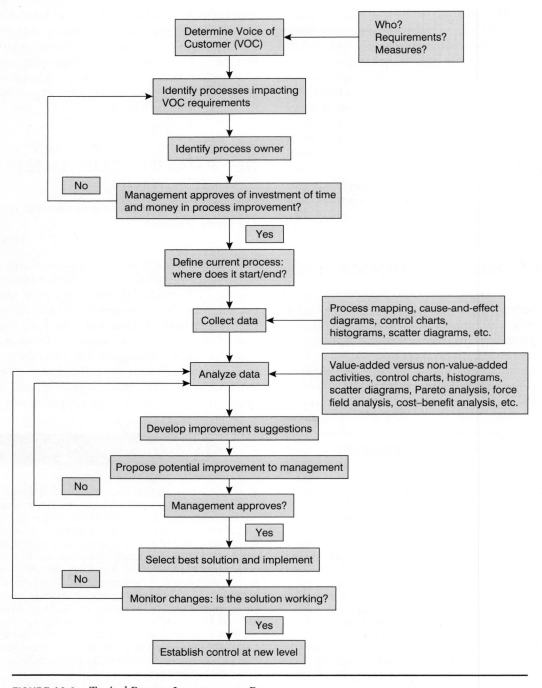

FIGURE 10.2 Typical Process Improvement Process

Regardless of the specific steps taken, in order to effectively improve processes it is critical to:

1. Determine the objective of the process as it relates to the customer.
2. Determine the boundaries of the processes as the customer sees them.
3. Involve representatives from each major activity associated with the process in the improvement effort. Identify where conflicts exist between the boundaries of the processes as they are related to existing functional departments.
4. Identify the process owner.
5. Create a process map by identifying all of the activities in the process.
6. Separate the non-value-added activities from the value-added activities.
7. Eliminate the non-value-added activities.
8. Identify, analyze, and eliminate variation in the process.
9. Determine whether the remaining value-added activities are truly the "best practice."
10. Redesign the process using the knowledge gained in the first nine steps.

The process should be studied from the customer's point of view. True process improvement comes from a knowledge of what adds value or meaning for a customer. Not looking at the process from a customer viewpoint often leads to a narrow definition of the process. This narrow definition limits improvement efforts because it fails to study the customer experience. For example, if a hotel views the check-in process internally—that is, not as a customer—it will probably identify check-in as the actual process of being assigned a room and receiving a key. Customers may take a broader view of where check-in begins, seeing where it begins as the moment they reach the front door of the hotel, and where it ends as the moment they are seated in their room with their shoes off. Improving simply the time spent receiving a room key fails to take into account the customer's overall impression or experience, thus missing the opportunity to increase customer perceived value.

Process ownership *refers to identifying who is ultimately responsible for seeing that a process is completed in a manner that results in customer satisfaction.* These individuals are in a position to make, and have the necessary power to make, changes to the process. It is important to involve individuals from the key activities in the process in the improvement effort. Doing so will create buy-in. Remember, the people who are going to have to live with the new process should be the ones who fix it. If they are involved in identifying, creating, and making the necessary changes, chances are very good that they will live with those changes and work to make them permanent. Involvement from all key activities also breaks down barriers between existing departments and provides everyone with a clearer understanding of how work gets done in the organization. Once the key processes have been identified, it is important to determine whether there are conflicts between the existing functional organization structure and the newly emphasized process approach. By identifying these conflicts, leadership can take steps to minimize the difficulties associated with the transition.

PROCESS MAPPING

One important tool used to study process performance is process mapping. Process mapping is a useful and simple tool that enables users to clearly define a process and reduce one of the five sources of variation identified in Chapter 1: process procedures. Knowledge gained during process mapping can be used to improve the process by removing non-value-added activities, creating clear work descriptions, and designing standard process procedures.

In most organizations very few people truly understand the myriad required activities in a process that creates a product or service. Process maps are powerful communication tools that provide a clear understanding of how business is conducted within the organization. Identifying and writing down the process in pictorial form helps people understand just how they do the work that they do. Process maps have the ability to accurately portray current operations, and they can also be used to evaluate these operations. In addition, a process map identifies the activities that have been added to a process over time in order to adapt older processes to changes in the business. Once changes have been proposed, process maps are equally powerful for communicating the proposed changes in the process.

Process maps are known by many names, including flowcharts, process flowcharts, and process flow diagrams. *A **process map** is a graphical representation of all of the steps involved in an entire process or a particular segment of a process.* An example of a process map is shown in Figure 10.3. Diagramming the flow of a process or system aids in understanding it. Flowcharting is effectively used in the first stages of problem solving because the charts enable those studying the process to quickly understand what is involved in a process from start to finish. Problem-solving team members can clearly see what is being done to a product or provided by a service at the various stages in a process. Process flowcharts clarify the routines used to serve customers. Problems or non-value-added activities nested within a process are easily identified by using a flowchart.

The construction of process maps is fairly straightforward. The steps to creating such charts are the following:

1. Define the process boundaries. For the purpose of the chart, determine where the process begins and ends.
2. Define the process steps. Use brainstorming to identify the steps for new processes. For existing processes, actually observe the process in action.

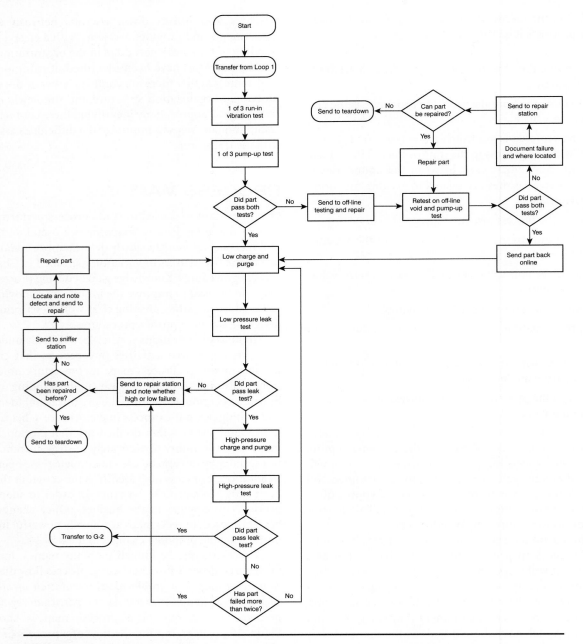

FIGURE 10.3 Example of a Process Map

3. Sort the steps into the order of their occurrence in the process.
4. Place the steps in appropriate flowchart symbols (Figure 10.4) and create the chart.
5. Evaluate the steps for completeness, efficiency, and possible problems such as non-value-added activities.

Because processes and systems are often complex, in the early stages of flowchart construction, removable 3-by-5-inch sticky notes placed on a large piece of paper or board allow creators greater flexibility when creating and refining a flowchart. When the chart is complete, a final copy can be made

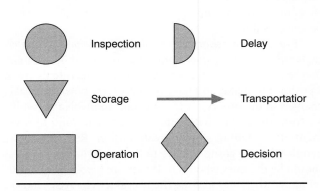

FIGURE 10.4 Process Map Symbols

Creating a Process Map

LEAN SIX SIGMA TOOLS at WORK

To help determine the current baseline performance of the process, the team at Plastics and Dashes investigating instrument panel warranty claims created a flowchart of the instrument panel assembly process.

Step 1. *Define the boundaries.* First the team examined the process from the customer's point of view. The team identified where the process begins and ends in the customer's experience

Step 2. *Define the process steps.* The team members brainstormed the steps in the assembly process. They double-checked their steps by observing the actual process. They wrote down each step on sticky notes.

Step 3. *Sort the steps into the order of their occurrence in the process.* After reconciling their observations with their brainstorming efforts, the team sorted the steps into the order of their occurrence.

Step 4. *Place the steps in appropriate flowchart symbols.* With the steps in the correct order, it was a simple task to add the appropriate flowchart symbols and create the chart (Figure 10.5).

Step 5. *Evaluate the steps for completeness, efficiency, and possible problems.* The team reviewed the finished chart for completeness. Several team members were unaware of the complete process. Because it creates a greater understanding of the process, this diagram will be helpful during later problem-solving efforts.

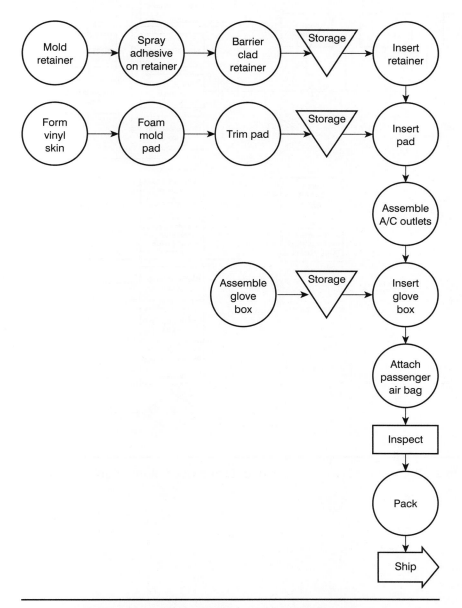

FIGURE 10.5 Plastics and Dashes: Glove Box Assembly Flowchart

utilizing the correct symbols. Either the symbols can be placed next to the description of the step, or they can surround the information.

A variation on the traditional process flowchart is the deployment flowchart. When a deployment flowchart is created, job or department titles are written across the top of the page, and the activities of the process that occur in that job or department are written below that heading (Figure 10.6). Flowcharts can also be constructed with pictures for easier understanding (Figure 10.7). When process flowcharts are used as routing sheets, they often appear as shown in Figures 10.8 and 10.9, which includes additional details such as process activities, operator self-inspection notes, and specifications.

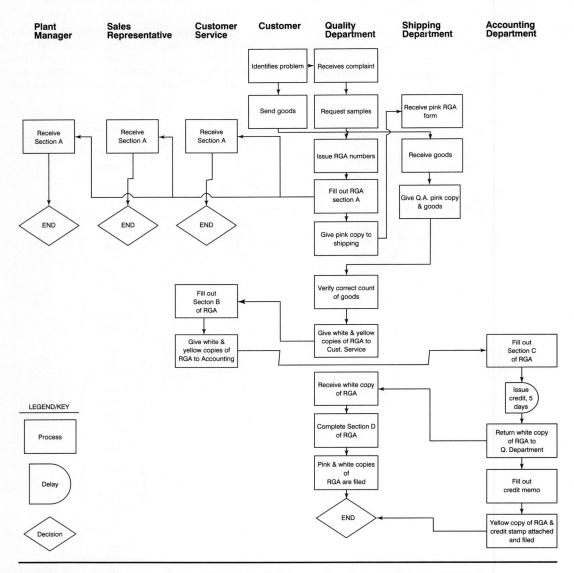

FIGURE 10.6 Deployment Process Flowchart for Return Goods Authorization (RGA) Forms

Value-Added Process Mapping 105

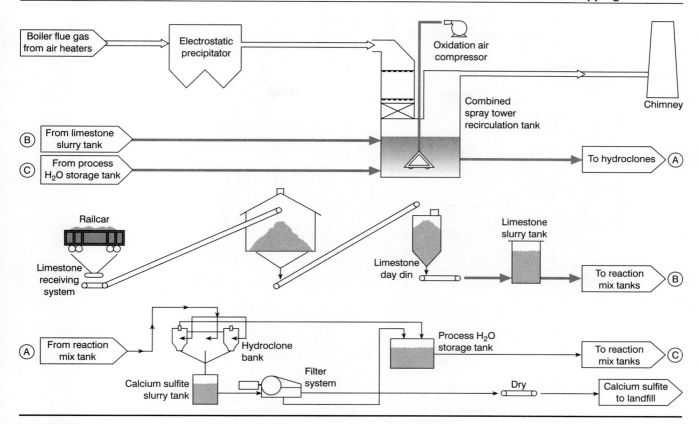

FIGURE 10.7 Flue Gas Desulfurization Process

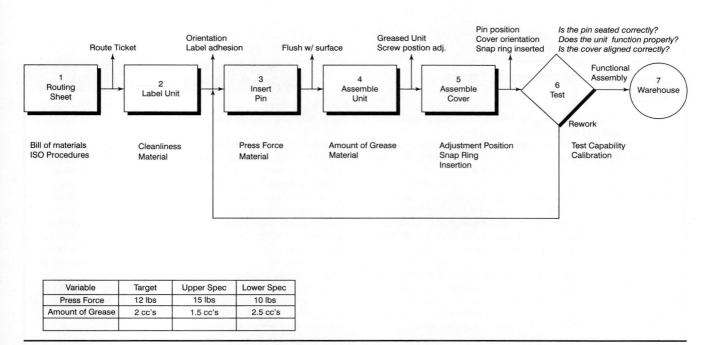

FIGURE 10.8 Process Flowchart with Instructions and Specifications

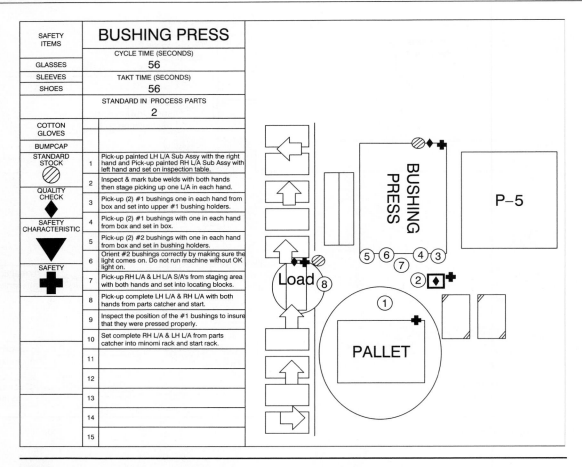

FIGURE 10.9 Bushing Press Flowchart with Instructions and Specifications

Improving the Quoting Process

LEAN SIX SIGMA TOOLS at WORK

Prior to new leadership at JF, the activity of turning a quote into a finished part had no uniform process. This led to incorrect quotes, lost quotes or orders, past-due orders, incorrect orders, improperly priced orders, and other problems. Knowing that standardized processes reduce the chance of errors, leadership formed a team with the objective of defining order processing. The boundaries of the process were set at customer requests quote to customer receives parts. The team included the order entry clerk, the scheduler, and the process engineer. The order entry clerk was designated the process owner since the clerk is responsible for a majority of the initial interaction with the quote/order.

By tracking the flow of an actual order through the shop, the team members developed a list of activities that must take place. As they studied the existing process, they used the five Ws and two Hs from Figure 10.10 to help them see the whole picture. Following a number of orders through the shop also helped. The original shop layout (Figure 10.11) resembled a spider web of parts moving here and there in the plant, so order tracking was not easy. Further complications arose when incorrect routers accompanied the orders. These were modified when necessary, before, it was hoped, incurring significant costs of quality (see Chapter 6) from producing the parts incorrectly.

When the team members reviewed their findings, the first thing that they noticed was that there were no checks or balances in the process. Quotes were rarely checked to ensure correct material costs, routings, or part print revisions. Those converting a quote into an order did not verify up-to-date material price, shop schedules, or delivery availability information. Promised due dates often did not correspond to true equipment uptime availability. This frequently was the cause of late orders (see Chapter 15). So, although the activities in the process were value added, they did not create value because of the variation in processing technique.

As the team members analyzed the process, they could clearly see that a better order processing method was needed. They readily considered the idea of computerizing the process; however, they recognized that computerizing a nonfunctional process failed to help company performance. Their efforts focused on redesigning the process. Figure 10.12 shows the new process map. The new steps in the process are shown in bold italics. Errors have declined significantly. Computerization of the new process speeds the quoting/order process even more.

	Questions to Be Asked		Questions to Be Asked
Who?	Who performs the process? Who is affected by this process?	Why?	Why does the company need this process? Why is this process important? Why must it be done?
What?	What is the purpose of this process? What are the steps in this process? What sequence should the steps take? What does this process accomplish? What is being done better? What could be done differently? What purpose does it serve?	How?	How does this process relate to other company processes? How is the work being done? How could it be done differently? How can it be changed to match or exceed the best? How will results get measured?
Where?	Where does the activity take place? Where does it need to take place?	How much?	How much does the old method cost? How much will the new method cost?
When?	When do the activities take place? When should the activities take place? When is the right time for the activities to take place?		

FIGURE 10.10 Five Ws and Two Hs

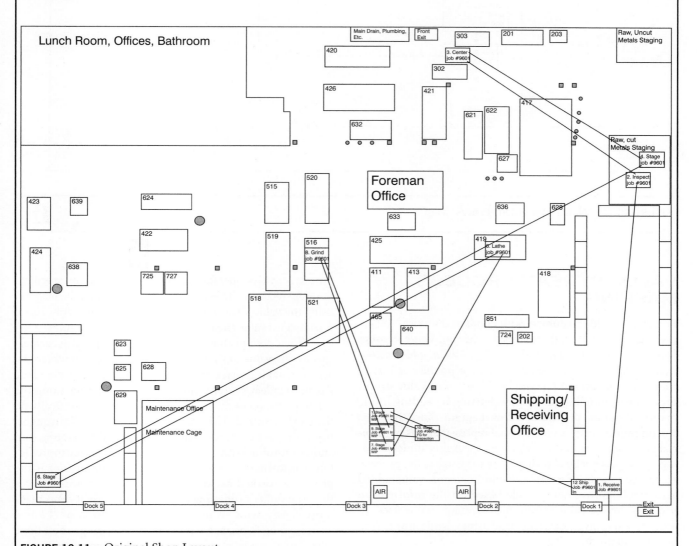

FIGURE 10.11 Original Shop Layout

(Continued)

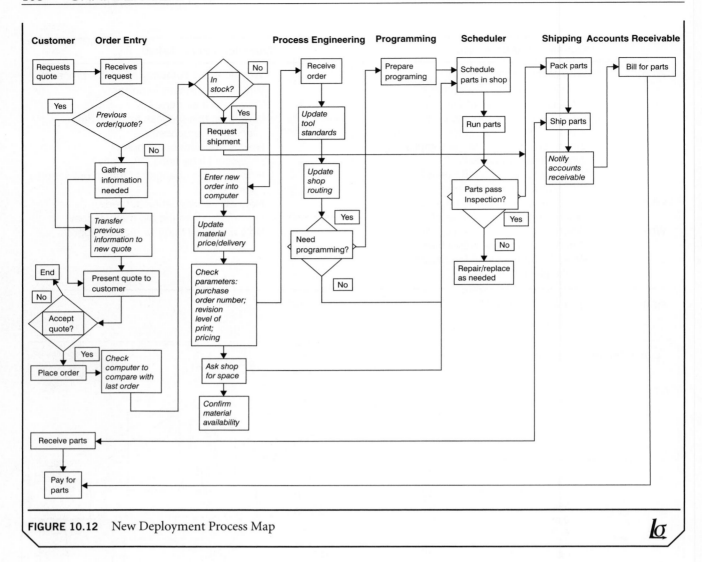

FIGURE 10.12   New Deployment Process Map

VALUE STREAM PROCESS MAPPING

Processes provide customers with goods and services. Wherever there is a process, there is a value stream. *Value streams, which include both value-added and non-value-added activities, are the actions required to create a product or service from raw material until it reaches the customer.* Value stream process mapping builds on flowcharting by adding information about the process to a flow diagram. Since the value stream process map contains visual representations of material and information flow, it provides a clearer understanding of the process by allowing users to visualize the process, recognize sources of waste, and eliminate non-value-added activities. Improvements made based on the information provided by the value stream process map enable users to reduce costs, inventory, and lead times while improving quality and productivity.

To create a value stream map, the steps are the same as a process map or flow diagram except that during the investigation, details such as cycle time, changeover time, uptime, process activities, operator self-inspection notes, and specifications are recorded. These details are displayed on the diagram using the value stream map symbols shown in Figure 10.13. A completed value stream map is shown in Figure 10.14.

When doing value stream process mapping, accurately specify the value desired by the customer. Go to where the action is taking place and identify every step in the value stream. Collect current information by walking along the actual pathways of material and information flow. Really do the work required. This may require several walkthroughs, beginning with a quick walk along the entire value stream and continuing with subsequent visits to gather more detailed information. Process mapping is often easier if the process is worked backwards. Working backwards reduces the probability of missing an activity because it takes place more slowly, without jumps to the conclusion of "I know what is going to happen next." Process mapping is best done with pencil, paper, and stopwatch. Though many computer programs exist to help with process mapping, it is difficult to get them to where the action is taking place. After all, the point of creating the map is to understand the flow of

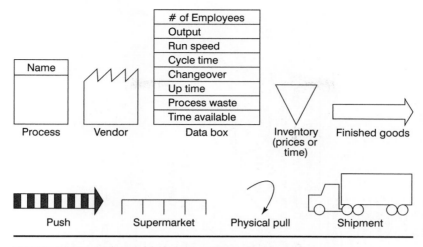

FIGURE 10.13 Value-Added Process Mapping Symbols

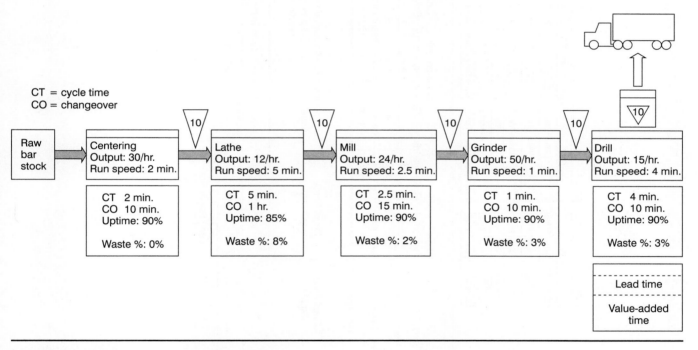

FIGURE 10.14 JF Value-Added Process Map

information and material, not in creating a map. Once the map is complete, remove the waste. Based on the requirements of the customer, make value flow from the beginning to the end of the process.

Typical data gathered during the value stream mapping process includes cycle time and changeover time information, machine uptime, production batch sizes, number of operators, lead times, number of products and their variations, pack sizes, working time, scrap rates, rework rates, and others. Cycle time refers to the actual time it takes to complete a product, part, or service. It is the time it takes for an operator to go through his/her work activities one full cycle.

Value creation time is rarely equal to cycle time. Value creation time is the time it takes to complete those work activities that actually transform the product into what the customer wants. Lead time is the time it takes to move one piece, part, product, or service all the way through the process. All of this information can be captured on the value stream map (Figure 10.15). When the value stream map is complete, it will show the areas that are in need of improvement. From this, create an improvement plan establishing what needs to be done and when, outlining measurable goals and objectives, complete with checkpoints, deadlines, and responsibilities laid out.

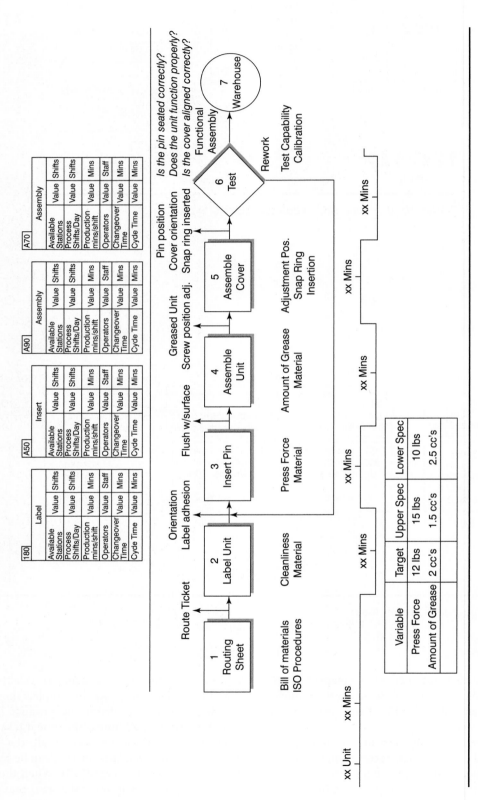

FIGURE 10.15 Value-Added Process Map

Value-Added Process Mapping

LEAN SIX SIGMA TOOLS at WORK

The chocolate candy manufacturing line at Yummy Chocolates is a fully automated line, requiring little human intervention. However, when a lean Six Sigma team reviewed cost of quality information related to production cost overruns, it found excessive scrap and rework rates, and high inspection and overtime costs. To tackle the cost overruns, the team members wanted to isolate and remove waste from the process, thus preventing defectives by removing the sources of variation. To enable them to better understand the production line activities, they mapped the process (Figure 10.16).

The process map revealed several non-value-added activities. Chocolate is melted and mixed until it reaches the right consistency, then it is poured into mold trays. As the chocolates leave the cooling chamber, two workers reorganize the chocolate mold

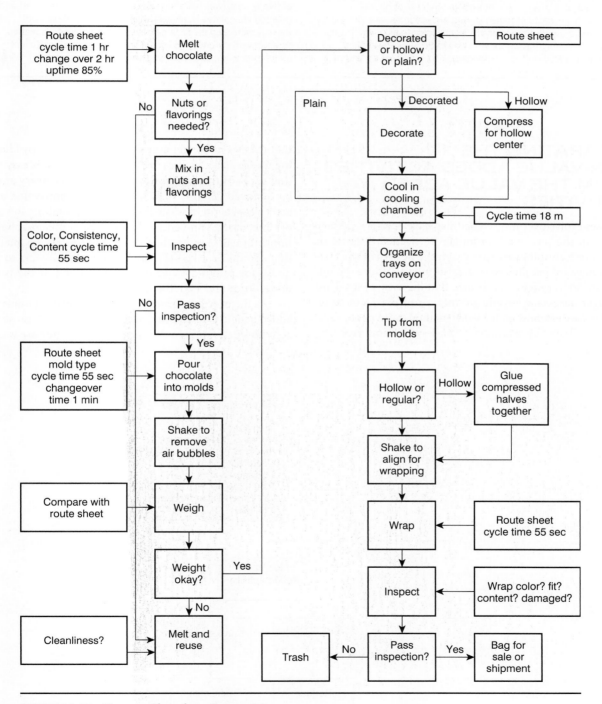

FIGURE 10.16 Yummy Chocolates Process Map

(Continued)

trays on the conveyor belt. This prompted the lean Six Sigma team to ask: why are the trays bunching up on the conveyor in the first place? This non-value-added activity essentially wastes the time of two workers. It also could result in damaged chocolates if the trays were to flip over or off the conveyor.

Yummy Chocolates prides itself on quality product. To maintain high standards, the company has four workers inspect nearly every piece of chocolate as it emerges from the wrapping machine before packaging. A full 25 percent of the chocolate production is thrown out in a large garbage can. Though this type of inspection prevents defective chocolates from reaching the customer, it is a very high internal failure cost of quality.

This huge waste of material, human resources, and production time prompted the lean Six Sigma team to ask: why are the chocolates being thrown away? The simple answer was that the chocolates didn't meet standards. They persevered and discovered that the rejected chocolates were improperly wrapped. Despite the large amount of chocolate being thrown away, no one had suggested that the wrapper machine should be repaired.

If the team hadn't studied the process carefully using value-added process mapping, these two enormous sources of waste and their associated costs would have gone unnoticed. Up until this study, Yummy Chocolates' employees considered these costs part of doing business. During a subsequent kaizen event, modifications made to the production flow through the cooling chamber and to the wrapping machine resulted in a personnel savings of five people (who were moved to other areas in the plant) and rejected chocolates went from 25 percent to 0.05 percent. Savings were evidenced in decreases in in-process inspection, scrap, rework, production cost overruns, overtime, inefficient and ineffective production, and employee lost time.

SEPARATING THE NON-VALUE-ADDED ACTIVITIES FROM THE VALUE-ADDED ACTIVITIES

Processes evolve over time. What may have been a necessary activity in the past may no longer be needed because of changes in technology or changes in the desires of the customer. Because people are so accustomed to doing the job that they do, very few even realize that these extra steps are no longer necessary. People get into a routine. Over time, traditions are established for work methods, and people will comment, "We've always done it that way," without realizing that there is no longer a need to do it that way. By identifying these extra activities on the process map, it is easy to see that some of these activities may not be necessary and can be removed (Figure 10.17). Eliminating non-value-added activities gets rid of waste in the process, resulting in savings of time, money, and effort. As with any problem-solving or improvement activity, the five Ws and two Hs are questions that should be asked (Figure 10.10). The answers to these questions provide a clearer understanding of the process under study and lead to improvements.

As the previous examples show, eliminating non-value-added activities saves time, money, and effort. Because these activities detract from the main business, they are wasteful

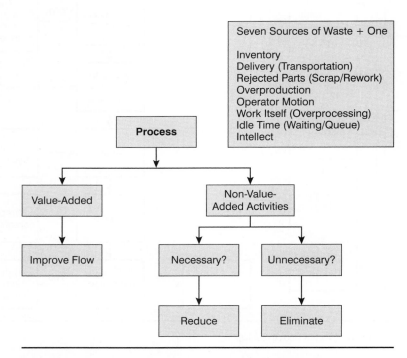

FIGURE 10.17 Value-Added versus Non-Value-Added Decision Tree

Value-Added Process Mapping

> **LEAN SIX SIGMA TOOLS at WORK**
>
> **We've Always Done It That Way**
>
> A local company decided to study its customer purchase record-keeping process to ensure that it accurately reflected customer activities. The team mapped the process. During this activity the team leader, an individual who was not familiar with the process, asked what the "white paper" was that everyone was referring to and what it was used for. The answers she received were, "We have to fill that out"; "It contains important information"; "We have one of those for each customer"; and "It takes a long time to fill out." Obviously these replies did not really answer her question. The fact that no one answered her question, combined with the fact that the white paper takes a long time to fill out, inspired her to be persistent and ask over and over again: "What is the white paper, and why is it important?" Finally, an individual with over 35 years of experience in the department remembered that the white papers were originally created when computers were first introduced. Due to the limited backup capabilities of the computers at that time, records were also kept on "white papers." For 35 years "white papers" had been dutifully filled out by successive generations of department personnel, a totally unnecessary activity in today's computerized world.

and should be eliminated. Eliminating non-value-added activities clarifies the process and allows it to focus on meeting the needs, requirements, and expectations of the customer. Process improvement methodologies, which can help identify and eliminate non-value-added activities, are covered in later chapters.

Removing the non-value-added activities in a process is just a start to process improvement. Following this step, it is critical to revisit the value-added process steps and check to see whether better ways to do the work do exist. The "we've always done it that way" mentality carries over into the value-added activities just as easily as it did the entire process. Determining best practices enables an effective organization to eliminate poor work methods, out-of-date practices, wasteful activities, and unnecessary steps. Doing so makes the organization highly competitive.

> **LEAN SIX SIGMA TOOLS at WORK**
>
> **Redefining the Value-Added Steps in a Process**
>
> JF has been investigating the process of returning defective raw material to suppliers. In the original process it took 175 hours, nearly four weeks, for defective raw material to be returned to a supplier. There were 14 steps to complete, none of which ran in parallel.
>
> A team comprised of engineers, purchasing agents, and shippers mapped the process and studied it to locate non-value-added activities. During four weeks of meetings, the team developed a new process. These improvements reduced the time from 175 hours to 69 hours. The new process included nine steps, two running in parallel (being completed at the same time).
>
> When the team presented its proposed process changes to upper management, the managers asked the team to begin the exercise again, starting with a clean slate, keeping only what absolutely had to take place to get the job done. The team was given one hour to report their new process. The new process, developed in one hour, has four steps and takes one hour to complete. This process was achieved by eliminating the remaining non-value-added steps and empowering the workers to make key decisions.
>
> Though they were able to develop a new, simplified process map for the defective raw material return process, while in their meeting it dawned on the team members that, in the big scheme of things, why were they allowing defective raw material at all?
>
> The team members learned two things from this experience: It doesn't pay to spend time fixing a badly broken process, and sometimes the reason for the process needs to be eliminated.

SUMMARY

Process improvement focuses on eliminating waste—the waste of time, effort, material, money, and human resources. It is the combined knowledge gained during the improvement efforts that enables an organization to develop its own best practices and reach a new level of performance, resulting in delight for its customers.

>
>
> **LEAN SIX SIGMA TOOLS at WORK**
>
> **Using a Process Map to Improve Work Flow and Increase Customer Satisfaction**
>
> Distribution centers pick, pack, and ship orders in one of two methods: wave (also called batch) and discrete. The newly appointed head of CH Distributors has been studying the flow of the pick, pack, and shipment of orders. To optimize this flow, Mr. H. considers the questions, How many times does a picker visit a particular location or area? How many items are picked? How much space is taken up? To better understand the flow, Mr. H. has created the process map shown in Figure 10.18.
>
> After creating the process map based on observing the workers doing their jobs, Mr. H. studied it. He was able to identify
>
> *(Continued)*

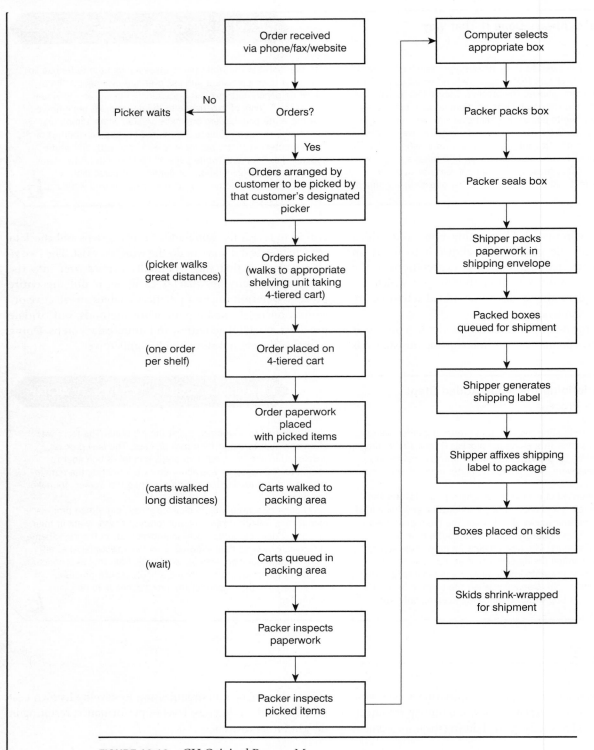

FIGURE 10.18 CH Original Process Map

where delays were occurring, where workers were idle, and where backlogs were created. Based on the process map and his observations, he made the following changes:

Allowed workers to pick orders in order of receipt, regardless of customer, eliminating specialized pickers for each customer. This change eliminated picker idle time, which occurred when no order for a particular customer existed. The change also balanced the workload among all of the pickers, eliminating bottlenecks.

Reorganized pack and ship areas. Previously spread throughout the entire distribution center, the pack and ship areas were combined into one layout. A 10-foot roller conveyor is used to move packages from the four packing stations to the two shipping stations. This change balanced the workload among packers and shippers. This change also allowed workers to interface more easily, enabling them to share process improvement ideas.

Reorganized the distribution center. The Pareto chart provided information concerning the number of customer orders generated by a particular customer. The distribution center was reorganized to place high-volume customers in close proximity with the pack and ship area. This change eliminated considerable walking and searching on the part of the pickers.

Cross-trained pickers, packers, and shippers. During peak times, all workers pick until queues have been generated at either the packing stations or the shipping stations. This change balanced the workload among all workers.

Moved the packaging of the order to the packing responsibility. This eliminated the bottleneck that occurred at the shipping station, while balancing the workload between the packers and shippers more effectively.

Allowed packers to select optimum box size. Since shipping cost is based on the size of the box, this often results in shipping savings for the customer.

Figure 10.19 shows the new process flow map.

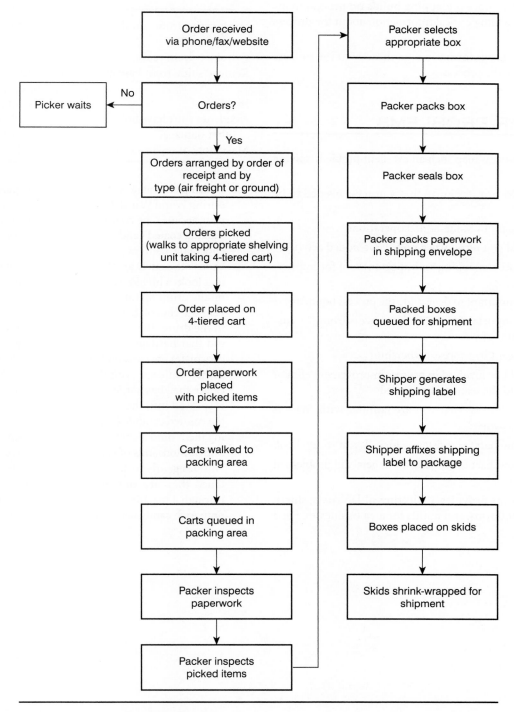

FIGURE 10.19 CH New Process Map

TAKE AWAY TIPS

1. Process mapping is an essential tool in process improvement.
2. Process mapping creates awareness of process activities.
3. Process mapping is used to standardize work methods and processes.
4. Process mapping identifies both value-added and non-value-added activities in a process.
5. Value-added process mapping builds on process mapping by providing critical information about the process activities.
6. To capture key information, go directly to the process and observe the process.

CHAPTER PROBLEMS

1. Why is a process map such an excellent problem-solving tool?
2. How would you recognize that a process needed to be improved?
3. What is meant by value-added operation?
4. How would you recognize a non-value-added activity?
5. Why is process mapping an important tool for improving processes?
6. Why is it important to determine the process boundaries?
7. Why is it important to study the process from the customer's point of view?
8. What is meant by process ownership?
9. Who should be involved in process improvement efforts? How should the team be structured?
10. Describe a process that you are familiar with. Who is the process owner?
11. Create a flowchart for registering for a class at your school.
12. Create a flowchart for solving a financial aid problem at your school.
13. WP Uniforms provides a selection of lab coats, shirts, trousers, uniforms, and outfits for area businesses. For a fee, WP Uniforms will collect soiled garments once a week, wash and repair these garments, and return them the following week while picking up a new batch of soiled garments. At WP Uniforms, shirts are laundered in large batches. From the laundry, these shirts are inspected, repaired, and sorted. To determine whether the process can be done more effectively, the employees want to create a flowchart of the process. They have brainstormed the following steps and placed them in order. Create a flowchart with their information. Remember to use symbols appropriately.

Shirts arrive from laundry.	Ask: Is shirt beyond cost-effective repair?
Pull shirts from racks.	
Remove shirts from hangers.	Discard shirt if badly damaged.
Inspect.	Sort according to size.
Ask: Does shirt have holes or other damage?	Fold shirt. Place in proper storage area.
Make note of repair needs.	Make hourly count.

14. Coating chocolate with a hard shell began with M&Ms during World War II. Coated candies were easier to transport because the coating prevented them from melting. Making coated candies is an interesting process. First the chocolate centers are formed in little molds. These chocolate centers are then placed in a large rotating drum that looks a bit like a cement mixer. Temperature controls on the drum maintain a low enough temperature to prevent the chocolate from softening. While rolling around in the drum, the chocolates are sprayed with sugary liquid that hardens into the white candy shell. Since the chocolates are constantly rotating, they do not clump together while wet with the sugary liquid. Once the white candy shell has hardened, a second, colored, sugar liquid is sprayed into the drum. Once the color coating dries, the colored candies are removed from the drum by pouring them onto a conveyor belt where each candy fits into one of thousands of candy-shaped depressions. The belt vibrates gently to seat the candies into the depressions. Once they are organized on the belt, they proceed through a machine that gently imprints a maker's mark onto each candy with edible ink. Map this process.

15. Create a flowchart using symbols for the information provided in the table below.

16. Create a process map for the milk processing example in Chapter 15.

Materials	Process	Controls
	Buy filter	
	Buy oil	
Wrench to unscrew oil plug		
Oil filter wrench		
Oil drain pan		
	Start car and warm it up (5 min)	
	Shut off car when warm	
	Utilize pan for oil	
	Remove plug	
	Drain oil into pan	
	Remove old filter with wrench when oil is completely drained	
	Take small amount of oil and rub it around ring on new filter	
	Screw new filter on by hand	
	Tighten new filter by hand	
	Add recommended number of new quarts of oil	
	Start engine and run for 5 min to circulate oil	
		Look for leaks while engine is running
	Wipe off dipstick	
		Check oil level and add more if low
	Place used oil in recycling container	
	Take used oil to recycling center	
	Dispose of filter properly	

JUST-IN-TIME AND KANBAN

The next process is the customer.
Kaoru Ishikawa

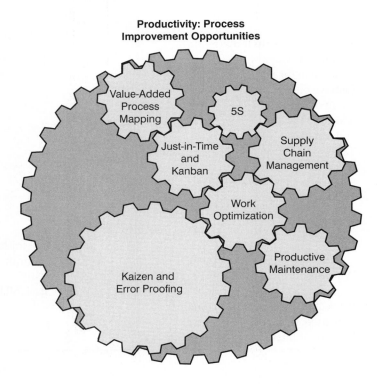

LEARNING OPPORTUNITIES

1. To introduce the concept of just-in-time
2. To introduce the concept of kanban
3. To introduce the concept of jidoka (autonomation)

JUST-IN-TIME

Henry Ford integrated the activities in his automotive empire so that the production from one area fed directly into the next step in the process. By studying his system, Taiichi Ohno, Shigeo Shingo, Ohno Toyada, and other employees of Toyota Motor Company were able to design an effective production system. Developed and refined during the 1950s and 1960s, Toyota's production system rests on two key pillars: just-in-time and jidoka. The key goals behind the system are to increase productivity, reduce costs, and enhance quality by improving the efficiency and effectiveness of processes. Taiichi Ohno (1912–1990) was the driving force behind the Toyota production system. Toyota's system has translated into lean manufacturing in the United States. Shigeo Shingo (1909–1990) focused on the importance of waste elimination to support the just-in-time (JIT) methodology. He was known for finding innovative ways to reduce process setup times and to prevent defects from inadvertent mistakes (poka-yoke).

Mr. Ohno based JIT on the U.S. supermarket concept. In 1956, he visited the United States to study the efficiency and effectiveness of the U.S. supermarket system. Consider what he witnessed. Customers go to a supermarket, select their items at the time they are needed in the amount they require. The supermarket replenishes these items as they are needed by the customer. The goal of the JIT system he designed is to have no buffer stocks in the system. In a manufacturing facility, *a JIT system means that incoming goods arrive, proceed directly to equipment for processing, become work-in-process, and through value-added activities become finished goods just-in-time for the customer to pick up.*

Just-in-time seeks to find the balance between excessive stocks and stock-out situations. Just-in-time, a pull system, is opposite a make-to-stock or push system. In make-to-stock environments, items are sent to the next step in the process as soon as they are completed, regardless of whether the workstation has a need for them. Items that cannot be used immediately are stored as inventory.

Utilizing a just-in-time system has many advantages including shorter lead times, reduced non-value-added time, and lower inventory levels. Just-in-time systems also help create a balance between different but integrated production processes. JIT requires three basic components:

A pull system
Continuous flow processing
Adherence to takt times

Pull systems are manufacturing systems requiring that products be produced only when needed by a customer. In a lean Six Sigma organization, the word *customer* refers to the next step in the process. Just-in-time is based on the idea expressed in the quote that opens this chapter. Because the customer is the next step in the process, items should not be sent to the customer unless the customer has requested them. In this system, the following workstations in a process draw from preceding workstations in the process (Figure 11.1). The customer, anyone who receives work from a step in the process, will not request defects so none should be sent. Just-in-time means that the exact number of defect-free items should be brought to the customer upon request and at the appropriate time. The ability to produce on demand is a pull system. Mr. Ohno and Mr. Shingo stated their JIT goal as follows: "Deliver the right material, in the exact quantity, with perfect quality, in the right place just before it is needed." Just-in-time enhances the efficiency of processes and enables an organization to respond quickly to market changes.

Continuous flow processing focuses on one-piece-at-a-time production. This means that stagnation of work-in-process inventory in and between processes must be eliminated. Inventory is often carried in preparation for problems such as defects, machine breakdowns, and raw material shortages. Just-in-time systems encourage workers to find and eliminate the inefficiencies of their work methods that lead to defects, machine breakdowns, and shortages. Muda (waste), mura (unevenness), and muri (overburden) are attacked and eliminated in order to create a balanced production process. Waste, according to Fujio Cho of Toyota, is "anything other than the minimum amount of equipment, materials, parts, space, and worker's time, which are absolutely essential to add value to the product."[1]

Seven sources of waste were identified by Taiichi Ohno, also of Toyota Motor Company:

Waste from overproduction
Waste from inventory
Waste in unnecessary transportation
Waste from producing defects
Waste in processes
Waste in waiting time
Waste in motions

Some people have modified this list to include intellect waste. Removing waste from the process occurs during 5S and kaizen activities (see Chapters 12 and 13). Mura or unevenness requires that a balance be achieved between the workstations in a process and between integrated processes. Standardization, level scheduling, setup time reduction, and reduced batch sizes support the elimination of mura or unevenness in processes. Overburden or muri occurs when

---

[1] Ohno, T. *Toyota Production System: Beyond Large Scale Production.* New York: Productivity Press, 1988.

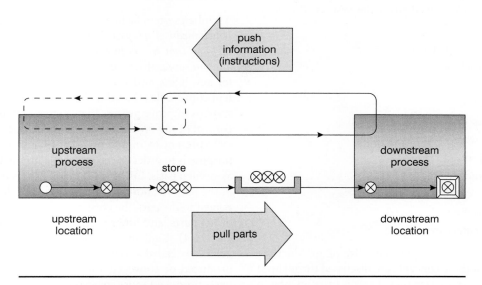

FIGURE 11.1 Pull Production System

> **LEAN SIX SIGMA TOOLS at WORK**
>
> ### Calculating Takt Time
>
> JF Inc. operates a single 8-hour shift five days a week. During the day, employees receive a 30-minute lunch and two 15-minute breaks. The total available working time per day is 7 hours. Customer demand for electric motor shafts is 90 per day. To calculate the takt time, the rate the process must produce the shafts in order to meet customer demand:
>
> $$\text{Takt time} = \frac{\text{Available working time per day}}{\text{Customer demand rate per day}}$$
>
> $$\text{Takt time} = \frac{420 \text{ minutes per day}}{90 \text{ per day}} = 4.67 \text{ minutes}$$
>
> A shaft must be produced every 4.67 minutes

one work area or worker's job is unbalanced when compared with the others. Line balancing occurs when all workers on the line perform their required work in the same amount of time. These concepts are covered in greater detail in Chapter 14. Continuous flow, or single piece processing, is not possible without removing the problems that exist in a process.

Just-in-time systems are based on a customer pulling needed items from a process. The rate at which the customer needs items can be compared with the available working time per day. The value calculated, takt time, is the rate a process must produce an item in order to meet customer demand. Takt time is defined as:

$$\text{Takt time} = \frac{\text{Available working time per day}}{\text{Customer demand rate per day}}$$

Takt times are based on customer requirements. It is the time allotted for the production of one complete item based on the customer demand rate. To eliminate the wastes of overproduction and overburden, a process should never produce more than what has been ordered by the customer. Takt times are used to balance line production with customer demand rates.

For JIT to function effectively, there must be a fundamentally sound system for it to function in. This means that problems have been removed from the system. Problems may include:

Equipment breakdowns
Equipment capability
Scheduling
Material variation
Excessive setup times
Nonstandard work
Waste in the process
Uneven demand
Layout restrictions
Unreliable suppliers
Employee absenteeism

Figure 11.2 presents a summary list of just-in-time principles. Many of these concepts are discussed throughout this text. Being able to manage these areas is instrumental to the success of a JIT system.

Encourage simplicity and flexibility
Eliminate waste
Reduce setup times
Reduce inventory
Standardize work
Practice error proofing
Implement 5S
Level schedules
Utilize jidoka
Use a pull system (kanban)
Achieve first-pass quality
Measure results
Implement preventive maintenance
Utilize visual controls
Maintain supplier relationships
Encourage employee involvement
Cross-train employees
Practice kaizen and continuous improvement
Utilize flexible equipment layout designs

FIGURE 11.2 Summary of JIT Principles

KANBAN

A kanban system facilitates just-in-time production. *Kanban, which means display card in Japanese, is normally a sign, card, or label that communicates what is needed and when.* The kanban serves three purposes: a label for the items, an order for new items, and a record of the items used. A kanban system coordinates the inflow of items to a work area. Kanban systems improve process management by focusing on visual control of the process. Whether the kanban cards are physically present or digital on a computer-based system, their purpose is to order the creation of a product or service on an as-needed basis. Information on the cards includes the name of the part, the number of parts to be produced, instructions related to the part, the creation date of the card, the due date of the parts, and other pertinent information (Figures 11.3 and 11.4). Kanban cards keep track of inventory. To be genuinely called a pull inventory system, parts must not be produced or conveyed without a kanban. The number of parts produced must correspond with the number of parts needed as listed on the kanban. Kanban is a production

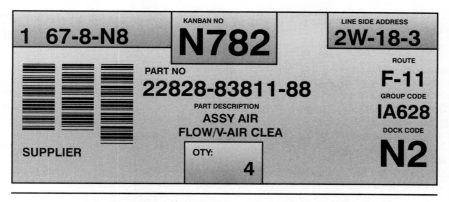

FIGURE 11.3 Bar-Coded Kanban Card

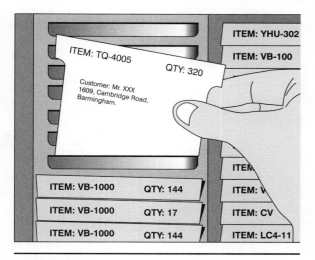

FIGURE 11.4 Kanban Card

control tool. It is not a full inventory control system, nor is it equal to just-in-time. Kanban works with other elements of a JIT system including level production and kaizen. Kanban works best when production schedules are level.

To understand how kanban works, consider the supermarket concept that forms the basis of JIT. A customer goes to the supermarket and selects the items he or she needs. At the checkout, the item is scanned using its bar code. This bar code serves as a kanban. It not only tells how much to charge for the item, but also alerts the system that the item has been sold. When all the kanban cards are collected, in this example the bar code readings compiled, the supermarket will know what has been sold during a certain period. This information triggers a pull signal to the suppliers that the goods need to be replenished (Figure 11.5). The rules for kanban, shown in Figure 11.6, are straightforward. Following these rules reduces the wastes associated with overproduction, unevenness, and transportation. Kanban serves as an excellent visual control system.

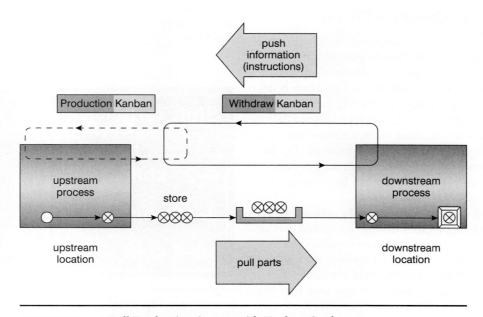

FIGURE 11.5 A Pull Production System with Kanban Cards

Using a Kanban System

LEAN SIX SIGMA TOOLS at WORK

During previous process improvement efforts, JF Inc. reduced the amount of in-process inventory significantly. What little remains still must be managed. A good inventory system enables users to easily identify what has been made and when. It also tells where it is and in what order it should be used.

JF's kanban inventory control system uses small discs the size of dog licenses. They have marked a small portion of their manufacturing floor with bright yellow squares matching the size of the carts used to hold inventory awaiting processing. Each square had a corresponding label painted in it: A-1, B-1, C-1, A-2, B-2, etc. For each of these squares, a corresponding disc labeled with the square location exists. When an inventory cart needs to be parked, a disc is selected from a storage rack that resembles a key holder. The number on this disc represents the parking location of the cart. After parking the cart in the appropriate square, the worker takes the disc over to a series of tubes. Each type of product is represented by a different tube. The disc is inserted into the top of the appropriate tube. Because the different tubes represent different types of inventory, the disc being placed in the tube matches the inventory on the cart. The more in-process inventory, the more discs in the tube, providing a visual representation of the in-process inventory for each type of product. This tube also serves to make sure that a first-in, first-out inventory control system is in place. When inventory is needed for processing, the operator pulls a disc from the bottom of the tube. This disc tells the operator which cart of inventory needs to be taken next.

---

1. Items should only be obtained from a previous step in the process when a kanban card is presented.
2. Only the quantity of items listed on the kanban card should be taken.
3. People in the process should only create items according to the information on the kanban card.
4. If there is no kanban card, no items should be produced or taken.
5. The kanban card should always accompany the item.
6. Defects should not be produced.
7. The number of items requested on the kanban cards should trend toward single piece flow.

FIGURE 11.6 Kanban System Rules

JIDOKA (AUTONOMATION)

Jidoka is the Japanese word for autonomation. Jidoka is one of the two main pillars of the Toyota production system. When a jidoka system is present, machines and equipment are designed to stop automatically when a problem is detected. Any operator on the line also has the power to stop production. Jidoka systems alert the worker to when a defective item is produced or a machine malfunction has occurred (Figure 11.7). This allows the problem to be dealt with immediately, preventing the production and passing of defects. Problems and the defects they cause can be more quickly localized, isolated, and corrected.

Visual controls often accompany jidoka systems. When a piece of equipment stops after having recognized that an abnormal condition exists, an andon or trouble light may come on. Andon means lantern in Japanese. Andons are lanterns that guide people to where there is trouble. When illuminated, this light alerts anyone in the area that the equipment has stopped and that there is an issue that must be dealt with. Because the andon alerts everyone in the area of the existence of a problem, troubleshooting can begin immediately. Corrective action can be taken quickly when problems are exposed. Andons can be lights, buzzers, alarms, or any method that immediately transfers information to the appropriate people without delay (Figures 11.8 and 11.9).

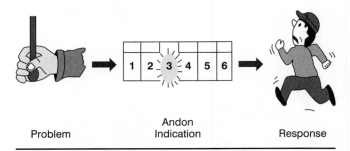

Problem — Andon Indication — Response

FIGURE 11.7 Jidoka System

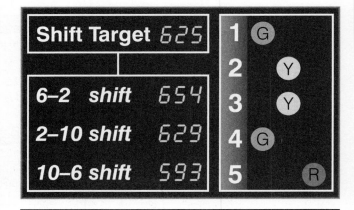

FIGURE 11.8 Andon Board

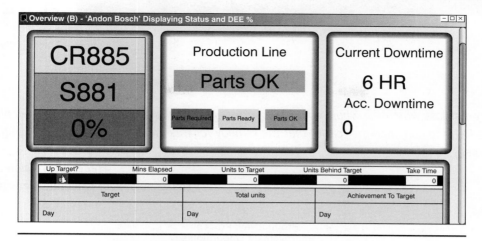

FIGURE 11.9 Andon Board Computer Screen

SUMMARY

JIT requires simplification, cleanliness, organization, visibility, consistency, and agility in order to be viable. As such, JIT is a philosophy, a method of managing that provides significant benefits to organizations that practice it. Lean Six Sigma organizations seeking to lower costs, increase quality, and improve productivity often have JIT as a supporting philosophy. In this text, future chapters discuss the tools and techniques that support JIT. Just-in-time affects and improves all aspects of a business including sales, marketing, engineering, customer service, and manufacturing.

TAKE AWAY TIPS

1. Just-in-time is one of the two main pillars of the Toyota production system.
2. Just-in-time refers to making what is needed, when it is needed, and in the amount needed.
3. Just-in-time relies on the pull system, continuous flow processing, and takt time.
4. Continuous flow processing refers to one-piece-at-a-time production.
5. Jidoka refers to the ability to shut down production lines.
6. Kanban is a key control signal for JIT systems.
7. Kanban serves as instructions for production and conveyance and visual controls over production.
8. Muda translates as waste. Waste can occur in overproduction, waiting, conveyance, processing, inventory, motion, and correction.
9. Mura refers to unevenness in a production schedule.
10. Muri refers to overburden.
11. Takt times are the time allotted for the production of one complete item.
12. Jidoka systems alert the worker to when a defective item is produced or a machine malfunction has occurred.

CHAPTER PROBLEMS

1. What are the two main pillars of the Toyota production system?
2. Describe each of the three basic components required by a just-in-time system.
3. What is the goal of a just-in-time system?
4. How does waste affect just-in-time systems?
5. Calculate the takt time for a company that works two 8-hour shifts per day. Each shift receives 30 minutes for a mid-shift meal and two 15-minute breaks. The customer would like to receive 4,000 tubes of toothpaste per day.
6. Calculate the takt time for a dairy that works an 8-hour shift each day. The shift receives 30 minutes for a mid-shift meal and two 15-minute breaks. The customer would like to receive 2,500 gallons of milk per day.
7. Calculate the takt time for a company that works two 8-hour shifts per day. Each shift receives 30 minutes for a mid-shift meal and two 15-minute breaks. The customer would like to receive 25 generators per day.
8. Create a diagram that describes how kanban systems work.
9. What does the word *autonomation* mean?
10. What are andon lights? How are they used?

5S

. . . the great loss which the whole country is suffering through inefficiency in almost all our daily acts . . . to try to convince the reader that the remedy for this inefficiency lies in systematic management . . .

Frederick Winslow Taylor
The Principles of Scientific Management

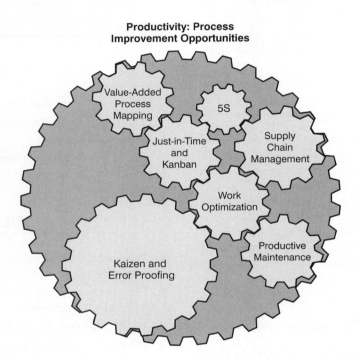

LEARNING OPPORTUNITIES

1. To familiarize the reader with the concept of 5S
2. To provide suggestions on implementing 5S

5S

Lean practitioners often consider 5S the foundation of a lean facility. *5S consists of activities that focus on creating orderliness in a facility*, thus supporting error proofing, setup time reduction, single piece flow, line balancing, visual management, and preventive maintenance. Many people would be surprised to learn that people have been practicing the concepts of 5S for years in a wide variety of situations. Consider the renowned chef Julia Child. By writing *Mastering the Art of French Cooking Volumes I and II*, Julia Child made French cooking accessible to cooks in the United States. In 1961, after years of living abroad, Paul and Julia Child moved to Cambridge, Massachusetts. Here Julia set about creating a kitchen designed to her requirements. After years of cooking in awkward, inadequate, poorly organized, dysfunctional kitchens, Julia carefully designed and organized her ninth kitchen. While she mapped out the functional requirements, Paul set about arranging the kitchen's elements. As they apply to cooking, 5S concepts can be seen in her 14' × 20' kitchen (Figure 12.1). Her kitchen walls were covered with pegboard. On these boards, Paul inscribed the outlines of the utensils, tools, pots, and pans that ought to be hung at each location. While she cooked, Julia Child did not have to waste time, motions, energy, or thought because she knew where things were kept. Donated in 2001, Julia Child's cleverly organized kitchen is now on display at the Smithsonian's National Museum of American History in Washington, D.C.

124

FIGURE 12.1 Julia Child's Kitchen
(**Source:** Julia Child's Kitchen at the Smithsonian National Museum of American History. Photographer Hugh Talman.)

The motto "A place for everything and everything in its place" neatly describes the 5S concept. First described by *5S* author Hiroyuki Hirano, the five Ss are words that serve as a reminder that process improvement lies in the basics. Each word refers to a specific step in the organizational process.

Seiri	Separate or sort
Seiton	Set in order, simplify the arrangement
Seiso	Shine, sweep, cleanliness
Seiketsu	Standardize
Shitsuke	Sustain, self-discipline

Inefficiencies in work methods include muda (waste), mura (unevenness), and muri (overburden). Proper arrangement also includes having work surfaces and equipment designed for ease of use. 5S attacks waste at its source. Whether the waste is from processing, motion, or waiting, when 5S is applied to a workstation, these wastes are eliminated. 5S attacks mura, too, by smoothing out the unevenness in a process. When needed items can be found easily, work pace becomes more balanced. When work processes are arranged according to 5S, workers can expect to have everything they need to perform their work right at their fingertips. In a 5S environment, whether it is a receptionist needing pen, paper, and a computer program to take a phone message or a surgeon needing particular scalpels for an operation, their tools are right at hand, ready for use. Workers no longer experience muri or overburden.

SEIRI

Seiri refers to separating or sorting needed tools, parts, and instructions from unnecessary items. When asked to practice seiri, most people merely tidy up. In this tidy-up process, obvious trash is removed from the work area, tools and other items are stored, and things in general are placed in a neater order. This type of clean-up process does not equal seiri. The objective of seiri is to get rid of all unnecessary items. This means that anything that will not be used frequently must be removed. If the item will never be used and fulfills no useful function elsewhere, it should be disposed of properly. Items that are needed, but only occasionally, should be stored appropriately.

Workplace layouts evolve over time, usually without very much planning or thought about how the work will be performed, what will be used during the process, and how work will flow. Seiri can be applied to existing or newly planned work areas to improve the overall layout and arrangement of items. The team working on 5S not only does a careful study of what is necessary, but also can explore how to improve work flow and communication. The tools and equipment present should be placed to support key work processes and flow.

To define an unnecessary item, consider the 24-hour rule. If an item is not used or touched during a 24-hour period, it should be stored elsewhere. Remember, seiri is not just making things neat; it is a concentrated effort to determine what is actually used and what is not. When performing seiri, allow the person who will perform the work to identify and decide which items will stay in the work area and which will not. Practicing seiri requires a person to make decisions about an item's usefulness. Typically, 40 percent to 60 percent of the items found in a work area end up being removed. At the end of seiri, a clean, uncluttered work area will exist. Clutter, excess and unused equipment, supplies, and papers will no longer be there.

> Look around your room or office or look at the lap drawer of your desk, how many of those items do you actually use on a daily basis? Seiri encourages the appraisal of each item and determining its usefulness. When practicing seiri, keep nothing that is not immediately useful.

A Place for Everything and Everything in its Place

LEAN SIX SIGMA TOOLS at WORK

In an April 1, 2005, article, "Boeing, Airbus Look to Auto Companies for Production Tips," the *Wall Street Journal* described many process improvement techniques including one involving 5S.

At an Airbus factory in Wales, where it builds wings, production teams used to walk far to the stockroom for bags of bolts and rivets, and frequently left them scattered about—a wasteful and unsafe practice—because they lacked nearby storage. Using work-analysis methods developed by the auto industry, project teams studied which fasteners were needed where, and when, and then organized racks on the shop floor. Now carefully labeled bins contain tidy sets of supplies needed for specific tasks. The change has sped up work and saved over $100,000 in rivets and bolts at the Welsh factory alone.

SEITON

Seiton refers to setting things in order, creating optimal boundaries and locations for each item in a work area. Whether it is surgical instrument trays or Julia Child's kitchen, seiton ensures that the items necessary to perform the work are placed in a sensible order or arrangement based on their use. For instance, in Julia Child's kitchen, the oil and vinegar were kept close to the stove. Sixteen baking sheets were stored vertically in a rack located next to the dishwasher. The goal of seiton is to arrange the key items in the work area so that they are easy to locate and use. When Paul Child drew outlines of pots and pans on pegboard, he was performing seiton by making key cooking items easy for Julia Child to locate and use.

Seiton should never be performed before seiri. It just does not make sense to organize unnecessary items. No work processes can be performed well if the work area is in disarray. Work centers utilizing 5S often have taped, painted, or clearly labeled locations for everything that is needed to perform the work (Figure 12.2). After each use, the operator returns the item to its convenient and marked location on the shadow board.

Seiton incorporates the principles of motion economy espoused by Frank and Lillian Gilbreth (1868–1924 and 1878–1972, respectively). The Gilbreths (Figure 12.3) studied the work activities of people in a wide variety of industries. Their aim was to find ways to increase productivity while making the job easier to perform. Essentially, this is what seiton encourages practitioners to do. For example, Frank Gilbreth began his working life as a bricklayer. He

FIGURE 12.3 Frank and Lillian Gilbreth
(**Source:** CORBIS–NY.)

noticed that different bricklayers used different methods and motions when laying brick. Because he was paid by the number of bricks he was able to lay in a day, Mr. Gilbreth sought to remove unnecessary and inefficient motions from his work. He quickly designed a variety of improvements, including a scaffold that could be easily adjusted so that the height of the working platform would enable the bricklayer to stand at the right height at all times. The design of the scaffold included a shelf for the bricks and mortar. This feature saved time and effort because it was no longer necessary for the bricklayer to bend down and pick up each brick. To further speed the process, bricks were brought to the bricklayer by lower paid workers. These bricks were stacked with the best side out and the end of each brick always in the same position. This eliminated the wastes of reaching, searching, and prepositioning. The scaffold's careful design meant that the bricklayer could pick up a brick with one hand and mortar with the other. Mr. Gilbreth's improvements reduced the number of motions made in laying a brick from 18 to 4 1/2. In terms of bricks, using Gilbreth's methods, a bricklayer could lay 2,600 bricks a day versus the traditional 500 bricks per day.

After removing unnecessary items from the workplace (seiri), 5S practitioners study the activities taking place while the person works (seiton). Often, this study takes the form of a time and motion study developed by the Gilbreths. During a time and motion study, the actions a worker takes are recorded in detail. Filming creates a record of the operation being performed multiple times. Designers of a work area can determine how items are used as well as their frequency of use. From this information, they can select a location for the item. Items are carefully placed to eliminate the wastes associated with long reaches, awkward motions, holds, searches, and transports. The new work

FIGURE 12.2 A Place for Everything and Everything in its Place

LEAN SIX SIGMA TOOLS at WORK

Bricklaying

Scaffolds and work methods based on Gilbreth's original designs still exist today. Annual bricklaying contests are held throughout the United States. The current world record holder, who can lay an average of 861 bricks per hour, works at McGee Brothers in North Carolina. McGee Brothers, which employs 600 masons, regularly involves its employees in sessions designed to use the 5S concept to improve the speed, efficiency, and effectiveness of bricklaying.

At MQ Homes, Max has read about the bricklaying contests. He has decided to teach his bricklayers how to work, not just lay brick. To support this mission, he has adapted a variety of different workable improvements into his processes:

- Special truck beds to make sure the masons have what they need when they need it
- Special jigs and fixtures to set up and align bricks easily
- Special scaffolding to reduce the amount of bending on the job
- Special mixing devices for the mortar to save time and effort
- Special products to offer the customer a customized look while saving its employees time and effort during the installation
- Special forklifts designed to transport and load material in the brickyard more easily
- Special materials trailers to hold all necessary materials for a job, including a special storage area for scaffolding
- A company store that sells everything and anything its employees might need on the job, including tools, jigs, fixtures, brackets, nails, materials, paper towels, and Gatorade

Many of these changes are based on motion and time study analyses that were used to study the jobs for performance improvements from productivity, human factors, and safety aspects. These studies focused on such areas as the best scaffolding heights for individual bricklayers. One result of these studies is the creation of an easily adjustable scaffolding that allows for two-inch increments. It is operated by a foot pedal. For the laborers who prep the job sites for the bricklayers, other studies investigated the optimum brick orientation, distance apart, and number per stack for the best bricklaying speeds.

Besides the changes to the equipment, Max took a look at his bricklayers. He divided them into crews of five people, three masons and two laborers, for each job. These crews are lead by the lead mason. These lead masons understand that the job site must be left "broom clean" upon completion of the job. This means that the job is considered complete only when the brick crew has cleaned up the materials from the job site.

Based on these 5S inspired changes, even on complex jobs, Max's crews can lay an average of 2,500 bricks per day. If the wall is uncomplicated, they can lay on average four bricks per minute. This performance level is 30 percent above the national average. Due to safety training and hard-hat requirements, the incident rate of accidents or injuries is very low, 75 percent below the national average. To spread improvements throughout MQ Homes, Max documents work processes and shares improvements and productive ideas among all crews. Max makes sure that his lead masons enforce the discipline to utilize the more efficient, effective, and safer methods.

---

area designs are tested for convenience and modified if necessary.

Workstations that have been designed with seiton in mind emphasize convenience. Their design places key items within arm's reach (Figure 12.4), keeps body motions to a minimum, and places items in their sequence of use. This search for one best way to perform a job makes performing the tasks more effective and eases worker fatigue. By filming people while working or by testing movements on their 12 children (Figure 12.5), the Gilbreths described hand motions as combinations of 18 basic motions called therbligs (Figure 12.6). The principles of motion economy focus on eliminating waste (muda) related to motions such as searching, reaching, holding, transporting. Eliminating wasted motions makes work performance faster and easier. The Gilbreths studied the activities of surgeons in the operating room because wasted time, motions, energy, or thought could have significant consequences for the patient. They developed and championed a system of arranging the tools

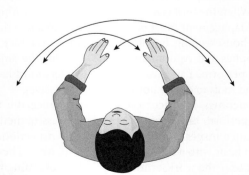

FIGURE 12.4 Everything Within Arms' Reach

FIGURE 12.5 The Gilbreth Family
(**Source:** AP World Wide Photos.)

Search	Find
Hold	Release load
Select	Use
Grasp	Inspect
Assemble	Disassemble
Transport loaded	Transport empty
Preposition	Position
Unavoidable delay	Avoidable delay
Plan	Rest

FIGURE 12.6 Therbligs

a doctor would need for a particular type of surgery on a tray (Figure 12.7). They also proposed that an assistant pass the doctor the tools upon request. In another arena, the Gilbreths devised the techniques used by soldiers to assemble and disassemble their weapons in the dark or blindfolded.

The Gilbreth's motion economy work is considered the foundation for the modern study of ergonomics. Thanks to the efforts of Purdue University's Dr. James H. Greene, department chair of Industrial Engineering, the Gilbreth's papers on motion economy now reside at Purdue University's Engineering Library. The Gilbreths are also the subject of several books including *Cheaper by the Dozen*. Dr. Lillian Gilbreth is considered one of the world's great industrial and management engineers. In 1984, she became the first woman industrial psychologist to be shown on a U.S. postage stamp. Late in her life, Lillian Gilbreth summarized their philosophy by saying, "When it comes to the questioning method, of course he shared with all the scientific management group the belief in the value of questions and the need to ask these questions over and over determining how the thing was to be done and why it was done and how the betterment could be brought about." She continued with "The things which concerned him more than anything else were the what and the why—the what because he felt it was necessary to know absolutely what you were questioning and what you were doing or what concerned you, and then the why, the depth type of thinking which showed you the reason for doing the thing and would perhaps indicate clearly whether you should maintain what was being done or should change what was being done" (*IW/SI News*, Issue 18, September 1968, "Pioneers in Improvement and Our Modern Standard of Living," 37–38).

These are thoughts to keep in mind when performing seiton on a work area. It is important to understand the whys and whats related to performing a job. To make improvements that make the work more effective and efficient, the work itself must be clearly defined and carefully thought out. As with Gilbreth's scaffold, the workspace can then be designed to support the work.

SEISO

Seiso is cleanliness. Some refer to this step as shining or sweeping. Here the focus is on removing grime and dirt from the work area. *Seiso means that every day the workplace should be cleaned, straightened up, and restored to order*. Tools and equipment should be cleaned at the end of each use. Anything unnecessary should be disposed of properly. Begin seiso by scrubbing down the work area. Clean machines, equipment, tools, and work areas enable unsafe or broken items to be identified. Unsafe conditions should be immediately reported and rectified. Any spills, leaks, or other sources of fluids should be cleaned up and repaired. Areas covered with dirt and grime are wasteful. Workers who feel unsafe will not be productive. Part of their attention will be rightfully focused on keeping themselves from harm.

Following the initial cleaning (seiso), it is appropriate to set up a housekeeping schedule with assigned responsibilities. Each person must understand the importance of maintaining a clean, organized workspace. The housekeeping schedule can detail when time should be allotted during the day for cleaning. Some companies chose to clean at the end of a work shift. This may or may not be appropriate, depending on the type of work performed. For instance, a machine setup that occurs in the middle of the day will require specific tools. When the setup is complete, it is appropriate to return the tools to their marked storage area. In other situations, it is appropriate to return the tool or item to its marked storage area immediately after use. This way it can be immediately accessible when needed again. Some companies install devices that suspend tools from the ceiling on retractable cords. When the tool is not being used, it hangs within easy reach. Management must be sure not only to allow time for regular housekeeping, but also to supply the necessary cleaning supplies and equipment. Seiso is not for the janitorial crew to perform. This means that the workers must have the skills and tools to do the housekeeping themselves.

Sometimes, in order to achieve compliance, it is necessary to perform inspections. One very successful method of compliance inspection is to conduct the inspections on a random basis with a rotating team of employees. These employees tour the workstations with a housekeeping review form listing targets to be met. The team makes appropriate notations about the cleanliness and organization of the

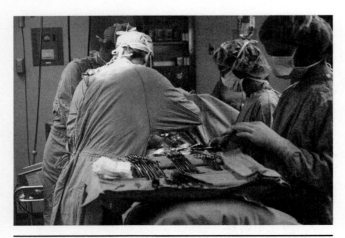

FIGURE 12.7 Nurse with Surgery Tray
(**Source:** G. B. Fry/Pearson Education/PH College.)

workstations. On a regular basis, prizes are awarded to employees who consistently keep their workstations clean and organized. The use of a rotating team of employees who inspect at random times keeps anyone from becoming a disliked inspector or a favorite. This type of visual control is often necessary to make the idea of cleaning and organizing a routine activity.

SEIKETSU

Once a work area has been sorted (seiri), organized (seiton), and now cleaned (seiso), seiketsu must be implemented. *Seiketsu reminds people to conduct seiri, seiton, and seiso at frequent intervals in order to maintain the workplace in pristine condition.* This fourth step focuses on standardizing the work processes put in place in the previous three steps. The sorting, organizing, and cleaning of the work areas should become a habit. After the initial push to sort, organize, and clean, smaller efforts need to be made on a daily basis to maintain the new level of order. Basically, workers in all areas of the company need to realize that this is the expected way of behaving. The tasks of sorting, arranging, and cleaning should be part of their daily job description. In other words, keeping their work area in order should be a habit, a way of life.

SHITSUKE

Shitsuke refers to sustaining. This step focuses on ensuring that the organization has the discipline to sustain the habit of sorting, organizing, and cleaning. The habit of sorting, arranging, and cleaning should become a daily routine. Can the new level of order be maintained or will the work areas return to their former levels of disorganization? It takes discipline and self-control to sustain a change of this magnitude.

 5S   **LEAN SIX SIGMA TOOLS at WORK**

JF management applied 5S to the entire facility within a few weeks of the purchase.

Seiri: Separate

One Monday, large trash cans appeared at every workstation. Employees were rather taken aback when at a short meeting all employees were given two directives: make a list of what you absolutely must have at your workstation and throw the rest out. This meant filling their trash cans by the end of the week.

The owner quickly dispelled disbelief by filling his own trash can with a multitude of unnecessary files from his own office. Slowly, others did the same. Soon, front-office desktops contained only what was needed for their function. As they dug out from the mounds of unnecessary files and papers, the office seemed bigger and lighter. Some separating was easy, like throwing out a phone book from 1998 or old lunch wrappers. Other separating took time, determining which files to keep and which to throw out or place elsewhere. Items in the work area were tagged red, yellow, or green. Only the green-tagged items remained in the work area (Figure 12.8).

In the shop, employees held many discussions concerning the optimum places to store tools and tooling (Figure 12.9). There weren't enough tools for each workstation, so setups frequently involved walking around looking for the right tool. Management invested in the needed tools so that each workstation had a complete set. To ensure that these tools didn't migrate around the plant, maintenance constructed boards drawn with outlines of each essential tool to hang vital tools and tooling on for each workstation (Figure 12.10). Involving all the operators, supervisors, setup, and maintenance employees in preparing these boards meant that their acceptance and use is ensured.

Seiton: Set in Order, Simplify, or Straighten

The most visible examples of straightening also took place in the inventory area and in tooling. Cooperative education students were hired to create a tooling area complete with labeled bins for all tools, gages, and spare parts. They also created tooling sheets listing all the tools and gages required to start a job. Setups used to be a process of having to walk around the shop looking for

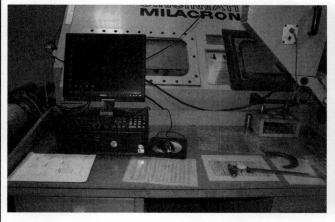

FIGURE 12.8 Workstation Following 5S

FIGURE 12.9 Messy Workstation

(Continued)

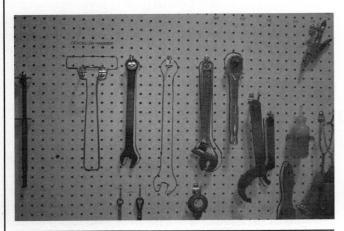

FIGURE 12.10 5S Workstation Boards

things. Now, the employee prints the job information, takes a basket, and walks around the small room gathering what is needed. Strict rules are in place for replacing what has been used. One individual has been assigned the task of daily receiving new tooling inventory and keeping the room organized (Figure 12.11).

The inventory storage area received special attention from the co-ops. Outdated and unneeded material was sold for scrap. New shelving units got material up off the floor and out of harm's way. Changes made to material ordering policies prevent overstocking. These changes resulted in a net space gain of 50 percent.

Seiso: Shine or Scrub

Making metal shafts means making metal chips. Chips attached themselves to shoes and got tracked everywhere. Chips overflowed in work areas, sometimes damaging parts during processing. Chip control became a key focus of cleanup efforts. Investigation revealed that chips, when compressed into units the size of a hockey puck, could be sold quite profitably for reuse. This fit nicely into management's concern for the environment, the need to recycle, and the need to eliminate waste. Management invested in a chip collection system. At each machine, vacuums placed on 55-gallon drums collect the chips. When full, these barrels are taken to a machine that compresses the chips into pucks. Once a week, a scrap dealer collects the chips. The $50,000 system paid for itself in four months. The shop is cleaner, and the root cause of many quality problems and machine maintenance headaches has been removed.

As machining areas employed the five Ss successfully, the machines in those areas were painted. Lathes sport a blue and white paint scheme. Mills are gray, and grinders are burnt orange. This visual statement also served as an incentive for other machinists to separate, straighten, and scrub their work areas.

Seiketsu: Standardize

Maintaining known and agreed-on procedures and conditions takes effort. The new tooling boards support faster setups, but more improvements could be made. The co-ops videotaped a number of different machine setups, and then the employees discussed a variety of ways to standardize the setups. The new methods are more effective, reducing some setup times by as much as 80 percent. In other areas, standards have also been set. Shipping procedures changed to ensure safer travel of the shafts to customers. The challenge in shipping has been to keep the area clean and organized with the increased volume of parts flowing through.

Shitsuke: Sustain or Systematize

Discipline to maintain the new environment is key. This discipline must occur on a personal level. Each employee must agree to follow standards and procedures necessary to maintain organizational neatness and cleanliness (Figure 12.12). JF management has put into place rewards and recognition that help support the employees as they make a habit of the five Ss.

FIGURE 12.11 Organized Tooling Storage

Housekeeping/Safety Audit
Area: _____ Shift: _____ Date: _____

1. Work area is clean and orderly.
2. All charts and information are organized and up to date.
3. Tooling, spare parts, and material storage areas are properly identified and stored in a safe and orderly manner.
4. Storage locations and 5S boards are being used consistently.
5. Floors are clean of oil, dirt, debris, and grime.
6. Machines, conveyors, work surfaces, and other surfaces are clear of hand tools, spare parts, and other material.
7. No accumulation of dirt, grease, or oil is on any surface.
8. Racks, baskets, pallets, and parts are properly identified. Nothing is stored on the floor.
9. Scrap hoppers are free of gloves, rags, and other nonrecyclable material.
10. Trash containers are emptied regularly and are not overflowing.

FIGURE 12.12 5S Audit Sheet

SUMMARY

5S is a five-step process of creating order out of chaos. Well-organized and well-designed workstations result in highly productive people and machines. Companies that pursue improvements in quality, cost, and productivity recognize that the foundation of effective and efficient work areas lies in organization. The five Ss remind us of the steps necessary in creating organized work areas. Seiri encourages the separation and sorting of items necessary to the work processes from those that add clutter and no value. Seiton focuses on simplifying activities in the work area by setting things in order. Seiso requires that the work areas be swept and shined on a regular basis. Seiketsu is standardization of the improved methods. Shitsuke reminds us that it takes discipline to sustain the changes made in the previous steps. This final step asks questions such as the following: What happens when no one is looking? Does disorder return? Does the work area get dirtier? Do unneeded items creep back in? 5S can fail if there is no effort to make a habit of the first four Ss. To achieve improvements in quality, cost, and productivity, management must be sure through audits and reward systems that the five Ss become ingrained in the organization's work practices.

TAKE AWAY TIPS

Seiri	Separate or sort
Seiton	Set in order, simplify the arrangement
Seiso	Shine, sweep, cleanliness
Seiketsu	Standardize
Shitsuke	Sustain, self-discipline

CHAPTER PROBLEMS

1. Describe seiri. How would you do it?
2. Describe seiton. How would you do it?
3. Describe seiso. How would you do it?
4. Describe seiketsu. How would you do it?
5. Describe shitsuke. How would you do it?
6. Study the activities that take place as a typical customer checks out at a grocery store. Study the checkout station itself. Once you have assessed the situation, perform a 5S on a typical checkout station. Be sure to include diagrams or pictures that enable you to clearly describe the existing conditions and the changes you would make. If you choose to visit a particular grocery store, be sure that you have written permission from the manager. To make this easy for the manager to grant, prepare a short document requesting permission that the manager can sign.
7. 5S your own personal work area. Don't skip any steps or keep unnecessary items. How hard was it to do? How long did it stay neat and organized?

KAIZEN AND ERROR PROOFING

There were many things that Scotty taught me. . . . Have respect for your tools. Have patience in the face of a medium that is not always cooperative. Most importantly, he taught me to always have pride in workmanship no matter how small the task. Strive to improve work procedures and think through solutions to simplify them.

Jay E. Minnich
Woodshop News, *December* **1997**

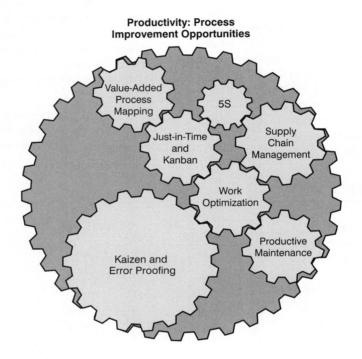

LEARNING OPPORTUNITIES

1. To introduce the kaizen concept
2. To introduce the concept of error proofing (poka-yoke)
3. To introduce the WHY-WHY diagram

KAIZEN

The writer of the quote that opens this chapter never met Dr. W. Edwards Deming, Henry Ford, or Masaaki Imai. Neither did his mentor D. Scotty Grau. They never had an introduction to lean either. Though Mr. Minnich celebrated his 90th birthday in 2009, the concepts taught to him by Scotty in the early 1930s are as relevant today as they ever were. "*Strive to improve work procedures and think through solutions to simplify them*" is a clear summary of the concept of kaizen. As this quote shows, the importance of process improvement and removal of waste from a process has been known for a long time. Lean Six Sigma practitioners use kaizen projects to achieve incremental improvements in quality, costs, and productivity.

Pronounced "k-eye-zen," the word *kaizen* is the combination of two words. In Japanese *kai* means "little," "ongoing," and "good." *Zen* means "for the better" and "good." Kaizen improvement efforts are little, ongoing, good improvements that make things better. *Kaizen events are short, highly focused projects that improve the activities in a work area.* Usually completed over a five-day period, these sessions begin with training followed by an analysis of a work area and the implementation of improvement ideas. The steps are straightforward. Begin by identifying that a problem exists such as defects, waste, or a process that is not working effectively. Next, generate improvement ideas. Review these ideas with management. Select and implement the most effective solution. 5S activities are often part of kaizen events. Kaizen and continuous improvement can be considered similar, if not the same thing. Both seek to improve processes with the goal of improving quality, reducing costs, and increasing productivity.

Continuous improvement efforts have as their foundation the teachings of Dr. Deming. Dr. Deming summarizes his philosophies on management involvement and continuous improvement in his 14 points. Several of those points directly relate to kaizen and continuous improvement. The first of his 14 points for management states: *Create a constancy of purpose toward improvement of product and service, with the aim to become competitive and to stay in business and to provide jobs.* This first point encourages leadership to constantly improve products or services through innovation, research, education, and continual improvement in all facets of the company. Dr. Deming's fifth point clearly states the kaizen and continuous improvement concepts: *Constantly and forever improve the system of production and service.* An organization cannot remain truly competitive unless it strives to continually enhance its business processes that provide the products and services its customers want. In the chapter's opening quote, to "have pride of workmanship no matter how small the task" is reminiscent of Dr. Deming's 12th point: *Remove barriers that rob people of their right to pride in workmanship.* Barriers are any aspect of a job that prevents employees from doing their jobs well. By removing them, leadership creates an environment supportive of employees and the continuous improvement of their day-to-day activities. In lean Six Sigma organizations, employees work patiently to continually improve the processes they work with. Dr. Deming's sixth point, *Institute training on the job*, also supports continuous improvement. Continual education and training creates an atmosphere that encourages the discovery of new ideas and methods. This translates to continuous improvement and innovative solutions to problems.

Consider the Japanese proverb: *If a man has not been seen for three days, his friends should take a good look at him to see what changes have befallen him.* Once again, it is apparent that change is expected. For kaizen practitioners, that change positively affects costs, quality, and productivity. Kaizen was first presented as a single concept in 1986, in Masaaki Imai's text, *Kaizen, The Key to Japan's Competitive Success.* In this text, he discussed the concepts, tools, and techniques that form the basis of kaizen. He defines kaizen as a separate concept from innovation. Kaizen, like continuous improvement, refers to small incremental improvements. Innovation focuses on significant breakthroughs in technology, inventions, or theories. Kaizen is often less high tech than innovation. It is also a people-oriented concept. The benefits kaizen provides are numerous. They include improvements in quality, throughput, safety, productivity, costs, and customer service.

Process improvement projects can take many forms. Most fall into two categories: long-term projects and kaizen projects. Long-term projects are just what the name suggests: large, complex improvement projects that will require significant time and cross-functional team effort to solve. Team members involved in this type of project integrate project work with their regular day-to-day activities. The duration of the project depends on its scope. Kaizen projects are normally well-defined from the beginning. The scope, objectives, and boundaries of the project are clearly defined and understood. Because these projects focus on obvious sources of waste, it is easier to take action and achieve results. Kaizen projects can usually be completed rapidly, often in one week or less. Teams consist of people who have been pulled from their regular work to focus intensely on solving a particular problem. The kaizen team consists of people closest to the problem. Support areas, such as maintenance, engineering, and information technology, are alerted to the need for quick response to the team's requests. These team members are trained in basic problem-solving tools and techniques.

Both types of projects utilize the problem-solving approach of define, measure, analyze, improve, control (DMAIC) methodology presented in Chapter 9. Sources of waste can be identified and verified through process mapping and data collection (Chapters 10 and 17). Long-term projects, due to their large scope, should use the project management steps presented in Chapter 8. The tools and techniques utilized to gather information and analyze the process will depend on the nature of the problem.

Kaizen events are typically concentrated improvement sessions that last about a week. Teams meet on the first day for training and problem identification. Kaizen practitioners go to the actual work area, work with the actual part or service, and learn the actual activities required in the work situation. From there they analyze the process by documenting activities, processes, and cycle times. This information enables them to discuss process improvement options. These improvement options are implemented and tested for their ability to solve the problem. This may involve rearranging existing equipment, creating mock-ups, or going through the motions of the newly designed process. The improvements are refined, tested again, and the results presented to management. During kaizen events, lean and Six Sigma tools and techniques blend together, enabling participants to optimize process, product, or service performance. Six Sigma tools and techniques emphasize root cause analysis through the use of a standardized problem-solving technique (PDSA or DMAIC) in combination with statistical analysis and performance measures. Kaizen improvements must plan a corrective action (Plan), implement it (Do), confirm its success or failure (Study), and determine whether more actions are needed (Act). Lean uses lean tools to locate and eliminate sources of waste. Many of the examples used in this text are from kaizen events. Kaizen activities use a variety of tools and techniques, many of which are covered in chapters throughout this text.

	Questions to Be Asked
Who?	Who performs the process?
	Who is affected by this process?
What?	What is the purpose of this process?
	What are the steps in this process?
	What sequence should the steps take?
	What does this process accomplish?
	What is being done better?
	What could be done differently?
	What purpose does it serve?
Where?	Where does the activity take place?
	Where does it need to take place?
When?	When do the activities take place?
	When should the activities take place?
	When is the right time for the activities to take place?
Why?	Why does the company need this process?
	Why is this process important?
	Why must it be done?
How?	How does this process relate to other company processes?
	How is the work being done?
	How could it be done differently?
	How can it be changed to match or exceed the best?
	How will results get measured?
How much?	How much does the old method cost?
	How much will the new method cost?

FIGURE 13.1 Five Ws and Two Hs

People practicing kaizen seek to improve work procedures, often by simplifying the activity taking place. Kaizen's guiding words are *combine, simplify, eliminate*. Kaizen seeks to standardize processes while eliminating waste. Waste can be considered any activity that consumes resources that do not add value to a product or service. Remember, value must be defined from the perspective of the customer. Waste in a process or system can be removed by combining steps or activities, simplifying steps or activities, and eliminating any waste in a system or process. Kaizen activities may take two forms: flow kaizen focusing on value stream improvement and process kaizen focusing on the elimination of waste.

Flow kaizen events study the value stream associated with providing a product or a service. The kaizen team studies the process by actually following the process from start to finish. Sometimes the team takes part in the process itself in order to better understand what the customer needs, wants, and expects. Properly performed, the study of a process will include determining the answers to key questions. These questions are easy to remember as the five Ws and two Hs (Figure 13.1). These questions help those studying the process understand it more clearly. Understanding leads to being able to differentiate between value-added and non-value-added activities. The predominant tools and techniques the team members use are described in Chapters 9 and 10.

Process kaizen focuses on the elimination of waste. This is not a new idea. Henry Ford (Figure 13.2) wrote about the problems associated with waste in his book *Today and Tomorrow*, published in the 1920s. Ford's commitment to lowering costs while maintaining quality caused him to seek out sources of waste in a process. His productivity improvements that eliminated waste include introducing the moving assembly line concept to his plants in 1913 (Figure 13.3). He also simplified the design of automobiles, creating the Model T in 1908. Process improvements enabled him to drop the price of the car from $825 in 1908 to $360 ($7,020 in 2008 dollars) by 1916. By 1918, half the cars in the United States were Model Ts. By 1927, the final year the Model T was made, 15,007,034 cars had been manufactured. Another of his

FIGURE 13.2 Henry Ford
(**Source:** Getty Images, Inc.–Hulton Archive Photos.)

FIGURE 13.3 Henry Ford's Assembly Line
(**Source:** Ford Motor Company.)

improved productivity, which enables a company to capture more of the market, which enables the company to stay in business and provide more jobs. Sound familiar?

In Japan, the methods of Henry Ford and W. Edwards Deming were carefully studied to determine how to increase productivity, reduce costs, and improve quality. Waste, according to Fujio Cho of Toyota, is "*anything other than the minimum amount of equipment, materials, parts, space, and worker's time, which are absolutely essential to add value to the product.*" Taiichi Ohno, also of Toyota Motor Company, categorized seven sources of waste:

- Waste from overproduction
- Waste from inventory
- Waste in unnecessary transportation
- Waste from producing defects
- Waste in processes
- Waste in waiting time
- Waste in motions

productivity improvement efforts served the United States well in World War II. He showed the U.S. government how to improve the production of the B-24 Liberation Bomber. Traditional aviation assembly techniques enabled the aviation industry to produce one B-24 a day under optimal conditions. At Ford's Willow Run facility in Michigan, B-24s rolled off the assembly line at the rate of one per hour! To achieve this incredible rate, workers at the plant constantly shared and implemented their process improvement ideas.

Henry Ford also recognized the importance of creating and maintaining an excellent workforce. High quality and productivity can be achieved only through an effective and efficient workforce. To hire and keep the best workers, Ford offered the highest wages available, beginning with the shockingly high $5 a day, first offered on January 5, 1914. Daily wages at that time were $2 or less. This wage and the highly motivated and skilled workforce it enabled him to create meant that labor turnover costs in his plants were so small he didn't even measure them. Ford pursued productivity improvements with zeal. He believed that productivity improvement is key to economic prosperity. He also understood that productivity gains may eliminate some jobs; however, he reasoned that this job loss would be balanced by the growth of new jobs elsewhere. This type of thinking is similar to Dr. Deming's economic chain reaction (Figure 13.4). Productivity improvements act as the catalyst necessary to start an economic chain reaction. Improving productivity and quality leads to decreased costs, fewer mistakes, fewer delays, and better use of resources, which in turn leads to

Overproduction waste is a common problem at both manufacturing and service industries. This waste is created every time too many products are produced for the market. Many reasons can be given for overproduction. Some industries require economies of scale when creating batch sizes for production. In Chapter 14, batch size reduction strategies are discussed. Smaller batch sizes are the result of flexible equipment and people. Flexible organizations are better equipped to deal with fluctuations of consumer demand.

Inventory waste carries with it a lot of costs. These costs include the cost of storing and managing the inventory. They also include the costs associated with having that inventory become obsolete or spoil. Excessive inventory hides problems, too. When a lot of inventory exists, quality problems often go unrecognized. It's easy to replace a defective item with one that is not. If a replacement is not readily available, attention is called to the existence of a quality problem. Manufacturing organizations have learned to maintain a lean facility, referring to their ability to maintain low levels of inventory. To prevent waste, inventory must be managed. Obsolete materials should be disposed of. Items that are not required in the short term should not be made or purchased. Manufacturing should be flexible and capable of making small lot sizes (Chapter 14).

Transportation waste is a non-value-added activity. Any time an item is handled or a customer is passed to another server, time and effort is wasted with no value to the process. How many times have you been passed from service representative to service representative hoping to find someone

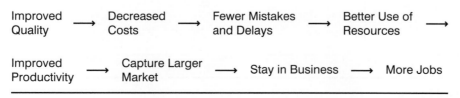

FIGURE 13.4 Dr. Deming's Chain Reaction

Improving Flow

LEAN SIX SIGMA TOOLS at WORK

Flow velocity is the movement of parts, people, or information through any system. At JF, flow velocity ran at a snail's pace. The chief problem related to how machines were placed in the plant. The machine arrangement arose over time; as new machines were added, the previous owners found places to put them without being concerned about material flow through the plant. This resulted in a confusing flow and lots of backtracking.

JF is a job shop making parts for specific customer orders. Using shop routings, cooperative education students performed an analysis of the flow of parts through the plant. They divided the parts into three part families: small, medium, and large. All raw bar stock is received and sent to be centered. Following centering, all bars proceed to the lathes to be turned to size. The next step is the mills, where keyways are cut. From there, differences arise. Approximately 40 percent of the shafts go to be drilled and tapped on their ends before going to the grinding area. The remaining 60 percent go directly to the grinders to be precision ground. Following this, the finished parts are inspected, packed, and shipped.

Changing the layout of a plant while product is being made is challenging. The realization that the parts traveled over 1,000 feet in a spiderweb pattern throughout the plant (Figure 13.5) encouraged management to take the plunge. Management and employees worked together to create a new layout. The new layout emphasizes U-shaped flow. It places all the lathes and millwork centers in one area. Each work center combines two lathes with a mill. In the new lathe/mill machining centers, one operator runs parts simultaneously on all three machines. Process studies showed that two shafts could easily be processed while the operator ran a part on the mill. For specialty products, two end mills have been combined in one work area staffed by a single employee. Having two loading docks enabled receiving and shipping to be separated. Additional floor space resulted from combining quality, maintenance, and scheduling offices together. A reorganization of the incoming materials area also gained floor space.

These changes significantly reduced material handling and therefore material damage. The new flow takes just 360 feet from receiving through the plant to shipping (Figure 13.6). Visual pull signals have been implemented because the work areas can see each other. Capacity planning and line balancing are simplified because productive maintenance and setup time reduction activities have made machines more capable. Changes have been made to make operators more flexible, too. Unlike the

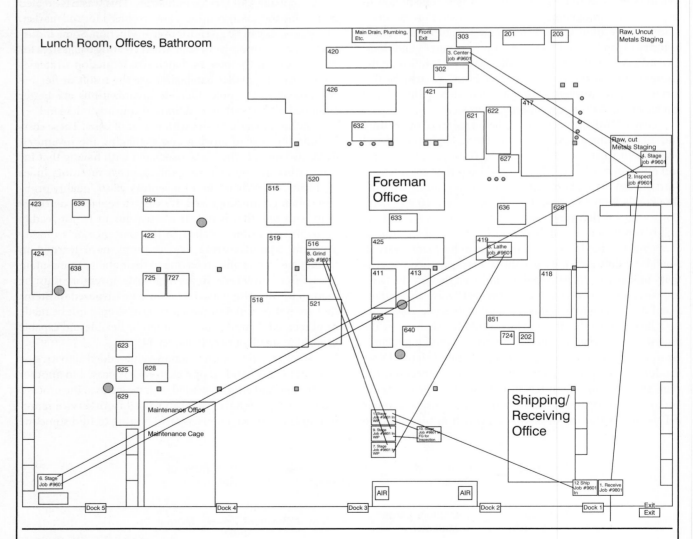

FIGURE 13.5 Original Layout

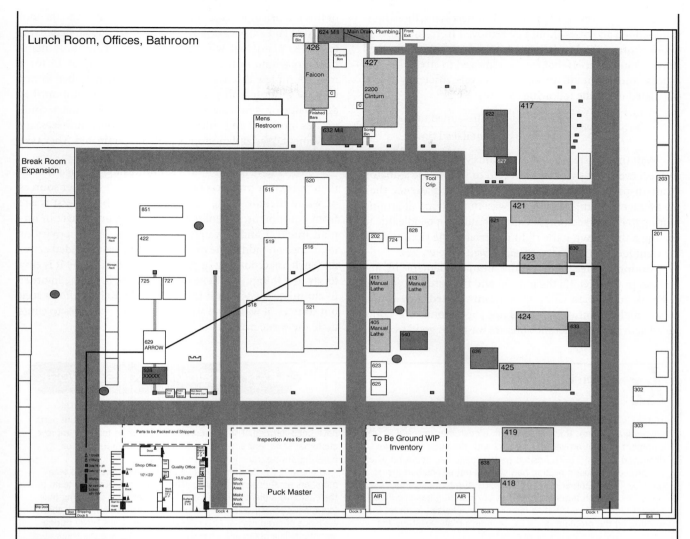

FIGURE 13.6 New Layout

previous system with one person to one machine, when work backs up in one area, cross-trained workers using standardized work procedures can move from work area to work area as needed. Closer tracking of customer orders using a computerized system has reduced batch sizes, inventory requirements, and work-in-progress. Many orders approach single piece flow. Visual management is easier, too, because the shop is cleaner and the more organized product flow more visible.

who can solve your problem? Transportation, the handling or passing that occurs, is waste. If the item is an actual product, there is the need to pick the item up, handle it, select where to store it, pick it up and handle it again later, and so on, until it is used. In any environment, items being processed often must be moved from one location to another. The distance and time an item travels should be kept to a minimum. How the item is moved should also be studied. Are conveyors or forklifts necessary? Or would carts suffice? Study should be made into the best way to move items and the optimum number of items to be moved at a time.

Defects waste is obvious. Doing something wrong or making something wrong wastes time, effort, and resources. Consider the title of the book *If You Haven't Got the Time to Do It Right, When Will You Find the Time to Do It Over?* by Jeffery J. Mayer. The book discusses the waste of time that results when people work ineffectively, creating defects. Regardless of their line of work, doing things wrong or producing defects results in scrap and rework that must later be corrected. As discussed in Chapter 6, these costs of quality must be identified and eliminated. Preventing defects is the most cost effective way to operate.

Process waste is sometimes more difficult to define. Processes develop over time, sometimes without rhyme or reason. The resulting processes are often carefully protected by the people performing them. It is not unusual to hear the words "but we have always done it that way." Processes need to be studied using value-added process mapping (Chapter 10) to identify and remove the non-value-added activities.

If a step in a process is ineffective or unnecessary, these non-value-added activities should be removed from the process. Often motion studies (Chapter 12) identify motions and activities in a process that can be removed to streamline the process and eliminate waste. Process cycle efficiency can be calculated using the following formula:

$$\text{Process cycle efficiency} = \frac{\text{Value-added time}}{\text{Total lead time}}$$

Wait time waste is another obvious waste. Consider how you feel every time you are expected to wait to see a doctor. Work-in-process inventory may have more patience than the average human, but it is still wasteful if items sit around waiting to be worked on. Machines should not be idle due to waiting either. Having the right material at the right time in the right location can prevent many wait time wastes.

Motion wastes refer to the human element in any process. In Chapter 12, the motion and time study teachings of Frank and Lillian Gilbreth were introduced. Any time a person makes a move that does not correspond to a value-added activity, his or her motions are wasted. A good example of this is a machine setup worker who must return to the tool crib to obtain a device critical to the setup. A much better system is to plan what will be needed and bring those items to the machine in the first place. Review Chapter 12 for a more complete discussion of this waste. Remember to not confuse motion with work. Work is a kind of motion that adds value or is necessary to add value. Motion that accomplishes nothing and adds no value is only unnecessary movement. Wasteful motions should be eliminated. They take precious time and do not add value.

Others have added to this list of waste or refined it to fit their particular products or service processes. Other sources of waste include waste caused by work-in-process; waste from equipment, expense, and planning; and waste in human intellect and resources. Simply put, if an activity or item does not add value or fill a key non-value-added activity (such as accounting or payroll), then it is waste. It is easy to say that things should be done right the first time in order to eliminate waste, but it is harder to do. Kaizen efforts seek out sources of waste and strive to simplify processes to eliminate the waste present.

Kaizen Event

LEAN SIX SIGMA TOOLS at WORK

JF Inc. is justifiably proud of its ability to ship just about anything anywhere. Recently, though, a customer rejected a shipment when it arrived at its docks with the ends of the box destroyed (Figure 13.7). Further part damage had occurred when the protective plastic netting had slipped out of position inside the box. The customer rejected the parts due to nicks and scratches. Unfortunately, this sort of box bursting has occurred several times before. At JF, a lean Six Sigma kaizen team is working on the problem using Dr. Deming's Plan-Do-Study-Act cycle, as well as a variety of quality tools.

PLAN
The team defined the problem as follows: Shipments are being rejected due to the boxes bursting during transit. The customers will not accept the shipment if the box is destroyed and the parts have been harmed. The problem statement is as follows: Improve customer shipment acceptance numbers by eliminating box bursting during transit.

Following the development of a problem statement, the team members established a baseline. They immediately sent a team to the customer's plant to obtain detailed information including photographs of the burst boxes. One of the boxes without straps shows evidence of strap marks. This means that the straps broke during shipment. The photo also shows the wide white tape the truck driver used to hold the box together when the straps came loose.

Back at JF, the rest of the team members watched the operator pack boxes. With this information they mapped and

FIGURE 13.7 Broken Box of Parts

photographed the key steps in the process (Figures 13.8 and 13.9). These photographs and diagrams helped them understand the process better.

The creation of a WHY-WHY diagram helped them determine the root cause of the problem (Figure 13.10). Analyzing the photos taken of the packaging process enabled them to see that

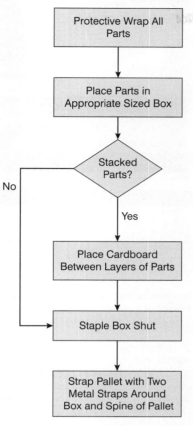

FIGURE 13.8 Packaging Process Map

FIGURE 13.9 Taped Box with Metal Straps Missing

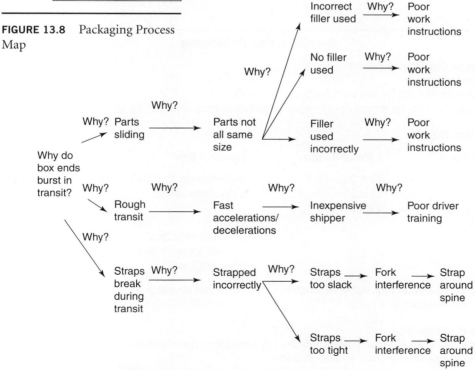

FIGURE 13.10 WHY-WHY Diagram for Packaging Difficulties

(*Continued*)

Interference

FIGURE 13.11 Incorrectly Strapped Box

the straps interfere with the forks of the fork truck during loading on or off a semitruck. The forks can not only cut the straps when picked up or moved by a fork truck, but also interfere during the packaging and strapping of the box to the pallet. Figure 13.11 shows that when strapping a box to a pallet that is held at a proper work height for the packer, the straps must be threaded down between the forks and around the spine. The strap is then pulled tight to secure the load. The problem occurs when the forks are subsequently removed from under the pallet, leaving slack in the straps. The slack straps provide no protection for the box during shipment. Their investigation enabled them to determine that the root cause of the burst boxes was the box strapping method.

DO
To prevent future box bursts, steps were taken to determine the best way to prevent improper box strapping in the future. The team developed a new method of strapping the box to the pallet. By strapping the box to the pallet using the top planks on the pallet, the interference from the forks will be eliminated. The new process is shown in Figure 13.12.

Standard Box Packaging

Limitations of Cardboard Boxes
- Length limit: 47 in.
- Do **Not** use cardboard for any part weighing over 250 lbs
- Boxes less than 100 lbs do not get strapped to pallets unless specified by customer

Standard Box Sizes

No Metal Straps	**2 Metal Straps**	**Wooden Box**
$8 \times 5 \times 5$	$24 \times 20 \times 10$	Build if parts do not fit within the limitations of a cardboard box
$12 \times 9 \times 3$	$36 \times 20 \times 10$	
$15 \times 10 \times 10$	$48 \times 20 \times 10$	
$20 \times 12 \times 4$		$W \times L$
$24 \times 6 \times 6$		69×42 Large Box
$36 \times 6 \times 6$		59×42 Medium Box
$48 \times 6 \times 6$		55×42 Small Box

Many small boxes headed for same destination may be stretch wrapped to a pallet.

Parts that do not fit box sizes must have a box made to fit.

Packaging Procedure
- Rust-paper, Naltex, or oil parts when required by packaging specification.
- Use packing peanuts for small individual parts.
- Parts layered in a box must have cardboard between layers with rust-paper when needed.

FIGURE 13.12 Standard Box Packaging

STUDY

The lean Six Sigma kaizen team monitored the new method and the shipping of boxes. No more broken or burst boxes occurred in the next two months.

ACT

To ensure permanence, new work procedures were created for the shipping department. These also included photos and line drawings to visually remind the operator about proper box preparation (Figures 13.13 and 13.14).

Standard Box Packaging

Strapping Procedure:

- Strap box to pallet with at least two straps

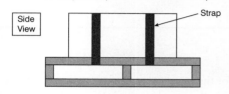
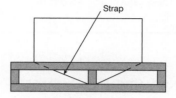

RIGHT
Truck forks do not interfere with straps

WRONG
Strap may break when lifted by forklift

- Always strap boxes as shown below

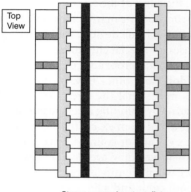

Straps secure box to pallet and parts together.

Straps secures box to pallet

- If more than 1 box on a skid, add more straps (up to 4 straps)

FIGURE 13.13 Strapping Procedure

Standard Box Packaging

Box Packaging Standard:

Skid used:
 – 34 inches wide

** Height of skid and boxes must NOT be higher than 14 inches together **
(Skids go through demagnetizing machine at Goulds)

** Strap skids same as **Standard Box Packaging** **

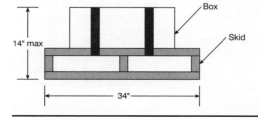

FIGURE 13.14 Box Packaging Standard

Kaizen and Error Proofing at Starbucks

LEAN SIX SIGMA TOOLS at WORK

Starbucks is known for its fresh coffee and baked goods. As reported in the August 4, 2009, *Wall Street Journal* article, "Latest Starbucks Buzzword: 'Lean' Japanese Techniques," during a recent kaizen event, a team discovered that carrying newly delivered baked goods from the delivery receiving area to the pastry case took as many as 40 round trips! Total time elapsed: one hour and 15 minutes. To speed up this process, rolling carts have been added. These carts enable workers to transport several trays at once.

Other changes made during the kaizen event centered on error proofing. Bins of coffee are kept on top of the counter for easy access. They are color coded so that a barista can find a particular roast without having to read a label. To make it easy to identify which product is which and eliminate errors, pitchers of soy, nonfat, or low-fat milk are labeled with different color tape.

ERROR PROOFING (POKA-YOKE)

Chapter 1 presented five sources of variation that Six Sigma seeks to eliminate. Poka-yoke or error proofing is used to counteract human variation or error. Kaizen events often focus on error proofing by developing simple methods of preventing human errors from occurring in a process. Coined by Shigeo Shingo, the Japanese word *poka-yoke can be translated as foolproof mechanism.* Error proof designs do not hinder worker performance; instead they eliminate the chance for error by putting mechanisms in place that prevent wrong action. Significant benefits of error proofing a process include the decrease in defects and the reduced inspection requirements. Poka-yoke ideas come from a variety of sources including the workers and engineers working most closely with the process. Poka-yoke designs and devices may be based on weight, dimensions, shape, procedure, sequencing, meters or counters, limit switches, contact switches, or other technology. Error proofing should be clever, logical, simple, and inexpensive.

The concept of preventive action known as error proofing or poka-yoke follows five principles: elimination, replacement, facilitation, detection, and mitigation. Elimination refers to the need to design work and processes that eliminate the potential for error. Replacement means replacing a faulty process with another, more reliable process, which has less of a potential for error. When a process is facilitated, it is easier for the operator to perform without error. Detection encourages the use of methods that enable an error to be easily spotted, either at the original workstation or at the very next operation. This prevents errors from compounding as the work progresses. Mitigation, the final choice, refers to minimizing the effect of the error if it does occur.

Error proofing may be accomplished through the design and use of fail-safe devices, counts, redundancy, magnifying the senses, or special checking and control devices. Fail-safe devices may foolproof an action by preventing the work from being done any other way. Mechanisms or work-holding devices may signal the operator when the work has been done correctly. Limiting mechanisms on tools may be used to prevent a tool from exceeding a certain position or amount of force. Counts or count-downs can help an operator keep track of where they are in a process. Redundancy can be effective when identifying parts, for instance, labeling a part with both color and a bar code. Double checks can be used to ascertain if the work has been completed correctly. Humans have five senses: seeing, hearing, feeling, tasting, and smelling. Tools that provide feedback in a variety of ways can alert an operator as to whether their process is operating the right way. Special checking and control devices of varying levels of complexity may be designed to help operators detect whether the work they performed is correct.

Poka-yoke

LEAN SIX SIGMA TOOLS at WORK

The quality specialist at JF Inc. received the following information concerning warranty claims.

Condition/Symptom Customers have been requesting warranty service for inoperative and incorrectly reading fuel gages for recently purchased automobile models R and Q.

Probable Cause Incorrect fuel gages or incorrect fuel pumps installed at the factory.

Immediate Corrective Action Replace fuel units (pump, gages, tank) with correct parts depending on model.

Market Impact To date there have been 39 warranty claims associated with fuel systems on models R and Q. Each claim costs $3,000 in parts and labor.

During a kaizen event, the quality specialist visited the Fuel System Creation workstation to study its layout (Figure 13.15). He created a process flow map (Figure 13.16). He was able to determine that the creation of the fuel system requires that the operator visually identify the handwritten designation for the fuel tank type and install the correct fuel system (pump and gage combination). Visiting the related workstations, he determined

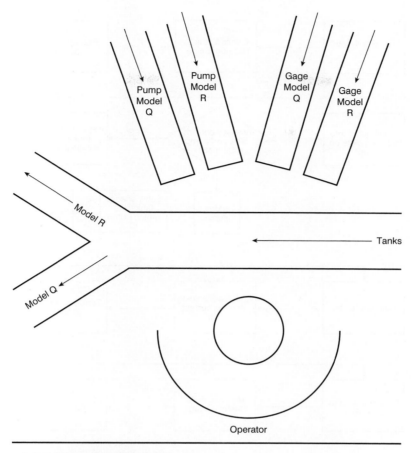

FIGURE 13.15 Workstation Layout

that the labeling of the tank is currently done by hand at the station where the part is unloaded from the paint rack. The part sequence, date, shift, and model type are written in a location visible to the person at the Fuel System Installation workstation, but not easily read by the operator at the Fuel System Creation workstation. The pumps and gages are not labeled, but arrive at the workstation on separate conveyors. The operator is responsible for designating the tank type and sequencing the tank into the correct production line for either model R or model Q. The unit then proceeds to the final assembly line.

Models R and Q are produced on the same final assembly line. Similar in size, they require fuel pumps, gages, and tanks that on the outside look remarkably similar. A variety of opportunities for error exist:

An R pump and gage combination may be installed in a model Q tank

A Q pump and gage combination may be installed in a model R tank

An R pump with a Q gage may be installed in a model Q tank

An R pump with a Q gage may be installed in a model R tank

A Q pump with an R gage may be installed in a model Q tank

A Q pump with a Q gage may be installed in a model R tank

The pump may be inoperable or substandard, repeating the same permutations

The gage may be inoperable or substandard, repeating the same permutations

With only 55 seconds in which to select and assemble a pump, a gage, and a tank, the possible errors are numerous. An operator might misread the handwritten identification on the fuel tank, select the wrong pump, select the wrong gage, or place the completed fuel system on the wrong assembly line.

Countermeasures To eliminate errors and foolproof the process, the following countermeasures were put into place:

Label all components with a barcode

Install barcode readers on the conveyors feeding the Fuel System Creation workstation

Install gates on all incoming and outgoing part conveyors

Utilize computer software that reads the type of tank and releases only the correct fuel pump and gage from material storage; this system also selects the appropriate conveyor line leaving the workstation

This kaizen activity prevents future errors by more clearly identifying the incoming parts and limiting the choices the operator can make.

(Continued)

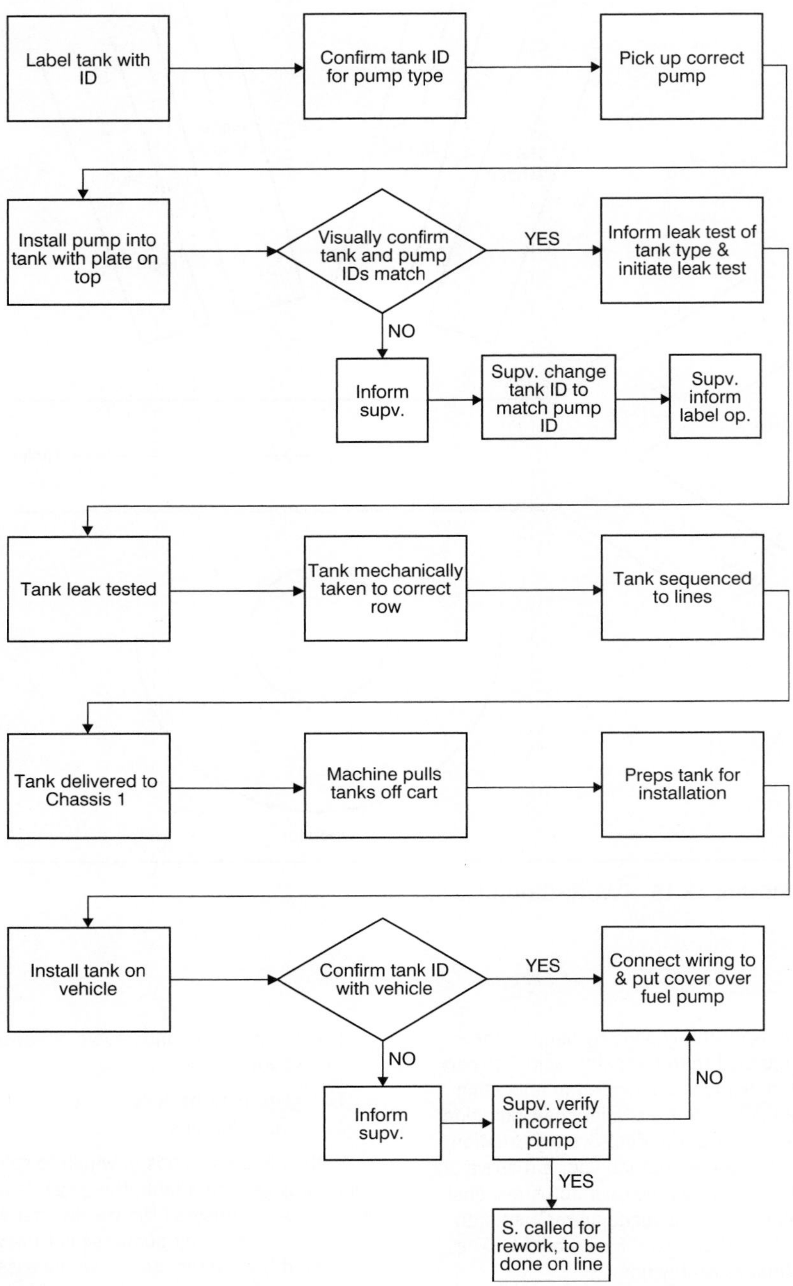

FIGURE 13.16 Process Flow Map

SUMMARY

Toyota Motor Company relies on kaizen to generate process improvements. It believes that the kaizen concept empowers its employees to make improvements to their work environment. These improvements can focus on quality, costs, safety, environmental issues, and productivity. Kaizen refers to the need to make continuous improvement through small incremental changes. Management support through the provision of training, tools, and resources results in a cultural attitude change that emphasizes day-to-day improvement. Incremental kaizen improvements accumulate and will ultimately lead to significant gains over the competition.

TAKE AWAY TIPS

1. Kaizen's guiding words are as follows: combine, simplify, eliminate.
2. Kaizen seeks to standardize processes while eliminating waste.
3. Waste can be considered any activity that consumes resources that do not add value to a product or service.
4. Kaizen activities may take two forms: flow kaizen focusing on value stream improvement and process kaizen focusing on the elimination of waste.
5. Most process improvement projects fall into two categories: long-term projects and kaizen projects.
6. Both kaizen and long-term projects should utilize the problem-solving approach of define, measure, analyze, improve, control (DMAIC) methodology.

CHAPTER PROBLEMS

1. Describe the philosophy behind kaizen.
2. Describe the kaizen concept.
3. How do Dr. Deming's teachings support kaizen?
4. List the steps for a kaizen project.
5. Describe each of the seven sources of waste.
6. What is meant by error proofing?
7. Describe an example of error proofing that you have seen or used.

CHAPTER FOURTEEN

WORK OPTIMIZATION

In 1808, Henry Maudslay created the first modern assembly line when he divided the work of creating blocks used in ship's tackle between 44 task-specific machines and 44 men to operate them. Formerly, all blocks were carved by hand. His system, which was used for the next 50 years without significant change, produced this key component of sailing ships in such vast quantities at such a high level of quality so quickly that England's Navy became a force to be reckoned with on the seas.

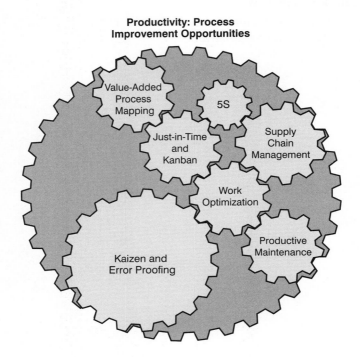

 LEARNING OPPORTUNITIES

1. To introduce the concept of line balancing
2. To familiarize the reader with the concept of setup time reduction
3. To present the concept of single piece flow
4. To describe the concept of schedule leveling
5. To present the concept of standardized work
6. To introduce the concept of visual management

Why is the making of blocks with machine tools important to modern workplace optimization? A block and tackle device (Figure 14.1) provides a mechanical advantage that enables a person to haul or lift a load four or more times the size that he or she could without a block and tackle. This mechanical advantage is critical to manning a sailing ship (Figure 14.2). Besides dramatically changing how things were made, from handcrafted to machine work, Henry Maudslay also had to deal with the problems of optimizing a production system. His 44 workers at their 44 machines needed to work together to create the complex device known as a block (Figure 14.3). Note the complexity of the device: its inner workings are surrounded by an outer

Work Optimization 147

FIGURE 14.1 Block and Tackle System

FIGURE 14.3 Block

shell. Formerly, blocks were carved from a single block of wood. Imagine the time-consuming process of hand carving one moving device inside the shell of another. Maudslay designed and built the machines that made manufacturing it possible. He also developed new machine tools that helped him balance and optimize machine production rates, including the first machine for cutting screws and improvements to the lathe. He combined these two concepts into the first screw-cutting lathe. Maudslay went on to apply these same production optimization concepts at flour mills, saw mills, mints, and marine engines. From his early work stems much of modern understanding about the need for work optimization through the lean Six Sigma concepts of line balancing, setup time reduction, single piece flow, schedule leveling, standardized work, and visual management.

FIGURE 14.2 Block and Tackle on a Sailing Ship

Line Balancing and Takt Time Calculations

LEAN SIX SIGMA TOOLS at WORK

The greatest difference between manufacturing and financial services is that financial activities do not allow for a backlog of orders. Can you imagine waiting for weeks for your checks to be cashed? For this reason, banks are required to process all the work that is delivered for a particular day. Forecasting the amount of work is essential as is careful capacity planning. The variation of volume from day to day and week to week can be extremely large. Banks often use a metric of holdover to judge their performance. Holdover refers to the percentage of work that was not processed on the day it was received. Many banks target 10 percent or less holdover per day.

When you drop off a paper check at a bank to be cashed or deposited, it must be processed before your account can be credited. At a large state or regional bank, as many as two million checks may need to be processed per month, requiring 24 hour a day operations and a staff of several hundred. If the check you are cashing has not been drawn on the same bank that you are depositing it at, the paper copy of your check must be mailed to the origination bank. When these envelopes arrive in bulk at a bank, envelopes with checks must be opened, checks scanned for validity, checks imaged for record keeping, deposited through computer keystrokes that record the account number and amount of check, transmitted by entering the keyed data into the system to reconcile with accounts receivable, and the paper copy with its appropriate paperwork placed in a repository file. All of these steps are performed manually by workers at the bank.

At QRC Bank, checks are processed 24 hours a day on three 8-hour shifts. Time for meals and breaks totals 1 hour per shift leaving 21 hours of usable work time. The bank receives 30,000 checks per day on average to process. Its takt time is as follows:

$$\text{Takt time} = \frac{\text{Available working time per day}}{\text{Customer demand rate per day}}$$

$$\text{Takt time} = \frac{1{,}260 \text{ minutes per day}}{15{,}000 \text{ per day}} = 0.084 \text{ minutes or 5 seconds}$$

Checks must be processed at a rate of one every 5 seconds in order to meet customer demand. This means that each step in the process—envelopes opened, checks scanned for validity, checks imaged for record keeping, checks deposited through computer keystrokes that record the account number and amount of check, checks transmitted by entering the keyed data into the system to reconcile with accounts receivable, and the paper copy with its appropriate paperwork placed in a repository file—must be balanced so that each operation takes 5 seconds.

LINE BALANCING

When systems are in balance, the work is performed evenly over time with no peaks and valleys in demand placing undue burdens on employees or machines. Each machine and each operator makes what is needed timed to match when it is needed. Unbalanced lines are evident when workstations make more than is needed or have to wait for production from the previous work center to reach them. Overproduction represents significant waste. Making more than the customer wants or more than can be sold means that valuable resources of time, material, and money have been spent to no benefit for the organization. Schedule leveling and line balancing create a thread that links customer needs with production. Lean thinking focuses on getting processes to make only what is needed by the customer, either internal or external. All processes should be linked, from the raw material to the final customer. This smooth flow will provide the shortest lead time, highest quality, lowest cost, and least waste.

Value stream mapping and kanban systems will reveal where lines need to be balanced. Takt times, how often a single part should be produced, provide the starting point for line balancing.

$$\text{Takt time} = \frac{\text{Available working time per day}}{\text{Customer demand rate per day}}$$

Takt times (Chapter 11) are used to synchronize the pace of production to the demands of the customer. Because it is based on the customer requirements, this measure enables organizations to balance their lines. No operation should run faster than the takt time, just as no operation should produce more than what has been ordered by the customer.

Takt times do not leave a margin for error. To meet takt times, organizations have to remove sources of waste from their processes. The organization must be able to respond quickly to problems and eliminate possible causes of unplanned downtime.

Line Balancing

LEAN SIX SIGMA TOOLS at WORK

At JF Inc., a kaizen event studied the packaging of driver's side airbags. This three-step process takes a completed driver's side airbag, folds it properly for instantaneous expansion, wraps it for insertion into the steering wheel, and inserts it into a metal bracket that will later hold the airbag to the interior of the steering column. The existing design resulted in an unbalanced line as evidenced by the percent load chart (also called a Yamazumi chart) in Figure 14.4. Operators 1 and 2 are overburdened, and they are unable to keep Operator 3 busy.

Takt time calculations require that one driver's side airbag package must be created every minute. With this knowledge, the entire work area was redesigned to accommodate three operators for folding, two operators for wrapping, and one operator for insertion.

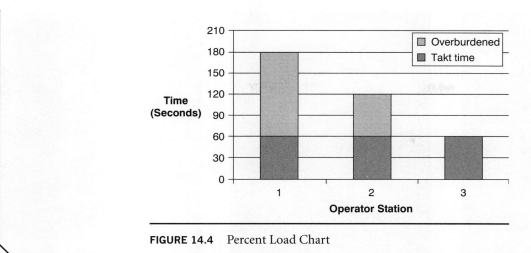

FIGURE 14.4 Percent Load Chart

SETUP TIME REDUCTION

Line balancing is a concept that makes good sense in both service industries and manufacturing. No one who has ever waited for a service or a product could disagree with that. Line balancing supports just-in-time, too. Unfortunately, line balancing isn't always easy to accomplish. Setup time reduction can help by taking the non-value-added time out of the process. In the broad sense, *setup time can be defined as the time between the production of the last good part in one series of parts and the production of the first good part in the next series of parts.* Setup times encompass taking an existing setup apart, preparing for the next setup, installing the next setup, and determining whether the new setup can create good parts.

Lean thinking encourages the reduction of setup time. Though necessary, setup times do not add value and therefore are a source of waste. In some processes, the act of changing from one fixture to the next in order to handle a different type of part takes a lot of time. Setup processes should be studied for wasteful activities. Focusing on quick changeovers is crucial to remove waste from a process. Single minute exchange of dies (SMED) calls for tooling and equipment designed to allow easy changes from one tool to another. Figures 14.5 through 14.7 show examples of a job order preparation cart and tooling carts designed to reduce setup times. Without quick changeovers, lead times remain long. Machines are not available for production for the length of the setup. Because the cost of changing over tooling is high, small batch sizes cannot be justified. Large batch sizes hide quality problems. Lengthy changeovers limit an organization's flexibility to respond quickly to customer needs. All of this points to the need to reduce the amount of time it takes to change from one tool to another. 5S and other lean Six Sigma tools and techniques can help.

Setup Time Reduction

A Florida orange juice producer has been studying the time it takes to prepare an independent-hauler truck for shipment. This process begins when the production department notifies the traffic department to arrange for an independent hauler for a juice shipment. When the truck arrives at the plant, the truck's information is entered into the company's computerized tracking system (10 minutes). This paperwork includes information on previous loads and tank wash records. Because contaminants must not be allowed in the juice, having a clean tank is of paramount importance. It takes 55 minutes to clean a tanker: 25 minutes to set up, 15 minutes to wash, and 15 total minutes to travel. There is often a line of trucks waiting to be washed. Once back at the plant with a clean truck, the paperwork for the truck is completed, requiring an additional 5 minutes. After that, the truck is directed to the bulk storage area. Here the average time is 50 minutes to fill the tank. The total time in the system: 120 minutes or 2 hours to process a truck!

LEAN SIX SIGMA TOOLS at WORK

A kaizen team has decided to tackle setup time at the wash station. The team members studied the operation and noted several problems that cause a lengthy setup time. One of the most critical is that different trucks need different hose couplings to feed cleansing fluid to the truck. The hose couplings are sterilized between uses and rarely replaced on the shelves designed to hold them. Employees must determine the size needed and then go and search for it in several locations. To cut down on search time, the kaizen team members came up with a sterilizer that can be placed nearer to the truck wash station. They also purchased multiple couplings so that some can be sterilized while others are ready to use. The couplings are now kept on a clearly labeled rack so that they can be easily stored, identified, and removed for use. These changes resulted in a savings of 15 minutes.

FIGURE 14.5 Prepicked Tooling Prepared for Quick Changeover

FIGURE 14.6 SMED Tooling Storage on Rolling Cart

FIGURE 14.7 SMED Stored at Crimping Machine

Reducing Setup Time to Increase Production Capacity

LEAN SIX SIGMA TOOLS at WORK

JF Inc. faces a dilemma. On the one hand, it is woefully short of the capacity required to take on a new, rather lucrative job from a customer. On the other hand, it is short of the funds needed to purchase a new stamping press vital to the new order. Management has approached the operators of the equipment to discuss these concerns. Together, they have decided to conduct a kaizen event focused on reducing the setup times on the existing machines. Nearly everyone, from engineers to operators, expressed concern that there was not any room for improvement at all.

On the first day, the first two hours of the morning is spent discussing the machines, how they operate, the importance of certain steps in the setup process, and the concerns of the operators and setup people. Following the meeting, the team members returned to their workstations with one small change. Each operator and setup person was videotaped while performing his or her normal duties. After a few moments of discomfort, they forgot they were on camera.

On the morning of the second day, each of the four videos were watched by the team. Two things were immediately apparent to all who viewed the films. Operators often stopped their machines to make minor adjustments and during machine changeovers the setup people were always walking off in search of tools. When questioned, the response was "that's the way we've always done it."

Quickly changes were made. Setup personnel identified tools necessary for their jobs. Management invested in tool carts to contain the necessary tools (Figures 14.8 and 14.9). Setup personnel also identified minor machine and tooling modifications

FIGURE 14.8 Tool Arrangement in Work Area

Work Optimization 151

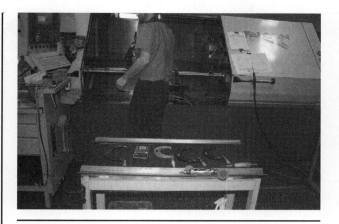

FIGURE 14.9 Machine Setup Cart

that could be made to greatly simplify their work. These changes, modest in scale, were rapidly implemented, especially when it was discovered that the improvements to setup would eliminate most, if not all, of the work stoppages for minor adjustments.

At the end of their kaizen event, the company was proudly able to report that the changes resulted in a capacity increase equal to one-third more production time. This exceeded the capacity increase from a new machine by more than two times. The company accepted the new order, rewarded its employees monetarily for their help, and scheduled future kaizen events for other areas of the plant.

SINGLE PIECE FLOW (REDUCED BATCH SIZES)

Reduced batch sizes result in shorter lead times, a shorter period of time between when a raw material arrives and is paid for and when a product or service reaches the customer and is paid for. Shorter lead times increase the number of inventory turns. Part of the ultimate goal of lean thinking is to create single piece or continuous flow wherever possible. In a continuous or single piece flow environment, as each piece is created, it flows immediately to the next activity with no delays, storages, or work-in-process inventories. Figure 14.10 shows a custom designed handcart that allows single pieces to be easily transported to the next operation. Kanban or pull systems work well with single piece flow because only what is needed is being made. Smaller lot sizes allow for items to be inspected in the sequence that they are made, revealing problems quickly (Figure 14.11). The process of going from large batches to single piece flow requires much problem resolution.

In some situations, it won't be possible to meet single piece flow; however, as downtimes due to changeovers are reduced and smaller in-line equipment utilized, batch sizes will get smaller and smaller, providing many of the benefits of single piece flow. Single piece flow does not work well when

FIGURE 14.10 Handcart Facilitating Single Piece Flow

processes are too unreliable to be linked closely to other processes. It can also happen when raw materials or components have to travel a great distance. Also, some processes, such as injection molding, stamping, or forging, achieve better economies of scale by batch processing. Still, where continuous single piece flow is not possible, batches should be linked with the same pull system, making only what is

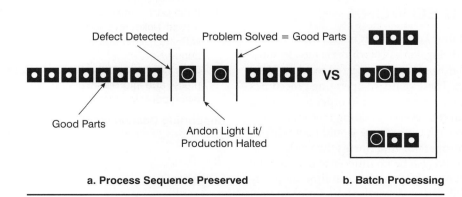

FIGURE 14.11 Process Sequencing

LEAN SIX SIGMA TOOLS at WORK

Single Piece Flow

In its April 1, 2005, article "Boeing, Airbus Look to Auto Companies for Production Tips," the *Wall Street Journal* reported that "Boeing, though, made one of the most dramatic production changes yet in 2001 when it began putting together planes on a huge moving line—a la Henry Ford and his Model T. The motion 'lent a sense of urgency to the process that we really didn't have when the planes were sitting still,' says Carolyn Corvi, the executive who oversaw the change. Ms. Corvi and other top Boeing executives made multiple visits to Toyota when they were first beginning to study how to convert the production process to a moving line. When many workers initially balked at the production line and unions filed complaints, Boeing took extra pains to win them over. The change paid off: Boeing halved the time it takes to assemble a single-aisle 737, and has started putting its other planes—including its oldest and largest product, the 747—on moving lines."

---

needed. Safety stock or buffers should be temporary, not a crutch to avoid improvement. Kaizen efforts, using the five Ss and other lean tools, should focus on working toward SMED.

SCHEDULE LEVELING

Much of the strength of a kanban system comes from level scheduling. Level scheduling creates a steady flow of material through the process. When a schedule is level, the same amount is produced each day or shift. A level schedule allows production people to focus their attention on their work without having to deal with sudden fluctuations in the amount of work. The need to expedite disappears because the organization is neither ahead nor behind. Though level scheduling becomes more difficult the closer the work is to the customer's final product or service, being able to predict the amount of work to be performed in a day reduces the demands on the process. Fluctuating customer demand can overwhelm the system's ability to respond appropriately. Uneven demand can result in waste associated with unpredictable downtime for setups or an excessive number of setups. Waste can also be found in the unpredictability in the mix and volume of customer demand. Level scheduling enables an organization to link the customer schedule with the pace of manufacturing, thus it requires cooperation from the customer. The drawback is that responding to customer requirement changes becomes more complicated. To remain flexible, lean organizations develop the ability to make every part every day.

STANDARDIZED WORK

Chapter 2 presented four key methods to sustain organizational gains. Standardizing work procedures is one. In some organizations, standardized work is referred to as standard operating procedures. These standard operating procedures refer to the activities that must happen in order to complete a process. They mean that everyone doing that job does it exactly the same way each time. There would be no difference between the way operator A performs the work versus operator B. There would also be no difference between the fifth time they did the work or the 1,000th time they performed the work. Well-designed standards and process procedures may include checklists, maintenance procedures, setup procedures, or other process steps. Whatever they are for, any written procedures should be visual, written in terms of expected behavior or actions, and must be auditable (Figure 14.12). Kaizen improvement teams often

---

First Piece Inspection Work Instruction Sheet

Purpose

To identify the procedure used by quality assurance inspectors for first piece inspection after initial job set up for a run of more than five pieces, for a new program, and whenever a new operator begins running an already set up job.

Procedure

1. Operator ready for first piece inspection informs quality assurance inspector.
2. Inspector reviews traveler for correct drawing revision, part number, and specific information related to the operation being inspected and that any required receiving inspections have been completed. If material/parts had been released without inspection, the inspector will perform required receiving inspection.
3. Inspector verifies that the operator's inspection/measuring tools are calibrated.
4. Inspector inspects the first piece part according to the characteristics as indicated on the part print appropriate to the specific operation completed.
5. If the part passes inspection, inspector signs the traveler and/or the SQC sheet as appropriate. Inspector marks part as a first piece.
6. If the part is nonconforming, inspector completes a Hold Ticket and marks the part as nonconforming with the ticket number and advises the operator and supervisor.
7. Operator corrects the setup, produces another first piece, and informs inspector, who inspects the new first piece per this procedure.

Supporting Documents/Records

Traveler

Hold Ticket

SQC sheet

FIGURE 14.12 Work Standards

Work Optimization

> **LEAN SIX SIGMA TOOLS at WORK**
>
> ### Standardized Work Through Well-Designed Work Instructions
>
> Well-designed work instructions produce quality work by preventing errors. Creating work instructions appears simple at first glance: just tell them what to do. However, it isn't that easy to make sure that everyone does what is expected of them. High-performance workforces begin with clear communication. Work instructions are an essential part of the communication process.
>
> When designing work instructions, consider the answers to two questions:
>
> What do you want them to do?
>
> What are the three most important things you want them to remember about the job that they must perform?
>
> The answers to these questions must be clearly defined. They will form the thrust of your communication and the focus of your work instructions. When designing your work instructions, think about how you will get across what you want them to remember.
>
> Industrial engineers in a regional shoe distribution warehouse noted the high number of price tags that were not properly placed on the shoes. They created a "Tagging Guide" to improve worker performance. They began by asking:
>
> What do you want them to do?
>
> Tag Hard Tags correctly on shoes and accessories (boxed and unboxed)
>
> Tag Sticky Tags correctly on shoes and accessories (boxed and unboxed)
>
> What are the three most important things you want them to remember about the job that they must perform?
>
> Hard Tags have particular placement locations on shoes with eyeholes and shoes without.
>
> Hard Tags have particular placement on accessories and women's shoes.
>
> Sticky Tags have particular locations on accessories and women's shoes.
>
> The tagging guide is actually a series of laminated signs that hang in the work area in strategic locations. Workers are visually reminded of what constitutes a properly tagged shoe and what is unacceptable. Since the hanging of these signs, errors and rework have dropped to zero from about 20 percent. The resulting savings is over $40,000 per month. People previously assigned to rework have been reassigned to other duties in the distribution center.

create standard operating procedures as they make improvements to an area. To become a learning organization, when changes prove to be useful in one work area, these same countermeasures should be adopted for other similar areas.

VISUAL MANAGEMENT

Visual management makes sense because people gain 80 percent of their information from their vision, 18 percent from their hearing, and just 2 percent from their other senses. In his book, *Out of the Crisis*, Dr. Deming used the word *transparency* to describe the concept of visual management. ***Transparency** refers to enabling anyone to have the ability to see, in real time, what is happening with a process. From the instantaneous information they gather, they should also be able to determine whether anything has changed or needs to be changed.* Note the first part of this definition refers to the word *visual*. The second sentence refers to *management*. As mentioned in Chapter 2, visual management is another of the four methods used to sustain gains. Visual management provides visual checks that allow anyone viewing an area to determine whether objectives are being met.

Establishing the visual methods comes first. The lean thinking concept of visual management focuses on the need to organize work areas, storage areas, processes, and facilities in such a way as to be able to tell at a glance if something is misplaced or mismanaged. The tools and techniques described in previous chapters include 5S outlines on a workstation work surface or a tool board marking locations for each tool used in the process (Figure 14.13). The white board shown in Figure 14.14 is an example of how information can be communicated visually. Another common example of this approach is the andon lights that flash at a workstation if it is unable to meet its production quota or takt time (Figure 14.15). The flashing light draws attention to the area experiencing problems. Visual management can also be seen when kanban cards are used to keep track of inventory. Visual management encourages the "a place for everything and everything in its place" mindset of 5S. Posting takt times and having andons that light when the takt time is not being met clearly communicate with the operator the performance expectations.

To use visual management to sustain the gains made in 5S and kaizen activities, management must pay attention to the form the visual aids take. Consider the bulletin board in

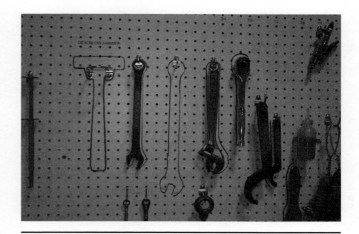

FIGURE 14.13 5S Tool Board

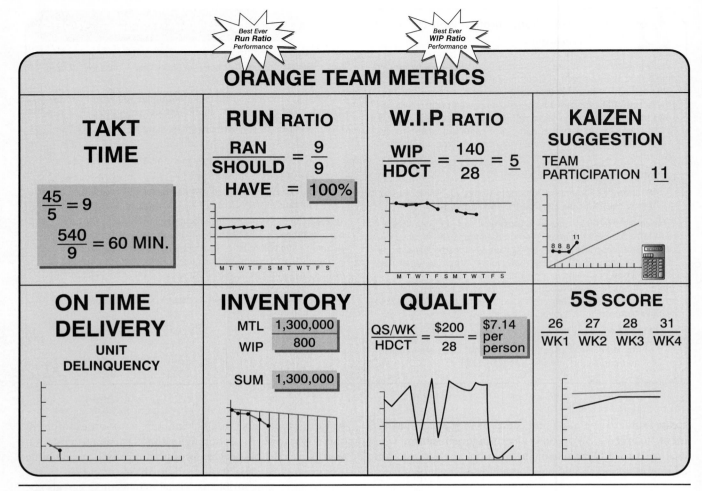

FIGURE 14.14 White Board with Takt Time

Figure 14.16. The papers on this board provide information about such things as monthly production rates and year-to-date sales figures. Good information to know, but not the information for making real-time, for today's process, types of decisions. When Dr. Deming discussed transparency, he was interested in making sure that leaders and those working with the process know how the process is performing right now. Useful information describes whether the process is currently meeting its takt time requirements or whether a machine is operational. Ultimately, transparency allows anyone, management or employees, to make instantaneous assessments of the current situation. From this information, if necessary, timely corrective actions or countermeasures can

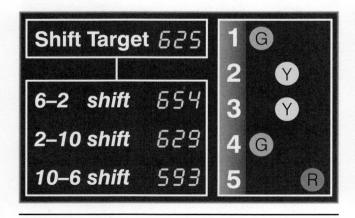

FIGURE 14.15 Andon Light

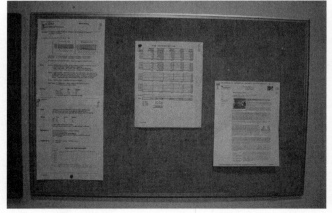

FIGURE 14.16 Bulletin Board

be implemented. Later, using the same transparent information source, the success or failure of the process improvements can be confirmed. Many of the figures in this chapter, including Figures 14.13 through 14.15, provide instant information to the viewer. A visual management tool must be able to allow viewers to quickly determine whether the process has changed or whether a countermeasure or action taken was successful. Visual controls focus attention on immediate process performance.

SUMMARY

Techniques such as line balancing, level scheduling, single minute exchange of dies, and others covered in this chapter enable lean Six Sigma organizations to be more responsive to their customers. Because these techniques focus on making the organization more flexible, customer requirements are easier to accommodate. The block and tackle device described at the beginning of this chapter helps sailors trim their vessels for optimal performance. The techniques presented in this chapter serve the same purpose for lean Six Sigma organizations by helping them optimize process performance.

TAKE AWAY TIPS

1. When line balancing occurs, the work is performed evenly over time with no peaks and valleys in demand that place undue burdens on employees or machines.
2. Takt time = $\dfrac{\text{Available working time per day}}{\text{Customer demand rate per day}}$
3. Setup time is defined as the time between the production of the last good part in one series of parts and the production of the first good part in the next series of parts.
4. Single minute exchange of dies (SMED) calls for tooling and equipment designed to allow easy changes from one tool to another.
5. In a continuous flow or single piece flow environment, as each piece is created it flows immediately to the next activity with no delays, storages, or work-in-process inventories.
6. When a schedule is level, the same amount is produced each day or shift.
7. Standardized work means that everyone doing a job does it exactly the same way each time.
8. Visual management means that everyone has the ability to see, in real time, what is happening with a process.

CHAPTER QUESTIONS

1. What tasks do you perform daily, either at home or at work? How would you go about reducing the setup times involved in these tasks? Select a task and make some changes to reduce the setup times. How much of a reduction did your changes provide?
2. What are the benefits of producing smaller batch sizes?
3. What is meant by the term *line balancing*?
4. What is meant by the term *takt time*?
5. What is meant by the term *level scheduling*?
6. What do users hope to accomplish by reducing setup time?
7. What is meant by the term *single minute exchange of dies*?
8. What is meant by the term *single piece flow*?
9. What is meant by the term *standardized work*?
10. What is meant by the term *visual management*? What types of tools are helpful for visual management?

PRODUCTIVE MAINTENANCE

If it is not broken, don't fix it.

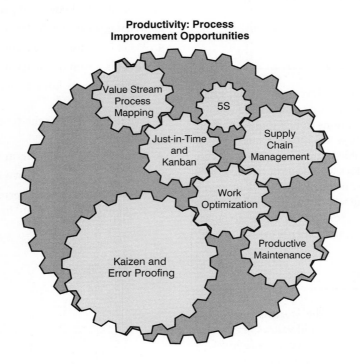

LEARNING OPPORTUNITIES

1. To introduce the concept of productive maintenance
2. To introduce the four major productive maintenance activities
3. To introduce the concept of overall equipment effectiveness
4. To introduce the concept of preventive maintenance
5. To introduce the concept of predictive maintenance
6. To introduce the concept of autonomous maintenance

PRODUCTIVE MAINTENANCE

Quality, costs, productivity. Equipment problems prevent an effective organization from optimizing these. Where do production losses come from?

Quality/nonconformance issues
Raw materials shortages
Machines or equipment availability
Cycle time losses

Productive maintenance deals with two of these four by improving the availability of machines or equipment while reducing cycle time losses. Lean Six Sigma organizations practice productive maintenance in order to reduce process equipment variation, one of the five sources of variation discussed in Chapter 1.

Every time equipment breaks down, work has to be stopped, production delayed, and completion times adjusted. *Productive maintenance programs optimize overall equipment effectiveness by focusing on total life cycle equipment management.* Optimization occurs when equipment is able to do what it needs to do when it needs to do it. Productive maintenance programs ensure that equipment functions well and runs reliably. A solid productive maintenance program will lead to reduced inventories; shorter lead times and setup times; lower accident and injury rates; and fewer breakdowns, jams, standbys, speed losses, and startup losses. The benefits of productive maintenance programs support lean

Six Sigma by enhancing customer satisfaction, reducing costs, increasing productivity, and improving quality. Reliable, productive equipment is required for lean and Six Sigma.

Complaints are often heard about the high costs of maintenance. Productive maintenance programs actually help reduce maintenance costs. Maintenance costs are often over budget because of the emergency response approach to maintaining equipment. This "ignore it until it fails" mode of operating is expensive. Organizations that engage in fire fighting instead of a prevention approach to maintenance endure high costs because they do not recognize the real costs of unavailable equipment. Malfunctions and unavailable equipment means lost time, lost productivity, poor quality, low morale, and decreased customer satisfaction. Productive maintenance programs build awareness of maintenance needs, monitor expenditures closely, and respond to difficulties quickly. Productive maintenance programs reduce costs by establishing policies and procedures that reduce equipment downtime and unplanned maintenance.

Productive maintenance, by preventing equipment problems, seeks to increase overall equipment effectiveness and machine reliability. To calculate overall equipment effectiveness (OEE), use the following formula:

Overall equipment effectiveness = Availability × efficiency × quality

where

Availability means that equipment works when it is needed

Efficiency means that the equipment performed at the expected speed

Quality means that the product produced meets customer requirements

Overall equipment effectiveness can run as low as 50 percent to 60 percent of the production time available. With a strong productive maintenance program, it can climb to 80 percent or more. This value will never reach 100 percent due to routine maintenance needs and setup time requirements. High overall equipment effectiveness, supported by an effective productive maintenance program, increases the time available for value-added activities. There are four components to a productive maintenance program:

Elimination of equipment losses
Preventive maintenance
Predictive or planned maintenance
Autonomous maintenance

ELIMINATION OF EQUIPMENT LOSSES

Productive maintenance programs focus on eliminating three key sources of equipment losses:

Equipment downtime
Speed losses
Quality defects due to variation in equipment performance

Calculating Overall Equipment Effectiveness

LEAN SIX SIGMA TOOLS at WORK

The production of milk is a lot more complicated now than the days when milk used to come straight from the cow (Figure 15.1). Raw milk has a fat content of 4 percent or more. Today's market requires lower fat alternatives such as 2 percent, 1 percent, or skim milk (< 0.1 percent fat). For whole milk, the fat content has been reduced to 3.4 percent. Fat-soluble vitamins A and D may also be added to any type of milk.

The production of milk involves many operations, including clarification, separation (for the production of lower fat milks), pasteurization, and homogenization. Equipment downtime can dramatically affect the spoilage rates of this delicate product.

WP Dairy carefully monitors overall equipment effectiveness because failure of any one operation in the process results in significant product spoilage. It keeps detailed records of equipment availability, efficiency, and milk quality. Regular downtime is scheduled for daily cleaning and minor maintenance. The OEE calculation for its separation unit is:

Availability = 95%

Efficiency = 99%

Quality = 99.98%

Overall equipment effectiveness = Availability × efficiency × quality
Overall equipment effectiveness = 0.95 × 0.99 × 0.9998 = 0.94

WP Dairy's separation unit is up and running 94 percent of the time.

FIGURE 15.1 Cows
(**Source:** Arkady Mazor/Shutterstock.)

LEAN SIX SIGMA TOOLS at WORK

Equipment Downtime

Before processing, milk is stored in tanks (Figure 15.2) after being pumped from tanker truck coming directly from the farms. Milk can be stored for a maximum of two days, after which it either must be processed (cleaned) or disposed of. While in the tank, the raw milk is tested for the presence of certain diseases including coli plating and aerobic bacteria (Figure 15.3). Tests are also run to determine butter fat percentages and taste. If a tank fails to maintain the required temperature for even a few moments, thousands of gallons of milk must be thrown out.

After passing the tests, the milk goes on to further processing to add flavor, fat, and vitamins. The milk is pasteurized, exposed to high temperature to destroy certain microorganisms (Figure 15.4). Lab tests are run to check alkaline phosphates levels, an indicator of pasteurization.

Low-fat milk is made by skimming the raw milk. Skimming can be partial or whole. Partial skimming reduces the raw milk to the desired fat content. Whole skimming completely removes the fat and then adds back in the appropriate amount of cream to achieve the desired fat content.

Homogenization is next. Raw milk settles over time and the cream rises to the top. In homogenization, milk is run through tiny tubes to break up the fat globules in order to distribute them throughout the milk. This process reduces the size of the fat molecules and disperses them more evenly throughout the milk. Creaming on the top of the milk does not occur with homogenized milk. Homogenization results in longer lasting milk. Equipment failure can again result in significant losses.

Following homogenization, an emulsification test is run to check the processed milk for fat and vitamin content. After it passes the tests, it leaves the holding tanks and flows (usually by gravity) directly into the milk jugs. Once capped, the jugs are put into plastic crates and stored for a maximum of one day. Final tests are run on the milk for quality purposes and then the milk can be shipped via trucks to local grocery stores.

Recently, the jug capping operation has been experiencing a number of difficulties. The automatic machine has been screwing some caps on too tight and others too loose. The torque force exerted by the machine has been inconsistent. Too tight and the jugs are damaged, resulting in broken jugs and spilt milk. Too loose and the milk spoils. Equipment downtime for adjustments and slower operating speeds combined with the downtime resulting from spilt milk cleanup and quality issues for the spoiled milk have caused WP Dairy's costs to skyrocket. A corrective action team immediately attacked this problem in search of a root cause for the variation in torque. Once the faulty part causing the problem on the capping machine was identified and replaced, these costs disappeared.

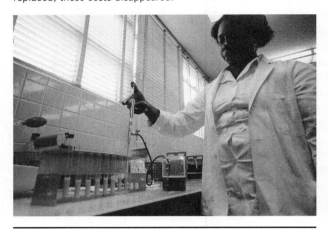

FIGURE 15.3 Milk Testing
(**Source:** U.S. Department of Health and Human Services.)

FIGURE 15.2 Milk Storage Tanks
(**Source:** Sonda Dawes, The Image Works.)

FIGURE 15.4 Pasteurization
(**Source:** James King-Holmes/Photo Researchers, Inc.)

Equipment downtime occurs whenever a piece of equipment fails to run or when the loss or reduction of a specified function causes production losses. Equipment does not run production all the time. There are necessary and planned downtimes and setup times. Unfortunately, OEE is further reduced by unplanned downtimes, idling, and minor stoppages for adjustments, malfunctions, speed losses, stops, jams, and standbys that occur randomly. Unscheduled machine downtime, whether from adjustments, reduced speeds, minor stoppages, or major breakdowns, represents wasted time and money. These losses can be reduced or eliminated by applying root cause analysis to search out and correct equipment problems.

Speed losses occur when the equipment cannot operate at design speeds. Designers theoretically calculate the optimum speeds at which a piece of equipment can operate; however, in the work environment, changes to the equipment or the product design may make it impossible to run at design speed. When equipment runs at a reduced speed, efficiency suffers. To aid planning, users of the equipment need to determine optimal production running speeds. They also study the equipment and correct inefficiencies in the original design, repair defective mechanisms or design weaknesses, and make adjustments to increase precision.

Quality is affected by variability in equipment performance. Reduced yields occur when equipment performance is suboptimal. Variability can come from a variety of sources, including the operator, as will be seen in the example on page 160. Productive maintenance teams work to find solutions to equipment malfunctions. They begin by measuring equipment effectiveness to establish a baseline for machine performance. Measures can take the form of counts, such as counting the number of occurrences of breakdowns, jams, and standbys. The duration of these breakdowns, jams, and standbys may also be measured and recorded. Because ineffective equipment can result in spoiled product, measuring spoilage rates is a good idea. Production rates and the mean-time-between-failures can be measured and then tracked on run charts (Chapter 9).

To resolve quality issues, teams use Six Sigma's DMAIC to prioritize improvement goals, set measurements to ensure reaching these goals, and work to resolve equipment effectiveness problems. Root cause analysis using Six Sigma's DMAIC can find and eliminate problems in the design of products, processes, and machines. By using the improvement tools presented in this text, such as Pareto diagrams, WHY-WHY diagrams, and cause-and-effect diagrams, organizations can reduce equipment variation.

Speed Losses

LEAN SIX SIGMA TOOLS at WORK

It is not unusual for a milk processing facility to make its milk jugs. At WP Dairy, a continuous flow of newly made plastic milk jugs flow directly into the filling area (Figure 15.5). Milk jugs are made just-in-time for use. Difficulties arise when the flow of the milk jugs is hindered in any way. Unfortunately at WP Dairy, that happens fairly regularly due to the design of the conveyor delivering the milk jugs to the filling area. The conveyor, designed to hold one-gallon milk jugs, winds its way from the injection molding machine to the milk jug filling station. Though the distance is only 50 feet, the path the jug takes is torturous. So many twists and turns take place before the jugs pass through the wall that the jugs often get hung up as they round a corner. This happens so often that the operators keep a broom stick at hand in order to poke at the trapped jugs until the line moves smoothly again. The jugs do not flow smoothly from one process to the next, a difficult problem in the just-in-time environment, because the flow of milk must be turned on and off whenever the jugs are jammed.

A kaizen team investigated the situation and determined that the operator in charge of monitoring the production of the jugs balances his time between three key activities: filling the milk jug injection molding machine with plastic pellets used to make the jugs, monitoring the temperature of the injection molding machine, and poking a stick at trapped jugs. Because the operators spend nearly a third of their time keeping the jugs flowing, the kaizen team members feel that action should be taken immediately to correct the problem. They studied the design of the conveyor and quickly determined that the reason for the twists and turns had been removed years ago. The conveyor had been designed to go around a large piece of equipment that had since been transferred elsewhere in the plant. The operators and their supervisors were so used to using a stick to keep the flow going that they never thought of changing the design of the conveyor once the obstruction was gone. They had even written the task into the job standards. The kaizen team had the conveyor modified to take a straight path from room to room and the problem disappeared.

FIGURE 15.5 Milk Jugs
(**Source:** Courtesy of Milk Promotion Services of Indiana, Inc.)

Tampering with the Process

LEAN SIX SIGMA TOOLS at WORK

The WP Dairy has been notified by its largest customer that WP Dairy will need to dramatically improve the quality level associated with the fat content of its whole milk. Currently the operation is unable to meet the specification limits set by the customer. WP Dairy has been testing the milk often, but it wanted to end this practice. The specifications are 3.750 ± 0.005 percent butter fat.

To determine the root causes of variation, the team is studying the skimming operation and the operator. The operator performs the process in the following manner. Every 18 minutes, he measures the fat content. The length values for the six consecutively produced jugs are averaged, and the average is plotted on $\overline{X}$ and R charts (Figure 15.6 and Chapter 18). Periodically, the operator reviews the evolving data and makes a decision as to whether the process mean (the fat content) needs to be adjusted. This process takes about 5 minutes and occurs fairly often.

The investigation team was quick to realize that the operator is adding variation to the process. The operator is overcontrolling (overadjusting) the process because he cannot distinguish between common cause variation and special cause variation. The operator has been reacting to patterns in the data that may be inherent (common) to the process. The consequences of this mistake are devastating to a process. Each time an adjustment is made when it is not necessary, variation is introduced to the process that would not be there otherwise. With each adjustment, quality is essentially decreased (made more variable) and production time is unnecessarily lost.

Compare the differences in Figures 15.6a and 15.6b, adjustments versus no adjustments to the process. In 15.6b, the process has stabilized because no unnecessary adjustments have been made. The method of overcontrol has proved costly from a quality (inconsistent product) and a productivity (machine downtime, higher scrap) point of view.

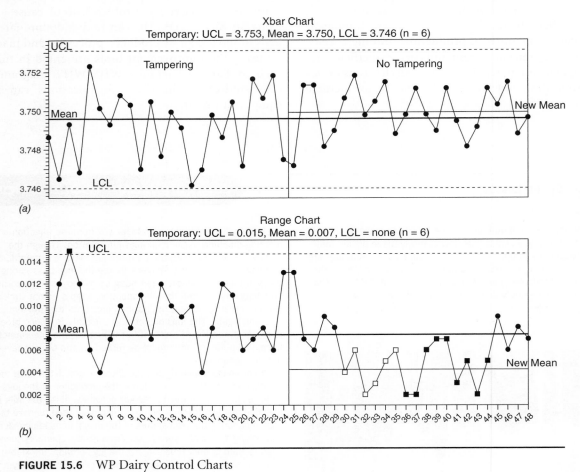

FIGURE 15.6 WP Dairy Control Charts

PREVENTIVE MAINTENANCE

Preventing the need for maintenance is another important goal of productive maintenance. *Maintenance prevention can be achieved through the design and acquisition of equipment that will be easy to maintain and operate.* The preventive maintenance component of a productive maintenance program attacks equipment failures as a source of waste. Just as homeowners change filters on their furnace or automobile owners change the oil in their car, organizations with sound preventive maintenance programs strive to maintain the equipment at the peak of its condition. The benefits of practicing preventive maintenance include reduced costs;

Productive Maintenance 161

> **Disaster Prevention** — LEAN SIX SIGMA TOOLS at WORK
>
> Productive maintenance could have prevented two disasters at British Petroleum (BP) facilities. In 2005, in Texas City, Texas, a refinery explosion and fire killed several employees. In 2006, a pipeline failure resulted in the leakage of more than 200,000 gallons of oil into an environmentally fragile area. Death, lost production, equipment and facility damage, wasted raw materials, and environmental impact—could these have been prevented by a sound productive maintenance program? According to a presentation made on May 16, 2007, in Washington, D.C., by the Chemical Safety Board (CSB) Chairperson Carolyn Merritt, the answer is yes. Chairperson Merritt told the House of Representatives subcommittee that "virtually all of the seven root causes identified for the Prudhoe Bay incidents have strong echoes in Texas City." These included the "significant role of budget and production pressures in driving BP's decision making—ultimately harming safety." The Prudhoe Bay pipeline suffered extensive corrosion due to lack of maintenance over several years. At the Texas City refinery, "abnormal startups were not investigated and became routine, while critical equipment was allowed to decay. By the day of the accident, the distillation equipment had six key alarms, instruments, and controls that were malfunctioning." Clearly, no one took the initiative to maintain, repair, or replace equipment as needed. As operating conditions deteriorated in both situations, the end result, disaster, was inevitable. A sound productive maintenance program could have averted disaster.

lower inventories; shorter lead times; fewer breakdowns, jams, standbys, speed losses, and startup losses; shorter changeover times; improved productivity; increased flexibility; and improved quality. Good preventive maintenance plans also result in good budgets. Because there is a distinct maintenance plan for each piece of equipment, there is also a budget in place to cover the costs. Typically, preventive maintenance programs can more easily forecast the costs associated with maintenance. Preventive maintenance prevents disasters, too. Or, as in the following example, lack of maintenance can be deadly.

One of the key steps of a preventive maintenance program involves establishing maintenance intervals for all vital equipment. This is done by:

- Periodically inspecting all equipment
- Scheduling regular maintenance
- Planning long-term maintenance
- Ensuring spare parts availability
- Lubricating equipment regularly
- Documenting maintenance

Periodic inspection of equipment is useful to detect conditions that might cause breakdowns. Follow-up action can be taken to reverse such conditions. Depending on the complexity of the equipment, periodic inspections can be done by the operator of the equipment (autonomous maintenance) or by maintenance personnel.

Preventive maintenance requires that maintenance of equipment be planned, scheduled, and then executed to performance standards. These performance standards should include the interval between planned maintenance, how the downtime will be scheduled, and what type of work is to be done. Will the equipment be maintained annually? Monthly? On a total number of hours used basis? The plan will also include information about how the downtime will be scheduled. Scheduled maintenance should start with high-priority equipment. The plan should be reviewed periodically.

Well-cared-for production equipment can have a long service life. *Long-term maintenance planning is critical to ensuring availability of older machines.* Schedulers will need to determine how much time it will take to perform the repair or overhaul. Repair-related downtime is always more significant than the actual repair time. Repair-related downtime includes finding or waiting for a repair person, diagnosing the problem, finding spare parts or tools, and testing the equipment following repairs. Knowledge of time requirements comes from realistic understanding of the equipment history including its mean time between failure, the time it takes to overhaul, operator experience with the equipment, and safety records.

Spare parts control is a critical part of productive maintenance. Carefully made plans will ensure that all necessary parts and materials have been procured before taking the equipment off-line. *Good spare parts control means that the necessary spare parts are available when needed.* Careful analysis should take place to determine the need, priority, storage method, restocking method, and cost of stock-out of key items. A preventive maintenance plan will include procedures to ensure that appropriate parts/materials have been procured and labor scheduled. Careful study should be made to determine which spare parts should be stocked.

Lubrication control is a key part of a preventive maintenance program. *Proper lubrication with appropriate oils, cutting fluids, greases, and solid lubricants prevents wear.* Lubrication control ensures that no leakage occurs. Deterioration or contamination of lubricants is prevented by daily inspection of fluid levels, time elapsed, and rate of usage.

Productive maintenance programs maintain good records. *Maintenance records document what maintenance has taken place, machine performance, inspections, and lubrications.* Good records provide a history that helps with planning maintenance. These records are used to assess equipment conditions, establish priorities, and deploy resources. Information can be focused on past history such as previous repair work as well as daily or periodic inspection records. This information can be used to help set maintenance priorities based on production importance, quality level, maintenance history, and safety.

Determining what records are needed and why and how they will be used is the first step in record keeping. Most organizations begin by simplifying and standardizing their current record-keeping procedures. These records may include, but are not limited to, routine inspection records, lubricant replenishment and replacement records, periodic inspection records, and repair and service reports. Preventive maintenance programs may require that a mean time between failure analysis (Chapter 20) be performed. Equipment logs, maintenance cost records, breakdown analysis records, and downtime occurrences and time are also good sources of information.

PREDICTIVE MAINTENANCE

Lean Six Sigma organizations think ahead and try to predict when maintenance might be needed. *Predictive maintenance monitors equipment performance over time by measuring, recognizing, and using signals from the process to diagnose the condition of the equipment and determine when maintenance will be required.* Many of the records kept for preventive maintenance are used in predictive maintenance. Predictive maintenance inspections take place while the equipment is running so as not to disrupt operations.

Many methods exist to monitor equipment performance over time. Depending on the type of equipment, monitoring systems may be vibration, oil, or acoustical analysis. Just as a car owner may check the cleanliness of the oil in the engine to predict oil life, these analyses allow users to determine how the equipment is wearing during normal use. To be effective, the cost of monitoring the equipment should be less than cost of repair or production losses.

AUTONOMOUS MAINTENANCE

One element of a productive maintenance program, autonomous maintenance, focuses on the idea that proper maintenance over time is a good defense. In autonomous maintenance, the user of the equipment is training in order to create a better understanding of the equipment. These efforts limit equipment downtime losses because those utilizing the equipment are aware of small changes in equipment performance. They are in the best position to eliminate the causes of equipment downtime.

Autonomous maintenance refers to the day-to-day regular maintenance performed by equipment operators. These daily cleaning, inspecting, lubricating, and bolt-tightening efforts prevent equipment deterioration. Autonomous maintenance activities transfer ownership to person running equipment. Autonomous maintenance includes six steps:

1. Initial cleaning
2. Countermeasures to eliminate sources of contamination
3. Establishing and following cleaning and lubrication standards
4. Autonomous maintenance standards
5. Regular inspections
6. Autonomous supervision

Autonomous maintenance begins with the thorough cleaning of all equipment. Some organizations choose to do this sort of cleaning area by area, whereas others prefer to clean the entire facility during a scheduled shutdown. Following the initial cleaning, future cleaning is made easier and faster by improving access to equipment; eliminating sources of contamination; establishing standard operating procedures; making sure proper lubricants, cleaners, and other necessary items are on hand; and teaching equipment operators the correct procedures, settings, and adjustments during basic training.

Basic training should also include teaching problem detection and resolution. Autonomous maintenance housekeeping rules include the following:

- Keep only necessary items at workplace
- Design locations for everything
- Keep workplace clean on a daily basis
- Ensure everyone participates
- Ensure continuous adherence

SUMMARY

Productive maintenance takes commitment. True productive maintenance programs are multiyear commitments requiring the understanding and commitment of management. The ultimate goal of a productive maintenance program is to perform required maintenance at a cost-effective time before the equipment loses its optimum performance level. Productive maintenance programs want to eliminate the unplanned and unexpected downtimes that disrupt production.

The benefits of practicing preventive and predictive maintenance include reduced costs; lower inventories; shorter lead times; fewer breakdowns, jams, standbys, speed losses, and startup losses; shorter changeover times; improved productivity; increased flexibility; and improved quality. What should be included in a productive maintenance program? Study overall equipment effectiveness. Eliminate sources of production loss. Prevent the need for maintenance. Use records and past data to predict the need for maintenance. Perform autonomous maintenance.

TAKE AWAY TIPS

1. Productive maintenance attacks two sources of production losses: machines or equipment availability and cycle time losses.
2. There are four components to a productive maintenance program:
    Elimination of equipment losses
    Preventive maintenance
    Predictive or planned maintenance
    Autonomous maintenance

3. Overall equipment effectiveness (OEE) = Availability × efficiency × quality
4. Productive maintenance programs focus on eliminating three key sources of equipment losses: equipment downtime, speed losses, and quality defects due to variation in equipment performance.
5. Maintenance prevention can be achieved through the design and acquisition of equipment that will be easy to maintain and operate.
6. Predictive maintenance monitors equipment performance over time by measuring, recognizing, and using signals from the process to diagnose the condition of the equipment and determine when maintenance will be required.
7. Autonomous maintenance refers to the day-to-day regular maintenance performed by equipment operators.

Productive Maintenance

LEAN SIX SIGMA TOOLS at WORK

Leaders at JF turned to productive maintenance to increase overall equipment effectiveness and machine reliability. In the past, as machines wore out, employees developed all sorts of workarounds. As a result, the machines in the plant were unreliable and prone to breakdowns. High costs of quality existed due to machines that were unable to hold tolerance and exhibited poor repeatability. One machine squealed so badly when operating that the sound level in the plant approached OSHA's 90-decibel maximum level.

The leaders at JF worked together to create a schedule of maintenance for the machines (Figure 15.7). It contains all four productive maintenance concepts. A new employee skilled in machine repair was hired. Money was spent to scrape the ways and replace the bearings, switches, and handles. A worn-out lathe was sold for parts and replaced with a significantly more capable machine. Because of its larger motor and newer CNC controls, this machine can handle shafts a foot longer.

Unscheduled machine downtime, minor stoppages for adjustments, speed losses, spoilage, and rework have all decreased significantly since the productive maintenance program was put in place. There are fewer breakdowns, jams, standbys, and changeover times. Training is easier, too. The quirks have been removed from the machines, making them easier to operate. These improvements have enabled the plant to increase production throughput by improving quality and productivity, increasing flexibility, and shortening lead times. The end result: JF has been able to double sales in two years without a major investment in new equipment, employees, or plant size.

Purpose
To identify the procedure for performing preventive maintenance on production equipment.

Procedure
1. Maintenance associates check oil; operator checks coolant level in machine. If not OK, operator adds fluid as needed. If level is very low, associate advises maintenance associate to check for leaks and maintenance associate repairs machine is needed. The maintenance associate completes the work order indicating what fluids and/or repairs are needed.
2. Maintenance associate reviews report from previous shift and checks any machine reported as having problems. If machine is inoperable, the manufacturing supervisor is informed.
3. Repair work priority is based on production needs. Maintenance supervisor schedules repairs and documents work performed on a work order. Machinery may be repaired by EICOM associates or contracted out.
4. Routine preventive maintenance is scheduled by the maintenance supervisor based on recommendations of the machinery manufacturers. Work orders are completed indicating date maintenance should be completed by. The manufacturing manager is advised of the work to be done so that production schedules are not impacted.
5. Maintenance requests initiated by associates are reviewed by manufacturing manager and scheduled by maintenance supervisor. When maintenance has been completed, the request is marked up and returned to the associate.
6. Maintenance associates perform scheduled maintenance as indicated on work orders and document work performed on the work order when completed.
7. Maintenance supervisor or an assigned associate logs completed work orders in the maintenance database. Paperwork is not retained.

Supporting Documents/Records
Maintenance request form
Maintenance electronic database

FIGURE 15.7 JF Productive Maintenance Procedure

8. Autonomous maintenance housekeeping rules areas follows:

    - Keep only necessary items at workplace
    - Design locations for everything
    - Keep workplace clean on a daily basis
    - Ensure everyone participates
    - Ensure continuous adherence

CHAPTER PROBLEMS

1. At a local machining company, the maintenance manager and the plant manager discussed the new moving assembly line and the new types of machines required to assemble their newest product. The maintenance manager suggested preparing a total productive maintenance program (TPM) plan. Based on what you learned while reading the chapter, discuss with the plant manager the need to prepare a TPM plan. Convince the manager that this is a great idea. Discuss why maintenance costs are often over budget. Answer such questions as: Why would the plant want to implement a TPM plan? What benefits would it receive?

2. Describe the critical components of a TPM plan. What are the components? How do they work together to become a single plan? What will need to be done to ensure that the critical components of the plan are actually performed?

3. Research an organization's TPM plan. Draw parallels between the plan presented in this chapter and any TPM/reliability information your research uncovered. Utilize your research as supporting documentation. In other words, comment on what TPM your research revealed, as well as, what parts of a TPM program you feel have been overlooked or you weren't able to find documentation for.

4. Discuss any difficulties you can see in implementing a TPM plan. How could these difficulties be overcome?

5. Outline a TPM plan. Provide details that you would be sure to include in a TPM plan for an entire assembly line.

6. Calculate the overall equipment effectiveness for a machine that is available 90 percent of the day, has an efficiency rate of 85 percent, and has a quality output of 95 percent.

7. Calculate the overall equipment effectiveness for a copy machine that has an availability rate of 93 percent, an efficiency rate of 97 percent, and a quality output of 90 percent.

8. Describe autonomous maintenance.

9. Describe the housekeeping rules used for a work area.

SUPPLY CHAIN MANAGEMENT

Companies don't compete against other companies; they compete against each others' supply chain.

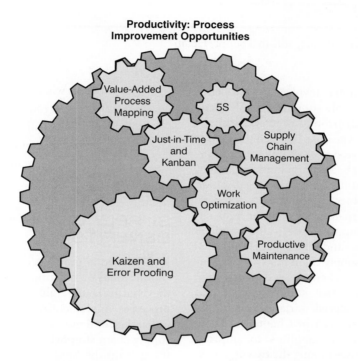

LEARNING OPPORTUNITIES

1. To become familiar with the concept of supply chain management
2. To understand the benefits of effective supply chain management
3. To become familiar with the elements of supply chain management
4. To understand the value of strategic partnering
5. To understand the importance of information exchange across a supply chain
6. To know the steps and challenges to creating an effective supply chain

SUPPLY CHAINS AS A PROCESS

Moving raw material through conversion to a product to the consumer takes effort. Consider what it takes to get a loaf of bread to a consumer. The wheat must be planted, cultivated, harvested, stored, and shipped to a mill. The mill refines the raw material, bags, and transports the flour to a bakery. The bakery combines a variety of other raw materials that have also come through a supply chain and bakes the bread. The bread is then wrapped, loaded onto a truck, transported to a store, unloaded, sorted, and shelved. Finally, it is available to the consumer for purchase. How many days does that take, from raw material to finished product? In the big scheme of things, the raw material cost is inconsequential to the total cost of traveling through the supply chain. Supply chains create value through two aspects: the physical movement of materials and exchange of information about those materials. As goods and services flow through the supply chain, various organizations add value to them. In today's global economy, a well-coordinated supply chain is a competitive necessity. Companies don't compete against other companies; they compete against each others' supply chains. Effective supply chains provide lean Six Sigma organizations with a strategic advantage.

A **supply chain is** *the network of organizations involved in the movement of materials, information, and money as raw materials flow from their source through production until they are delivered as a finished product or service to the*

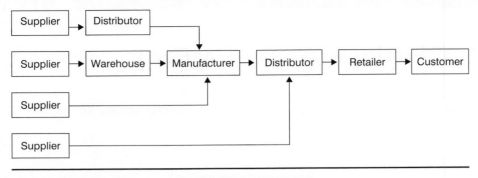

FIGURE 16.1 Participants in a Supply Chain

final customer. A supply chain may involve any number of organizations, including multiple suppliers, wholesalers, manufacturers, warehouses, distribution centers, retailers, and customers (Figure 16.1). The links between these units must be strong for an effective supply chain to exist. Supply chains are driven by three factors: materials, money, and information. Throughout the chain, each organization is both a supplier and a customer (Figure 16.2). A complete supply chain starts with raw material or parts or information flowing in from the supplier side, value is added through internal processes, and ends with the final customer (Figure 16.3). The supply chain links all the components together in order to meet customer demand. Lean Six Sigma organizations practice good supply chain management to reduce raw materials variation, one of the five sources of variation discussed in Chapter 1. These organizations work with their suppliers to improve their processes and the processes linking the two organizations. Raw materials variation can be reduced by having clearly defined raw materials requirements and specifications, correctly placed orders, and appropriate shipping protocols.

Lean Six Sigma organizations focus on supply chain integrity. This means they carefully manage the information, materials, and services that flow up and down the supply chain. A well-designed supply chain ensures that the right product or service is in the right place at the right time at an affordable price. Strong supply chains help lean Six Sigma organizations optimize quality, costs, and productivity. When designing a supply chain, they ask the following questions:

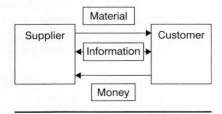

FIGURE 16.2 The Flow of Material, Information, and Money in a Supply Chain

What products will travel through the supply chain?
What services will travel through the supply chain?
What information will travel through the supply chain?
What is the structure of the supply chain?
How will each segment of the supply chain be managed and optimized?

SUPPLY CHAIN MANAGEMENT BENEFITS

As discussed in previous chapters, lean Six Sigma organizations focus on creating customer value. For this reason, they optimize the value-creation process by improving processes, implementing lean principles, optimizing asset utilization, and monitoring quality assurance. An effective supply chain supports the value-creation process. Everything from the kind of weather to market demand can affect a supply chain. To perform well, an effective supply chain must be designed with quality, cost, flexibility, speed, and customer service in mind. A well-designed supply chain ensures reliability, adaptability, reduced costs, and appropriate asset utilization. Effective supply chains are able to provide order fulfillment, on-time delivery, short response time, high value-added input per employee, high inventory turns, and short cycle times at a reasonable cost. Well-managed supply chains benefit effective organizations by enabling them to meet the challenges facing them. They are able to achieve their objectives through improved supplier quality, delivery, and price.

SUPPLY CHAIN MANAGEMENT ELEMENTS

Lean Six Sigma organizations manage their supply chains. They do so to improve business operations, manage inventories, and reduce costs. Global economic conditions have combined to force companies to carefully consider their supply chains. As transportation costs increase and outsourcing opportunities arise, companies are faced with decisions affecting their ability to move their product to market. The rise of e-commerce has changed the way many companies do business from both a buying and a selling

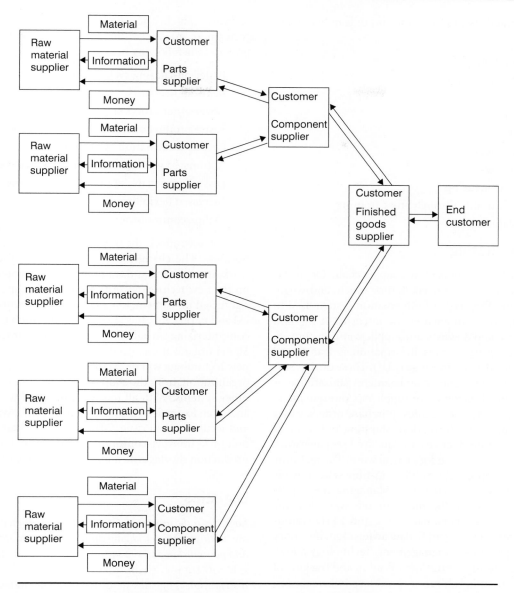

FIGURE 16.3 A Complex Supply Chain

point of view. All of these changes have resulted in more complex supply chains. Typical issues that face any organization include determining what products or services are needed, how many, and when; insourcing or outsourcing the manufacture or processing of goods or providing of services; managing the trade-offs between market demand and the cost of holding inventory; evaluating, selecting, communicating, and working with suppliers; monitoring supplier quality and quality assurance; and deciding how to best package, transport, and store materials. *The critical elements of supply chain management are inventory management, information sharing, e-commerce, logistics, and purchasing.*

Inventory Management

How much inventory should be held? Where should this inventory be stored? Is the inventory perishable? How much inventory needs to be held in order to satisfy the market and provide a good level of service? Organizations use a variety of systems to manage inventory: just-in-time, kanban, material requirement planning, enterprise resource planning, and others. For each of these systems, the goal is to meet customer demands through optimal use of equipment and people. Disruptions to the supply chain due to inventory shortages are costly. In some cases it may take weeks to recover. Supply chains in a global economy are vulnerable for a variety of reasons, including that organizations cannot identify or plan for all the risks involved. Many systems are not robust enough to adapt to different conditions, especially when unexpected. Even if a supply chain is working well, efforts are always under way to reduce inventory, improve customer service, reduce costs, and increase equipment and labor utilization. All of these factors make inventory difficult to manage. Lean Six Sigma organizations use a variety of methods to remain flexible when disruptions affect their supply chain. These may include elements and practices commonly found in lean Six Sigma organizations:

- Schedule leveling
- Schedule fixing
- Pull systems
- Short cycle times
- Safety stock
- Safety lead times
- Small batch sizes
- Flexible job sequencing
- Workforce flexibility through cross-training
- Capacity planning

Information Sharing

Information is a key component of a supply chain. Organizations need to know what is needed, how much, and when. Careful selection of the type of information system used to manage the supply chain ensures the integrity of the link between upstream and downstream suppliers and customers. At effective organizations, this information is transferred through electronic data interchange (EDI). These direct computer-to-computer interactions can be interorganizational or external to individual customers or suppliers. Common information exchanged includes quotes, purchase orders, order confirmations, pricing information, shipping notices, and more. EDI improves productivity and quality. Less paperwork is required, reducing the need for clerical work. The real-time exchange of information helps run just-in-time systems while reducing inventory and lead times. Managing the supply chain is also easier because the information is available at the touch of a button. Scanners, bar codes, and RFID (radio-frequency identification) simplify data acquisition, inventory control, and supply chain management. Technology allows real-time data gathering for the identification and tracking of products. Information about a product's whereabouts, contents, and physical status are quickly available. Software applications such as enterprise resource planning (ERP) systems also aid in managing the supply chain.

Information systems and networks provide increased data accuracy and reduced data transfer time. Unfortunately, these systems are vulnerable to equipment failure, power outages, and security threats. Effective organizations are careful to maintain information reliability. They recognize that the ability to rapidly disseminate information and capitalize on the knowledge it generates is a driving force in organizational success. Effective organizations design and manage their information management systems, make technological improvements, and measure system performance to ensure data accuracy, integrity, reliability, timeliness, security, and confidentiality.

E-Commerce

E-commerce is the use of electronic technology to aid business transactions. E-commerce allows organizations to reach global markets with little effort. Their products and services can be accessed through the Web, allowing consumers to make their choices. E-commerce often provides the following benefits:

- Faster purchasing cycle times
- Reduced inventory
- Fewer order processing people
- Faster product or service search and ordering
- Greater information availability
- Automated validation and preapproved spending
- Less printed material such as order forms or invoices
- Increased flexibility
- Online communication

E-commerce affects the way lean Six Sigma organizations do business. The changes it requires changes the way people work. It takes time and training to bring people up to speed on the new technology. All of this new technology requires significant investment, too. Manual processes need to be studied and improved; otherwise, the organization runs the risk of computerizing an already poor manual process. Lean Six Sigma organizations need to decide just how pervasive the new technology will be. Many choose to integrate the entire organization, from the shop floor to accounting, payroll, and other systems. With all the systems linked within the plant and often to outside customers and suppliers, system integrity and security are an issue. Effective supply chains utilize technology to improve communication. This reduces the risk of production shortages, lost materials, or stock-outs.

Logistics

Logistics is the process of determining the best methods of procuring, maintaining, packaging, transporting, and storing materials and personnel in order to satisfy customer demand. This physical side of supply chain management is sometimes called *distribution*. Logistics or distribution includes the material handling, packaging, warehousing, staging, and transportation of equipment, parts, subassemblies, tools, fuels, lubricants, office supplies, information, and anything else needed to keep the organization functioning. This means logistics manages the flow of materials and information within a facility and incoming and outgoing from that facility. When preparing a logistical strategy, frequently asked questions include:

- What is being shipped?
- How will it be packaged?
- What type of handling care will be needed?
- When will it be needed?
- When will it be shipped?
- What is the best way to ship it? Rail? Road? Air? Water? Pipeline? Satellite?
- What are the optimal shipping practices?
- What will it cost?

In effective organizations, this movement should be smooth and expedient. In some organizations, time and effort is lost because of poor internal logistics. These problems can also be evidenced between organizations. Effective supply chains safeguard products while they are in transit. Though it adds time and cost to the process, security is vital as products cross multiple borders and great distances.

Purchasing

Materials, parts, supplies, and services are all purchases necessary for operating a business. In most companies, these costs comprise 40 percent to 60 percent of the final cost of the finished goods. ***Purchases*** *are driven by several factors,*

Logistics and a Tsunami

LEAN SIX SIGMA TOOLS at WORK

On December 26, 2004, an earthquake of magnitude 9.3 centered in the Indian Ocean near the northwest coast of the Indonesian island of Sumatra triggered a tsunami that killed nearly 230,000 people and destroyed homes throughout that part of the world (Figure 16.4). Caused by a slip of the Indian Plate under the Sumatra Plate, the earthquake deformed the ocean floor, creating the tsunami (Figure 16.5). Forming a wave thought to be 80 feet high, it is estimated that nearly 200 billion gallons of water were dropped on the land in a single minute. In comparison, seismic activity for the 1906 San Francisco earthquake was 7.8 and that of Northridge, California, in 1994, 6.7. Hardest hit were Indonesia (167,739 dead or missing), Sri Lanka (35,322), India (18,045), and Thailand (8,212). The biggest challenge facing the humanitarian relief agencies responding to the situation was logistics. How can water, food, medicine, relief workers, and other critical supplies be brought to an area where much of the infrastructure had been damaged or destroyed? Normal supply chains simply did not exist.

Facing this logistical nightmare, relief organizations relied on those who are best in the business. Supply chain experts from around the world gathered together to help. The involvement of a number of firms enabled relief workers to set up temporary supply chains and use sea, air, and land assets of these organizations to move needed supplies from mainland warehouses to temporary warehouses built in the disaster areas. Within days, trucks, helicopters, planes, ships, landing craft, and ferryboats were mobilized to help cope with the disaster. As people and organizations from around the world contributed water, food, consumer products, tents, medicine, equipment, time, and money, this influx also needed to be managed. Logistics became a strategic weapon in the fight against further death and destruction. Logistics firms and their employees handled tons of supplies. Working 14- to 16-hour days, they coordinated the handling, stocking, warehousing, distributing, and shipping of lifesaving materials. Disaster response efforts lasted six months. Quick response to the crisis was possible because the chief relief organization, the United Nations World Food Programme, has long partnered with logistics organizations such as TNT, Logistics. The strategy to respond to such a natural disaster was in place two years before disaster struck. One phone call activated a well-coordinated logistical effort to aid people in need.

Before earthquake

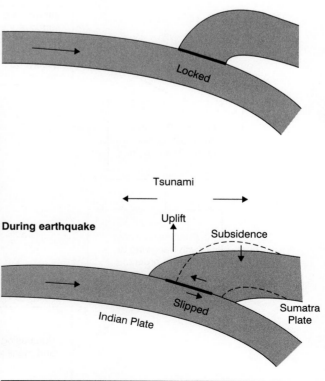

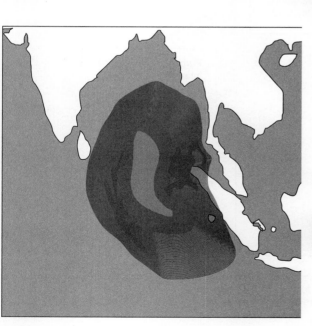

FIGURE 16.4   Earthquake Impact

FIGURE 16.5   Slipping Plates Cause Tsunami

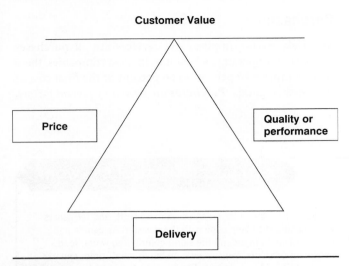

FIGURE 16.6 The Purchasing Triangle

including quality of the goods or services, timing of the deliveries of the goods or services, and the costs associated with purchasing the goods or services (Figure 16.6). The optimization of all three rarely occurs; effective purchasing happens when these three are balanced in a way that creates value for the customer. As business has become more global, purchasing has had to evolve. Originally, companies focused on finding suppliers of goods and services. This developed to include managing deliveries using logistics. Now they are buying a more complete package, the supply chain. And supply chains must be improved, optimized, and managed. Business-to-business relationships are crucial when managing the supply chain. Purchasing has had to adapt to Web sources, global suppliers, improvement efforts, and lean production.

The purchasing department is responsible for finding out what is out there, what the options are, how much it costs, and when it can be delivered. This means that purchasing agents not only order goods and services but also select and manage suppliers. The purchasing cycle is summarized in Figure 16.7. Purchasing agents negotiate contracts and act as the liaison between the organization's internal departments and the external suppliers. Understanding the entire supply chain enables them to make effective purchases. To ensure that what they order is correct, they work closely with the internal departments making the requests for goods or services. These departments focus on the quality of the item purchased and the timeliness of the delivery. They provide purchasing with specifications and required delivery dates. In many cases they may also have an idea about where the items may be obtained. Purchasing adds value to the process by working to balance the costs associated with buying the goods and services. Efforts to reduce the material costs associated with providing a product or service often weigh heavily on the purchasing department.

Effective organizations seek to optimize the sourcing of their needed material, equipment, and services. To maintain an effective supply chain, these organizations develop a

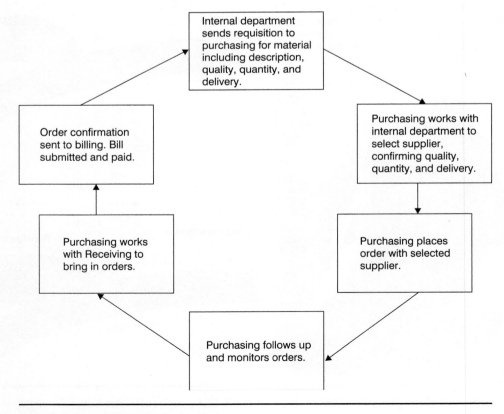

FIGURE 16.7 The Purchasing Cycle

complete and accurate profile of their supplier base. As they judge supplier performance, they consolidate and grade suppliers, granting preferred status to those suppliers who perform best on meeting quality, quantity, performance, cost, and delivery requirements. Effective organizations develop and manage suppliers using several approaches. They evaluate suppliers using quality assurance systems like ISO 9000, the Malcolm Baldrige National Quality Award (MBNQA) criteria, or their own criteria. They run supplier certification or qualification programs to aid supplier development. These are followed by supplier audits to ensure compliance. Effective organizations ask questions such as the following:

How are suppliers selected?

What will it take to become a preferred supplier?

Who are the preferred suppliers?

Why are these suppliers preferred?

How does supplier interaction occur?

How does communication with suppliers occur?

Where is the supplier located?

How will the supplier's location affect delivery performance?

Suppliers work closely with the purchasing department to determine material or service specifications, including quality, quantity, delivery date, and cost. Choosing reliable and trustworthy suppliers is crucial to maintaining the supply chain. The following are key performance factors to consider when choosing a supplier:

Type of product (new technology versus mature product)

Quality

Quality assurance

Flexibility

Reliability

Location

Price

Delivery capability

Lead times

Financial stability

Length of time in business

Relationship strength

Other factors, depending on the business

Supplier performance can be monitored through audits and certifications such as ISO 9000. Effective organizations typically audit and review the following:

Quality assurance activities

Quality planning process

Process flow diagrams for key processes

Key product characteristic capability

Key process capability

Process planning

Process performance measures

Continuous improvement based on performance measures

In-process inspection and testing

Operator instructions

Visual factory

First-pass quality performance

Supplier assessment

Customer communication

Workforce stability or employee turnover

Value added per employee

On-time delivery performance

Quality certifications (such as ISO 9000)

Supplier partnerships are pursued by many companies. The partnerships go beyond a price contract and focus on cooperation. The benefits associated with a partnership relate to improved operations. These long-term relationships help in maintaining the supply chain through improved delivery and quality, lower inventory, and lower costs. Computerized interaction on set schedules helps reduce purchasing, inventory, and transportation costs. These same suppliers also provide information about item improvements, new products, or other market data. Strategic partnerships may also provide the confidence an organization needs to invest resources in a particular product or service line. When deciding to partner, both organizations study the cost–benefit trade-off. They ask questions such as the following:

How critical to quality is what we are buying?

What are we buying through this partnership? Performance? Delivery? Savings?

How much is being bought?

What will the time investment be?

What will the dollar investment be?

What benefits will we receive through this partnership?

As with any partnership, there can be drawbacks. Though the relationship can be mutually beneficial, often investments in computer systems or dedicated equipment are necessary and costly. If the supply chain is complex, not all costs associated with a firm contract can be controlled. Sometimes, even the culture of the two businesses can be conflicting. Effective supply chain managers understand that people cannot be standardized and that there may be cultural differences. They standardize processes to reduce this difficulty. When trying to understand organizational and cultural needs, expectations, and differences, clear communication is essential. Product requirements start the supply chain process, but communication enables it.

There are advantages to having a single supplier and advantages for multiple sources. Single-supplier relationships are stronger, each side showing a greater commitment to the other. There are economies of scale present with a single supplier. Naturally, having only one other organization to communicate with simplifies information exchange and makes it easier to keep that information confidential. There are sound reasons for picking multiple suppliers, too. Having several suppliers minimizes the risks associated with having only a single source. If something affects a particular supplier's ability to deliver, such as a strike or a flood, then the supply chain is frozen until another source can be found. Multiple suppliers prevent this from happening. Multiple suppliers also encourage competition, which can bring about lower prices or better delivery schedules or other benefits. Multiple suppliers also provide more flexibility when demand increases unexpectedly. They also provide access to a wider variety of goods and services.

Selecting a supplier requires making an informed decision. Careful purchasing agents find out as much as they can about potential suppliers before making their selection. If the item being purchased is critical to quality or involves significant funds, it is not unusual for the purchasing agent to make a site visit. They come prepared to audit the supplier in order to determine whether the supplier is a good match for their organization. Suppliers that do not offer a strategic advantage are not chosen. Desperate manufacturers or service providers are also avoided. Their willingness to quote any part or service at any price when they may not have the expertise, capacity, or systems in place to support the demand can signal significant trouble later. An effective organization seeks suppliers who have the same mind-set that they do. They choose companies that want and value their business. Audits and site visits reveal whether a supplier has a focus on continuous improvement, value creation, and quality assurance. Effective organizations take time to ensure that the supplier they choose will be a good fit.

Working with suppliers requires a lot of negotiating. Negotiation is the process of making joint decisions when the people involved have different preferences. Simply taken, negotiating involves presenting the facts, discussing the facts, discussing the viewpoints on the facts, and reaching a rational agreement. Anyone who has participated in a negotiation will tell you that, in reality, negotiating is tough. There are at least two sides to the situation. This means that there are at least two sets of desires, interests, objectives, and agendas. The type of product or service involved can make negotiating more complex. If the product or service is highly technical, then the negotiation will take on more of a consulting nature. In this situation, the supplier provides information showing its in-depth knowledge of the product or service and its willingness to share it. This type of negotiating is different than one for a commodity product.

The outcome of any negotiation is significantly influenced by the amount of advance preparation on both sides. To conduct successful negotiations, prepare thoroughly in advance. Good preparation, including determining the other side's expectations, means fewer surprises. Outside the negotiation, when the pressure is off, information can be more clearly evaluated. While preparing in advance, good negotiators generate a variety of scenarios and alternatives to see how each one relates to their desired result. They come to the negotiation knowing which aspects of the negotiation must absolutely go their way and which ones are less essential.

Well-prepared negotiators have their thoughts organized. Advance preparation means that during the negotiation they can be factual rather than emotional. They have learned to focus on the situation and not on the attitudes

Supply Chain Management

LEAN SIX SIGMA TOOLS at WORK

The leaders at JF would like to achieve a two-week lead time from order placement to order receipt. They can't do this without managing their complete supply chain. Raw material must be available just in time for new orders. The shop must be prepared to handle the incoming orders. Shipping must deliver the parts quickly and safely. At JF, the chief material needed is steel bars made to specification and cut to length. JF uses a variety of steel, including 1045, 1144, 12L14, 300 series stainless, and exotics such as Monel and Hastelloy. JF makes small batches of parts to order and is too small to impact the purchase price of its steel raw material. For this reason, JF partners with suppliers who can create value for JF in other ways. Raw material suppliers regularly visit JF. Because they are in touch with lots of other shops, they provide information about where the industry is going. They have data on the industry that JF uses to improve and adapt to the market.

Because JF does not make large purchases, it shops every order among its six suppliers. It does not have specific contracts because of fluctuations in steel prices. There are six suppliers because JF recognizes that some suppliers are better at producing certain steel than others and chooses accordingly. Once the supplier meets quality assurance requirements, it is chosen on the basis of delivery and price. JF's inventory system is a pull system; raw material orders are triggered by customer order. For many customers, their orders repeat monthly. For these customers, a blanket raw material purchase order is in place with a single vendor.

Consistent mistake-free communication is crucial to ordering supplies and shipping parts. Improvements focused on modernizing JF computer capabilities. The new system standardizes order processing and links all computers in the plant. Order processing efficiency increased dramatically since automated spreadsheets allows quotes to be adjusted quickly to respond to customer requests. When accepted by the customer, quotes are turned into orders, complete with shop traveler, invoice, packing list, and material purchase order. When the material purchase order is sent via computer to the steel vendor, an e-mail reply notifying receipt of the order is requested. Confirmation of the order to the customer is sent via e-mail. Job progress through the plant can be tracked as the parts move from operation to operation. Shipments are also tracked. Taken together, the new order processing system supports the entire supply chain by tracking and confirming the order process from customer quote to final part delivery.

of the people involved. They also can recognize where the two sides have matters of common interest and agreement, smoothing the negotiating process. Good negotiators are good listeners, too. They take the time to listen carefully to proposals and counterproposals. This means that they don't hurry the negotiations along, and they remain as flexible as possible. They allow people to save face and don't ignore the needs of others. They conduct their negotiations in an atmosphere of trust and confidence. This gives them a reputation for being fair but firm. Once the negotiation is complete, good negotiators take the time to set everything in writing. They avoid problems later by being sure that the signed agreement conveys the intent of both parties.

There are false perceptions about the negotiating process. Good negotiators realize that negotiation is never a win-lose confrontation. Each side must be able to feel good about the end results. Too big an imbalance, and future transactions may suffer. There is an old adage that what goes around comes around. Care should be taken to strike a mutually beneficial bargain because each negotiation is rarely an isolated transaction. Another false assumption about negotiating is that it is all about obtaining the lowest possible price. As this next Lean Six Sigma Tools at Work

Negotiations

LEAN SIX SIGMA TOOLS at WORK

The leaders at JF decided to shop around and compare their shipping company with others in the market. Their current shipping company has performed well on price and delivery. However, one of the performance measures that JF tracks, product damage during shipment, has been creeping up over the past several months (Figure 16.8). While recognizing that their heavy product is relatively easy to damage, the leaders at JF would like to keep the number of shipments damaged to one shipment damaged in any three-month time period, the level of the previous year's performance. Even better would be to reduce this damage. They are also concerned with the responsiveness of the shipper to their requests for meetings concerning damaged product. The company has been evasive and has not dealt with insurance claims quickly. Damaged parts are a significant burden on the JF manufacturing facility. Because JF and most of its customers operate on a just-in-time basis, damaged parts seriously disrupt the supply chain. Customers have been sympathetic, knowing that JF did not release damaged parts to be shipped; however, they have production to get out and need good parts. It is JF's responsibility to provide good parts ready for use when needed. If JF cannot manage its part of the supply chain, then the customers will look around for someone who will.

Because the leaders at JF already have a clear idea of what they would like—maintain current price and delivery times while reducing shipment damage—they know what they need to discuss when they meet potential new suppliers. They have narrowed their candidates by preferring suppliers with the following characteristics:

Quality (low damage rate)
Quality assurance (ability to monitor and improve shipping process)
Flexibility (able to deliver to a wide variety of customers)
Reliability (on-time deliveries regardless of weather or other problems)
Price (competitive)
Delivery capability (able to manage shipments of JF size and weight)
Lead times (able to make regular deliveries in reasonable times for distances)
Financial stability (large enough to have and maintain effective shipping equipment and employees)
Length of time in business (not new to business or new to products of this type)

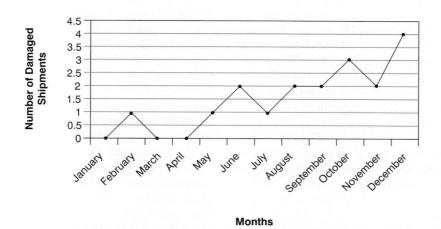

FIGURE 16.8 Damaged Shipments

(Continued)

> Several companies were selected for on-site visits. There, leadership from JF studied the companies' activities and processes in order to verify the following:
>
> Quality assurance activities (how does the organization capture the voice of the customer or JF's shipping needs?)
> Quality planning process (a plan for future support of customers, including improvements)
> Process flow diagrams for key processes (how are shipments handled?)
> Key process capability (what is their performance level with other similar shipments?)
> Process performance measures (how do they measure their performance?)
> Continuous improvement based on performance measures (what improvements are they planning in the near future?)
> Operator instructions (what are their drivers told to do, and how are they trained?)
> Visual factory (what does their equipment, trucks, look like?)
> Customer communication (how will information be shared?)
> Workforce stability or employee turnover
> On-time delivery performance
>
> JF leaders made careful study of the results of their supplier visits. In the final analysis, two companies stood out. They invited each to visit their plant to discuss their requirements. The negotiation moved quickly past the pricing question because both companies were very competitive. On the critical performance measures—on-time delivery rate and damage during shipment—both had similar delivery performance, too. When negotiations focused on performance, the two companies differed widely. One of the companies negotiated its involvement in the packing process. It had clearly prepared for the negotiation in advance. Its in-depth understanding of JF's real expectations became obvious. It clearly evaluated the situation and generated several alternatives focused on JF's desired result: no damage during shipment, on-time delivery, and competitive pricing. It understood what was essential, and this enabled the company to prepare a package detailing packaging changes that could better protect the parts. The package it presented even ended up saving JF money on packaging materials.

feature shows, supplier negotiations usually have three aspects: quality, delivery, and price.

Purchasing plays a key role in organizational effectiveness. It should not be taken lightly. Purchasing agents must do their homework when selecting suppliers, outsourcing activities, or developing partnerships. These relationships must be assessed and audited for both new and existing suppliers. Site visits ensure that suppliers are able to meet their commitments. Measures of performance apply to external as well as internal activities. Negotiations with chosen suppliers ensure value creation by balancing delivery, price, and quality. Effective organizations choose their suppliers carefully, knowing that they are not all created equal.

SUPPLY CHAIN MANAGEMENT CHALLENGES

Effective organizations treat their supply chains as processes that must be managed and improved. Process improvement techniques support a quality management focus that ensures reliability and high quality delivered to the customer. Because suppliers play a large role in providing a quality product or service to the customer, effective organizations ensure that suppliers share their philosophy. This means that the suppliers they chose conform to standards such as ISO 9000. Quality assurance audits confirm or deny good suppliers. Effective organizations appraise the performance of the supply chain. Performance parameters are negotiated in advance and are measured and tracked. Supply chains are continuously improved and made lean.

When a supply chain lacks synchronization, problems arise. The longer the supply chain, the greater the effects of changes in demand. Disruptions to any one of the organizations on the supply chain seriously affect its performance capabilities. These problems can be reduced by communication and information sharing throughout the supply chain. Information availability also helps to reduce lead times and supports smaller batch sizes.

SUMMARY

Supply chains play a strategic role in organizational success. Effective supply chains can be judged by a variety of metrics, including its ability to provide order fulfillment, on-time delivery, lead time, value added per employee, inventory turns, stock-outs, quality, response time, and cycle times. A supply chain must be designed with quality, cost, flexibility, speed, and customer service in mind. In this global economy, supply chains know no boundaries. To be effective, supply chains must be designed with the organization's strategy in mind. The supply chain should correspond to the organization's views on outsourcing and insourcing. Effective organizations insource tasks and processes related to their core competencies and outsource tasks that are not critical. For instance, at JF, its core competencies revolve around its ability to machine parts. It outsources payroll operations because an organization whose primary function is payroll generation and tracking will perform it more efficiently and cost effectively.

Effective supply chains are really networks (Figure 16.3). The functions cannot be managed as separate activities but as one in a series of events. Decisions concerning the supply chain must be carefully considered for their effects on other aspects of the chain. For this reason, effective organizations chose their supply chain partners carefully, knowing that effective supply chains are built on relationships. Business partners in the supply chain maintain clear communication channels and have a distinct interest in each other's success. Whether these organizations are worldwide or local, they display traits that strengthen relationships even when difficulties occur. Suppliers have a reputation for providing quality products or services on a timely basis. They are reliable and consistent on an ongoing basis.

TAKE AWAY TIPS

1. Supply chains create value through two aspects: the physical movement of materials and exchange of information about those materials.
2. A supply chain procures material, performs value-added processing, and distributes products or services to customers.
3. A supply chain supports organizational success by ensuring reliability, adaptability, low costs, and appropriate asset utilization.
4. The elements of a supply chain are inventory management, information sharing, e-commerce, logistics, and purchasing.
5. Information management and maintaining data integrity and security are crucial to the supply chain.
6. E-commerce refers to the electronic integration of information into the supply chain.
7. Distribution or logistics management refers to material handling, warehousing, packaging, and transportation.
8. The role of purchasing is to procure materials needed by an organization.
9. Negotiation is the process of making a joint decision about quality, delivery, and pricing.

CHAPTER PROBLEMS

1. What is a supply chain?
2. What are the benefits of a well-managed supply chain?
3. What are the objectives of effective supply chain management?
4. What organizations might be involved in a supply chain?
5. What do organizations involved in a supply chain transfer back and forth?
6. Why is information so critical to supply chain management?
7. What role does information play in a supply chain?
8. What role does e-commerce play in supply chain management?
9. What are the challenges to creating an effective supply chain?
10. Describe the role that purchasing plays in the supply chain.
11. Describe the purchasing cycle.
12. What is meant by logistics?
13. Why does logistics play a key role in supply chain management?
14. From a firm you have visited or worked with, describe the logistical activities there.
15. From a firm you have visited or worked with, describe supply chain activities, including who is involved and what they provide.

STATISTICS

If things were done right just 99.9 percent of the time, then we'd have to accept

- *One hour of unsafe drinking water per month*
- *Two unsafe plane landings per day at O'Hare International Airport in Chicago*
- *16,000 pieces of mail lost by the U.S. Postal Service every hour*
- *20,000 incorrect drug prescriptions per year*
- *500 incorrect surgical operations each week*
- *22,000 checks deducted from the wrong bank accounts per hour*
- *32,000 missed heartbeats per person per year*

Original source unknown

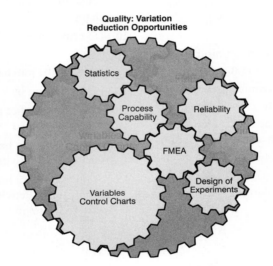

LEARNING OPPORTUNITIES

1. To review basic statistical concepts
2. To understand how to graphically and analytically study the variation present in a process by using statistics
3. To know how to create and interpret a frequency diagram and a histogram
4. To know how to calculate the mean, median, mode, range, and standard deviation for a given set of numbers
5. To understand the importance of the normal curve and the central limit theorem in quality assurance
6. To know how to find the area under a curve using the standard normal probability distribution (Z tables)
7. To understand how to interpret the information analyzed

Statistics, the collection, tabulation, analysis, interpretation, and presentation of numerical data, provides a viable method of supporting or clarifying a lean Six Sigma project. The lean Six Sigma methodology uses correctly collected, analyzed, and interpreted statistical information to understand and predict process behavior. The five aspects of statistics—collection, tabulation, analysis, interpretation, and presentation—are equally important when analyzing the variation present in a process. Once gathered and analyzed, statistical data are the mathematical measures that can be used to aid in decisions about making process changes or pursuing a particular course of action.

POPULATIONS VERSUS SAMPLES

Statistics can be gathered by studying either the entire collection of values associated with a process or only a portion of the values. A ***population*** *is a collection of all possible* elements, values, or items associated with a situation. A population can contain a finite number of things or it may be nearly infinite. As the size of a population increases, studying that population becomes unwieldy unless sampling can be used. A *sample is a subset of elements or measurements taken from a population.* A doctor's office may wish to sample 10 insurance claim forms per week to check the forms for completeness. A manufacturer of toothpaste may check the weight of a dozen tubes per hour to ensure that the tubes are filled correctly. A sample will represent the population as long as the sample is random and unbiased. In a *random sample*, each item in the population has the same opportunity to be selected.

An example of biased sampling could occur on the manufacturing floor. If, when an inspector receives a skid of goods, that inspector always samples from the top layer and takes a part from each of the four corners of the skid, the sample is biased. The parts in the rest of the skid have not been considered. Furthermore, operators observing this behavior may choose to place only the best-quality product in those corners. The inspector has biased the sample because it is not random and does not represent all of the parts of the skid. The inspector is receiving an incorrect impression about the quality in the entire skid. Data should be gathered in a random yet systematic fashion. Its collection should not rely on convenience or deception.

Unbiased samples depend on other features besides randomness. Conditions surrounding the population should not be altered in any way from sample to sample. The sampling method can also undermine the validity of a sample. A sound data collection plan will ensure the validity of a sample by asking the following eleven questions:

How was the problem defined?
What was studied?
How many items were sampled?
How was the sample taken?
What was its source?
How often?
Who collected the data?
Have conditions changed?
How will the data be used?
What will the data be used to predict?
How will the data be displayed?

Taking a Sample

LEAN SIX SIGMA TOOLS at WORK

An outlet store has just received a shipment of 1,000 shirts sealed in cardboard boxes. The store had ordered 800 white shirts and 200 blue. The store manager wishes to check that there actually are 20 percent blue shirts and 80 percent white shirts. He doesn't want to open all of the boxes and count all of the shirts, so he has decided to sample the population. Table 17.1 shows the results of 10 random samples of 10 shirts each.

A greater number of blue shirts is found in some samples than in others. However, when the results are compiled, the blue shirts comprise 19 percent, very close to the desired value of 20 percent. The manager of the outlet store is pleased to learn that the samples have shown that there are approximately 20 percent blue shirts and 80 percent white.

TABLE 17.1 A Sampling of Shirts

Sample Number	Sample Size	Number of White Shirts	Number of Blue Shirts	Percentage of Blue Shirts
1	10	8	2	20
2	10	7	3	30
3	10	8	2	20
4	10	9	1	10
5	10	10	0	0
6	10	7	3	30
7	10	8	2	20
8	10	9	1	10
9	10	8	2	20
10	10	7	3	20
Total	100	81	19	19

DATA COLLECTION

Two types of statistics exist: deductive and inductive. Also known as *descriptive statistics*, **deductive statistics** *describe a population or complete group of data*. When describing a population using deductive statistics, the investigator must study each entity within the population. This provides a great deal of information about the population, product, or process, but gathering the information is time consuming. Imagine contacting each man, woman, and child in the United States, all 300 million of them, to conduct the national census!

When the quantity of the information to be studied is too great, inductive statistics are used. **Inductive statistics** *deal with a limited amount of data or a representative sample of the population*. Once samples are analyzed and interpreted, predictions can be made concerning the larger population of data. Quality assurance (and the U.S. census) relies primarily on inductive statistics. Properly gathered and analyzed, sample data provides a wealth of information.

Two types of statistical data can be collected. **Variables data**, *those quality characteristics that can be measured*, are treated differently from **attribute data**, *those quality characteristics that are observed to be either present or absent, conforming or nonconforming*. Although both variables and attribute data can be described by numbers, attribute data are countable, not measurable. Attribute data is collected in the form of counts, percentages, and labels such as good/bad, pass/fail, and yes/no.

Variables data tend to be continuous in nature. When data are **continuous**, *the measured value can take on any value within a range*. The range of values that the measurements can take on will be set by the expectations of the users or the circumstances surrounding the situation. For example, a manufacturer might wish to monitor the thickness of a part. During the course of the day, the samples may have values of 0.399, 0.402, 0.401, 0.400, 0.401, 0.403, and 0.398 inch.

Discrete data consist of distinct parts. In other words, when measured, **discrete data** *will be countable using whole numbers*. For example, the number of frozen vegetable packages found on the shelf during an inventory count—10 packages of frozen peas, 8 packages of frozen corn, 22 packages of frozen Brussels sprouts—is discrete, countable data. Because vegetable packages can only be sold to customers when the packages are whole and unopened, only whole-numbered measurements will exist; continuous measurements would not be applicable in this case. Attribute data, because the data are seen as being either conforming or not conforming to specifications, are primarily discrete data.

A statistical analysis begins with the gathering of data about a process or product. Sometimes raw data gathered from a process take the form of ungrouped data. **Ungrouped data** *are easily recognized because when viewed, it appears that the data are without any order*. **Grouped data**, on the other hand, *are grouped together on the basis of when the values were taken or observed*. Consider the following Lean Six Sigma Tools at Work feature.

LEAN SIX SIGMA TOOLS at WORK

Grouping Data

A company manufactures various parts for automobile transmissions. One part, a clutch plate, resembles a flat round plate with four keyways stamped into it (Figure 17.1). Recently, the customer brought to the manufacturer's attention the fact that not all of the keyways are being cut out to the correct depth. The manufacturer asked the operator to measure each keyway in five parts every 15 minutes and record the measurements. Table 17.2 shows the results. When managers started to analyze and interpret the data they were unable to do so. Why?

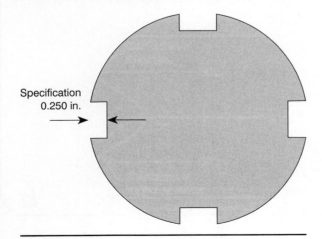

FIGURE 17.1 Clutch Plate

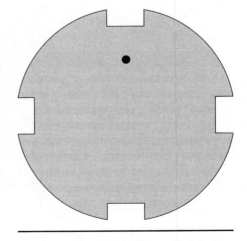

FIGURE 17.2 Clutch Plate with Mark

TABLE 17.2 Clutch Plate Ungrouped Data (in inches)

0.247	0.245	0.271
0.254	0.260	0.276
0.268	0.278	0.268
0.261	0.260	0.230
0.231	0.224	0.243
0.241	0.224	0.225
0.252	0.222	0.232
0.258	0.242	0.254
0.266	0.244	0.242
0.226	0.277	0.248
0.263	0.222	0.236
0.242	0.260	0.262
0.242	0.249	0.223
0.264	0.250	0.240
0.218	0.251	0.222
0.216	0.255	0.261
0.266	0.247	0.244
0.266	0.250	0.249
0.218	0.235	0.226
0.269	0.258	0.232
0.260	0.251	0.250
0.241	0.245	0.248
0.250	0.239	0.252
0.246	0.248	0.251

An investigation of this raw, ungrouped data reveals that there is no way to determine which measurements belong with which keyway. Which keyway is too deep? Too shallow? It is not possible to determine the answer.

To rectify this situation, during the stamping process, the manufacturer placed a small mark below one of the keyways (Figure 17.2). The mark labels that keyway as number 1. Clockwise around the part, the other keyways are designated 2, 3, and 4. The mark does not affect the use of the part. The operator was asked to measure the keyway depths again, five parts every 15 minutes (Table 17.3).

By organizing the data according to keyway, it could then be determined that keyway number 2 is too deep, and keyway number 4 is too shallow.

TABLE 17.3 Clutch Plate Grouped Data (in inches)

	Keyway 1	Keyway 2	Keyway 3	Keyway 4
Subgroup 1	0.250	0.261	0.250	0.240
Subgroup 2	0.251	0.259	0.249	0.242
Subgroup 3	0.250	0.258	0.251	0.245
Subgroup 4	0.249	0.257	0.250	0.243
Subgroup 5	0.250	0.262	0.250	0.244
Subgroup 6	0.251	0.260	0.249	0.245
Subgroup 7	0.251	0.258	0.250	0.241
Subgroup 8	0.250	0.259	0.249	0.247
Subgroup 9	0.250	0.257	0.250	0.245
Subgroup 10	0.249	0.256	0.251	0.244
Subgroup 11	0.250	0.260	0.250	0.243
Subgroup 12	0.251	0.258	0.251	0.244
Subgroup 13	0.250	0.257	0.250	0.245
Subgroup 14	0.250	0.256	0.249	0.246
Subgroup 15	0.250	0.257	0.250	0.246

ACCURACY, PRECISION, AND MEASUREMENT ERROR

The validity of a measurement not only comes from the selection of a sample size and an understanding of the group of data being measured, but also depends on the measurements themselves and how they were taken. Measurement error occurs while the measurements are being taken and recorded. **Measurement error** *is considered to be the difference between a measured value and the true value.* The error that occurs is one either of accuracy or of precision. **Accuracy** *refers to how far from the actual or real value the measurement is.* **Precision** *is the ability to repeat a series of measurements and get the same value each time.* Precision is sometimes referred to as repeatability. **Repeatability** *means being able to achieve the same result time after time.* Challenges to repeatability occur even when using the same person, or instrument, or machine, or setup, or environmental conditions. The real challenge is achieving reproducibility. **Reproducibility** *refers to being able to recreate the same results even when using different people, but the same instrument, or machine, or setup, or environmental conditions.* The variation associated with repeatability is attributable to the instrument, machine, setup, or environmental condition. With reproducibility, the variation is due to the person. Gage repeatability and reproducibility (Gage R&R) studies seek to reduce the variation present in instrument, machine, and setup, as well as operator variability. If the process is stable, all that is left is part variability.

Figure 17.3a pictures the concept of accuracy. The marks average to the center target. Figure 17.3b, with all of the marks clustered together, shows precision. Figure 17.3c describes a situation in which both accuracy and precision exist. The following Lean Six Sigma Tools at Work feature and Figures 17.4, 17.5, and 17.6 illustrate the concepts of accuracy and precision.

Measurement errors may contribute to the lack of accuracy and precision. Measurement errors are not always the fault of the individual performing the measuring. In any situation, several sources of error exist, including environment, people, and machine error. Environmental problems, such as with dust, dirt, temperature, and water, cause measurement errors by disturbing either the products or the measuring tools.

Significant figures and associated rounding errors affect the viability of a measurement. **Significant figures** *are the*

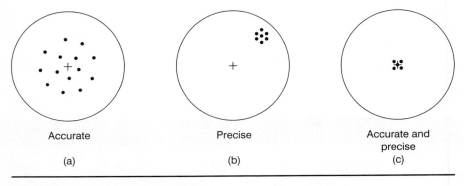

FIGURE 17.3 Accuracy and Precision

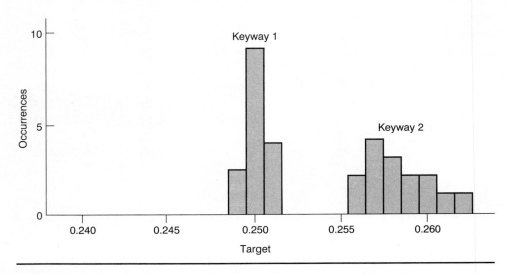

FIGURE 17.4 Data for Keyways 1 and 2 Comparing Accuracy and Precision

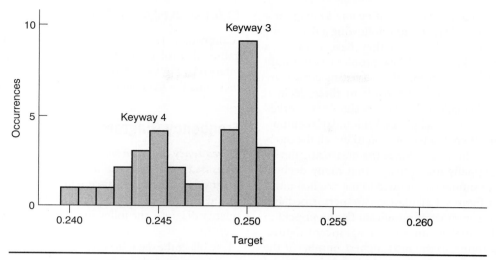

FIGURE 17.5 Data for Keyways 3 and 4 Comparing Accuracy and Precision

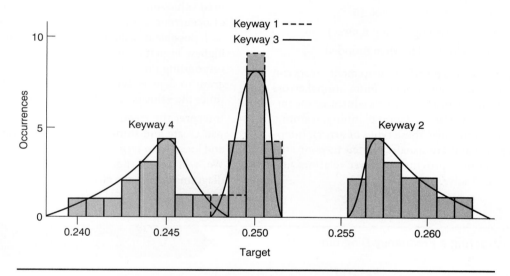

FIGURE 17.6 Data for Keyways 1 and 3 Are Both Accurate and Precise Whereas Data for Keyways 2 and 4 Are Not

Accuracy and Precision

LEAN SIX SIGMA TOOLS at WORK

Accuracy and precision describe the location and the spread of the data. Look at Figures 17.4, 17.5, and 17.6 showing the data from the clutch plate example. When compared, the difference in the keyways' accuracy and precision becomes apparent. The data for keyways 1 and 3 exhibit greater accuracy than that for keyways 2 and 4. Note how the data for keyways 1 and 3 are concentrated around the target specification of 0.250 inch. The data values for keyway 2 are greater than the desired target specification and those for keyway 4 are smaller. Notice, too, the difference in precision. Data for keyways 1 and 3 are precise, tightly grouped around the target. Data values for keyways 2 and 4 are not only far away from the target, but also more spread out, less precise. These figures show that more variation is present for keyways 2 and 4. Changes to this stamping process must be twofold, improving both the accuracy and the precision.

numerals or digits in a number, excluding any leading zeros used to place the decimal point. Zeros following a digit—for example, 9.700—are significant if they have truly been measured. When working a statistical problem, you should use only the number of digits that the measuring devices are able to provide. If a micrometer reads to three decimal places, then all mathematical calculations should be worked to no more than three decimal places. With today's computers and calculators, there is a temptation to use all the numbers that appear on the screen. Since the measuring device may not have originally measured to that many decimal places, the number should be rounded to the original number of significant figures. Calculations are performed and then rounded to the number of significant figures present in the value with the smallest number of significant figures. When rounding, round to the next highest number if the figure is a 5 or greater. If the figure is 4 or below, round down. Consider the following examples:

$23.59 \div 3.8 = 6.2$ (3.8 has fewest significant figures)

$3{,}456 \div 12.3 = 281$ (12.3 has three significant figures)

$3.2 \times 10^2 + 6{,}930 = 7.3 \times 10^3$ (3.2×10^2 has two significant figures)

$6{,}983 \div 16.4 = 425.79268 = 426$ when rounded

Human errors associated with measurement errors can be either unintentional or intentional. Unintentional errors result from poor training, inadequate procedures, or incomplete or ambiguous instructions. Good planning, training, and supervision can minimize these types of errors. Intentional errors are rare and are usually related to poor attitudes; they will require improving employee relations and individual guidance to solve.

DATA ANALYSIS: GRAPHICAL

A thorough statistical analysis of the data that has been gathered involves three aspects: graphical, analytical, and interpretive. A variety of different graphical methods exist, including the frequency diagram and the histogram.

Frequency Diagrams

A *frequency diagram* shows the number of times each of the measured values occurred when the data were collected. This diagram shows at a glance which values occur the most frequently as well as the spread of the data. To create a frequency diagram, the following steps are necessary:

1. Collect the data. Record the measurements or counts of the characteristics of interest.
2. Count the number of times each measurement or count occurs.
3. Construct the diagram by placing the counts or measured values on the x axis and the frequency or number of occurrences on the y axis. The x axis must contain each possible measurement value from the lowest to the highest, even if a particular value does not have any corresponding measurements. A bar is drawn on the diagram to depict each of the values and the number of times the value occurred in the data collected.
4. Interpret the frequency diagram. Study the diagrams you create and think about the diagram's shape, size, and location in terms of the desired target specification. We will learn more about interpreting frequency diagrams later in the chapter.

Constructing a Frequency Diagram

LEAN SIX SIGMA TOOLS at WORK

To respond to customer issues, the engineers involved in the clutch plate problem are studying the thickness of the part. To gain a clearer understanding of incoming material thickness, they plan to create a frequency diagram.

Step 1. *Collect the data.* The first step is performed by the operator, who randomly selects five parts each hour, measures the thickness of each part, and records the values (Table 17.4).

TABLE 17.4 Clutch Plate Thickness: Sums and Averages

						$\sum X_i$	$\overline{X}$
Subgroup 1	0.0625	0.0626	0.0624	0.0625	0.0627	0.3127	0.0625
Subgroup 2	0.0624	0.0623	0.0624	0.0626	0.0625	0.3122	0.0624
Subgroup 3	0.0622	0.0625	0.0623	0.0625	0.0626	0.3121	0.0624
Subgroup 4	0.0624	0.0623	0.0620	0.0623	0.0624	0.3114	0.0623
Subgroup 5	0.0621	0.0621	0.0622	0.0625	0.0624	0.3113	0.0623
Subgroup 6	0.0628	0.0626	0.0625	0.0626	0.0627	0.3132	0.0626
Subgroup 7	0.0624	0.0627	0.0625	0.0624	0.0626	0.3126	0.0625
Subgroup 8	0.0624	0.0625	0.0625	0.0626	0.0626	0.3126	0.0625
Subgroup 9	0.0627	0.0628	0.0626	0.0625	0.0627	0.3133	0.0627

TABLE 17.4 (continued)

Subgroup 10	0.0625	0.0626	0.0628	0.0626	0.0627	0.3132	0.0626
Subgroup 11	0.0625	0.0624	0.0626	0.0626	0.0626	0.3127	0.0625
Subgroup 12	0.0630	0.0628	0.0627	0.0625	0.0627	0.3134	0.0627
Subgroup 13	0.0627	0.0626	0.0628	0.0627	0.0626	0.3137	0.0627
Subgroup 14	0.0626	0.0626	0.0625	0.0626	0.0627	0.3130	0.0626
Subgroup 15	0.0628	0.0627	0.0626	0.0625	0.0626	0.3132	0.0626
Subgroup 16	0.0625	0.0626	0.0625	0.0628	0.0627	0.3131	0.0626
Subgroup 17	0.0624	0.0626	0.0624	0.0625	0.0627	0.3126	0.0625
Subgroup 18	0.0628	0.0627	0.0628	0.0626	0.0630	0.3139	0.0627
Subgroup 19	0.0627	0.0626	0.0628	0.0625	0.0627	0.3133	0.0627
Subgroup 20	0.0626	0.0625	0.0626	0.0625	0.0627	0.3129	0.0626
Subgroup 21	0.0627	0.0626	0.0628	0.0625	0.0627	0.3133	0.0627
Subgroup 22	0.0625	0.0626	0.0628	0.0625	0.0627	0.3131	0.0626
Subgroup 23	0.0628	0.0626	0.0627	0.0630	0.0627	0.3138	0.0628
Subgroup 24	0.0625	0.0631	0.0630	0.0628	0.0627	0.3141	0.0628
Subgroup 25	0.0627	0.0630	0.0631	0.0628	0.0627	0.3143	0.0629
Subgroup 26	0.0630	0.0628	0.0620	0.0628	0.0627	0.3142	0.0628
Subgroup 27	0.0630	0.0628	0.0631	0.0628	0.0627	0.3144	0.0629
Subgroup 28	0.0632	0.0632	0.0628	0.0631	0.0630	0.3153	0.0631
Subgroup 29	0.0630	0.0628	0.0631	0.0632	0.0631	0.3152	0.0630
Subgroup 30	0.0632	0.0631	0.0630	0.0628	0.0628	0.3149	0.0630
						9.3981	

Step 2. *Count the number of times each measurement occurs.* A check sheet, or tally sheet, is used to make this step easier (Figure 17.7).

Step 3. *Construct the diagram.* The count of the number of times each measurement occurred is placed on the *y* axis. The values, between 0.0620 and 0.0632, are each marked on the *x* axis. The completed frequency diagram is shown in Figure 17.8.

Step 4. *Interpret the frequency diagram.* This frequency distribution is nearly symmetrical, but there is only one occurrence of the value 0.0629. The engineers should definitely investigate why this is so.

```
0.0620   /
0.0621   //
0.0622   //
0.0623   ////
0.0624   //// //// //
0.0625   //// //// //// //// //// /
0.0626   //// //// //// //// //// ////
0.0627   //// //// //// //// //// //
0.0628   //// //// //// //// ///
0.0629   /
0.0630   //// //// /
0.0631   //// //
0.0632   ////
```

FIGURE 17.7 Clutch Plate Thickness Tally Sheet

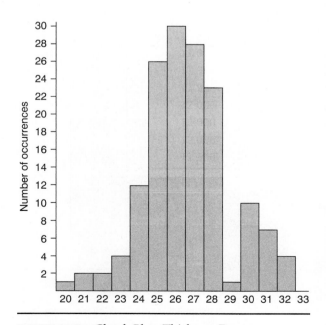

FIGURE 17.8 Clutch Plate Thickness Frequency Distribution (Coded 0.06)

Histograms

Histograms and frequency diagrams are very similar. The most notable difference between the two is that on a histogram the data are grouped into cells. Each cell contains a range of values. This grouping of data results in fewer cells on the graph than with a frequency diagram. The x axis scale on a histogram will indicate the cell midpoints rather than individual values. The following Lean Six Sigma Tools at Work feature details the construction of a histogram.

LEAN SIX SIGMA TOOLS at WORK

Constructing a Histogram

The engineers working with the thickness of the clutch plate have decided to create a histogram to aid in their analysis of the process. They are following these steps:

Step 1. *Collect the data and construct a tally sheet.* The engineers will use the data previously collected (Table 17.3) as well as the tally sheet created during the construction of the frequency diagram (Figure 17.7).

Step 2. *Calculate the range.* The **range**, represented by the letter R, *is calculated by subtracting the lowest observed value from the highest observed value.* In this case, 0.0620 is the lowest value and 0.0632 is the highest:

$$\text{Range} = R = X_h - X_l$$

where
  R = range
  X_h = highest number
  X_l = lowest number
  $R = 0.0632 - 0.0620 = 0.0012$

Step 3. *Create the cells.* In a histogram, data are combined into cells. Cells are composed of three components: cell intervals, cell midpoints, and cell boundaries (Figure 17.9). **Cell midpoints** *identify the centers of cells.* A **cell interval** is the distance between the cell midpoints. The **cell boundary** defines the limits of the cell.

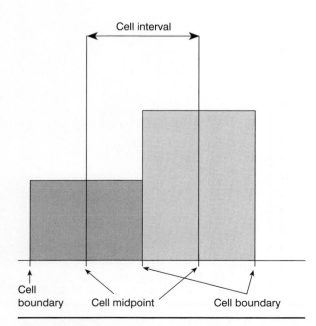

FIGURE 17.9 Histogram Cell Description

Cell Intervals Odd-numbered cell intervals are often chosen for ease of calculation. For example, if the data were measured to one decimal place, then the cell intervals could be 0.3, 0.5, 0.7, or 0.9. If the gathered data were measured to three decimal places, then the cell intervals to choose from would be 0.003, 0.005, 0.007, and 0.009. (The values of 1, 0.1, and 0.001 are not chosen for a histogram because they result in the creation of a frequency diagram.) For this example, because the data were measured to four decimal places, the cell interval could be 0.0003, 0.0005, 0.0007, or 0.0009.

Cell interval choice plays a large part in the size of the histogram created. To determine the number of cells, the following formula is used:

$$h = \frac{R}{i} + 1$$

where
  h = number of cells
  i = cell interval
  R = range

Since both i, the cell interval, and h, the number of cells, are unknown, creators of histograms must choose values for one of them and then solve for the other. For our example, if we choose a cell interval of 0.0003,

$$h = \frac{0.0012}{0.0003} + 1$$
$$h = 5$$

The histogram created will contain 5 cells.
For a cell interval value of 0.0005:

$$h = \frac{0.0012}{0.0005} + 1$$
$$h = 3$$

For a cell interval value of 0.0007:

$$h = \frac{0.0012}{0.0007} + 1$$
$$h = 3$$

As the cell interval gets larger, the number of cells necessary to hold all the data and make a histogram decreases. When deciding the number of cells to use, it is sometimes helpful to follow this rule of thumb:

For fewer than 100 pieces of data, use 4 to 9 cells.

For 100 to 500 pieces of data, use 8 to 17 cells.

For 500 or more, use 15 to 20 cells.

Another helpful rule of thumb exists for determining the number of cells in a histogram. Use the square root of n (represented as $\sqrt{n}$), where n is the number of data points, as an approximation of the number of cells needed.

For this example, we will use a cell interval of 0.0003. This will create a histogram that provides enough spread to analyze the data.

Cell Midpoints When constructing a histogram, it is important to remember two things: (1) Histograms must contain all of the data; (2) one particular value cannot fit into two different cells. Cell midpoints are selected to ensure that these problems are avoided. To determine the midpoint values that anchor the histogram, use either one of the following two techniques.

The simplest technique is to choose the lowest value measured. In this example, the lowest measured value is 0.0620. Other midpoint values are determined by adding the cell interval of 0.0003 to 0.0620 first and then adding it to each successive new midpoint. If we begin at 0.0620, we find the other midpoints at 0.0623, 0.0626, 0.0629, and 0.0632. If the number of values in the cell is high and the distance between the cell boundaries is not large, the midpoint is the most representative value in the cell.

Cell Boundaries The cell size, set by the boundaries of the cell, is determined by the cell midpoints and the cell interval. Locating the cell boundaries, or the limits of the cell, allows the user to place values in a particular cell. To determine the lower cell boundary, divide the cell interval by 2 and subtract that value from the cell midpoint. To calculate the lower cell boundary for a cell with a midpoint of 0.0620, the cell interval is divided by 2:

$$0.0003 \div 2 = 0.00015$$

Then, subtract 0.00015 from the cell midpoint,

$$0.0620 - 0.00015 = 0.06185, \text{the first lower boundary}$$

To determine the upper cell boundary for a midpoint of 0.0620, add the cell interval to the lower cell boundary:

$$0.06185 + 0.0003 = 0.06215$$

The lower cell boundary of one cell is the upper cell boundary of another. Continue adding the cell interval to each new lower cell boundary calculated until all the lower cell boundaries have been determined.

Note that the cell boundaries are a half decimal value greater in accuracy than the measured values. This is to help ensure that values can be placed in only one cell of a histogram. In our example, the first cell will have boundaries of 0.06185 and 0.06215. The second cell will have boundaries of 0.06215 and 0.06245. Where would a data value of 0.0621 be placed? Obviously in the first cell. Cell intervals, with their midpoint values starting at 0.0620, are shown in Figure 17.10.

Step 4. *Label the axes.* Scale and label the horizontal axis according to the cell midpoints determined in step 3. Label the vertical axis to reflect the amount of data collected, in counting numbers.

FIGURE 17.10 Cell Boundaries and Midpoints

Step 5. *Post the values.* The final step in the creation of a histogram is to post the values from the check sheet to the histogram. The *x* axis is marked with the cell midpoints and, if space permits, the cell boundaries. The cell boundaries are used to guide the creator when posting the values to the histogram. On the *y* axis, the frequency of those values within a particular cell is shown. All the data must be included in the cells (Figure 17.11).

Step 6. *Interpret the histogram.* As we can see in Figure 17.11, the data are grouped around 0.0626 and are somewhat symmetrical. In the following sections, we will study histogram shapes, sizes, and locations when compared to a desired target specification. We will also utilize measures such as means, modes, and medians to create a clear picture of where the data are grouped (the central tendency of the data). Standard deviations and ranges will be used to measure how the data are dispersed around the mean. These statistical values will be used to fully describe the data comprising a histogram.

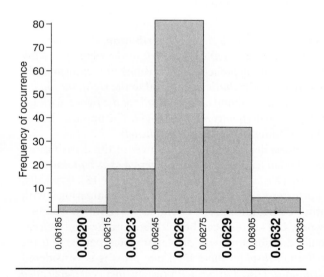

FIGURE 17.11 Clutch Plate Thickness Histogram

Shape, location, and spread are the characteristics used to describe a distribution (Figure 17.12).

Shape: Symmetry, Skewness, Kurtosis Shape *Symmetry, skewness, and kurtosis shape refer to the form that the values of the measurable characteristics take on when plotted or graphed.* Tracing a smooth curve over the tops of the rectangular areas used when graphing a histogram clarifies the shape of a histogram for the viewer (Figure 17.13). Identifiable characteristics include **symmetry**, or, in the case of lack of symmetry, **skewness** of the data; **kurtosis**, or *peakedness of the data*; and **modes**, *the number of peaks in the data*.

When a distribution is **symmetrical**, *the two halves are mirror images of each other*. The two halves correspond in size, shape, and arrangement (Figure 17.13). When a distribution is not symmetrical, it is considered to be skewed

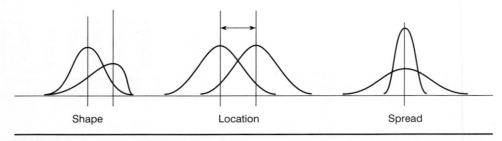

FIGURE 17.12 Shape, Location, and Spread

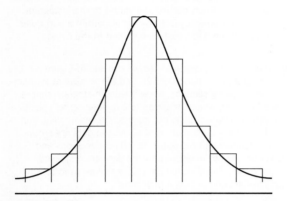

FIGURE 17.13 Symmetrical Histogram with Smooth Curve Overlay

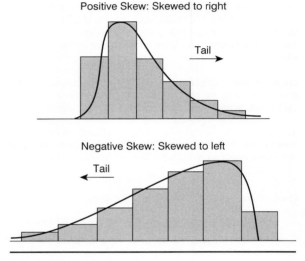

FIGURE 17.14 Skewness

(Figure 17.14). With a **skewed distribution**, *the majority of the data are grouped either to the left or the right of a center value, and on the opposite side a few values trail away from the center*. When a distribution is **skewed to the right**, the majority of the data are found on the left side of the figure, with the tail of the distribution going to the right. The opposite is true for a distribution that is **skewed to the left**.

Kurtosis describes the peakedness of the distribution. A *distribution with a high peak is referred to as* **leptokurtic**; *a flatter curve is called* **platykurtic** (Figure 17.15). Typically, the kurtosis of a histogram is discussed by comparing it with another distribution. As we will see later in the chapter, skewness and kurtosis can be calculated numerically. Occasionally distributions will display unusual patterns. *If the distribution displays more than one peak*, it is considered **multimodal**. Distributions with two distinct peaks are called **bimodal** (Figure 17.16).

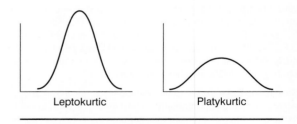

FIGURE 17.15 Leptokurtic and Platykurtic

Analyzing the Histogram

LEAN SIX SIGMA TOOLS at WORK

Analyzing Figure 17.11 based on the three characteristics of shape, location, and spread reveals that the clutch plate thickness data are fairly consistent. The **shape** of the distribution is somewhat symmetrical, though skewed very slightly to the right. The data are unimodal, centering on 0.0626 inch. Because we have no other distributions of the same type of product, we cannot make any comparisons or comments on the kurtosis of the data. **Location**, or where the data are located or gathered, is around 0.0626. If the engineers have specifications of 0.0625 ÷ 0.0003, then the center of the distribution is higher than the desired value. Given the specifications, the **spread** of the data is broader than the desired 0.0622 to 0.0628 at 0.0620 to 0.0632. Further mathematical analysis with techniques covered later in this chapter will give us an even clearer picture of the data.

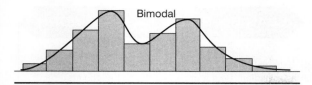

FIGURE 17.16 Bimodal Distribution

DATA ANALYSIS: ANALYTICAL

Though shape was easily seen from a picture, the location and spread can be more clearly identified mathematically. Location is described by measures of central tendency: the mean, mode, and median. Spread is defined by measures of dispersion: the range and standard deviation.

Location: Measures of Central Tendency

Averages, medians, and modes are the statistical values that define the center of a distribution. *Because they reveal the place where the data tend to be gathered, these values are commonly called the **measures of central tendency**.*

Mean The *mean of a series of measurements is determined by adding the data values together and then dividing this sum by the total number of data values.* When this value is calculated for a population, it is referred to as the mean and is signified by μ. When this value is calculated for a sample, it is called the *average* and is signified by $\overline{X}$ (X bar). Means and averages can be used to judge whether a group of values is accurate. To calculate the mean of a population, use the following formula:

$$\mu = \frac{X_1 + X_2 + X_3 + \cdots + X_n}{n} = \frac{\sum_{i=1}^{n} X_i}{n}$$

where

μ = mean value of the series of measurements
$X_1, X_2, \ldots, X_n$ = values of successive measurements
n = number of readings

The same formula can be used to calculate the average associated with a sample. To calculate the average of a sample, use the following formula:

$$\overline{X} = \frac{X_{s1} + X_{s2} + X_{s3} + \cdots + X_{sn}}{n} = \frac{\sum_{i=1}^{n} X_i}{n}$$

where

$\overline{X}$ = average value of the sample measurements
$X_{s1}, X_{s2}, \ldots, X_{sn}$ = values of sample measurements
n = number of readings

Median The *median is the value that divides an ordered series of numbers so that there is an equal number of values on either side of the center, or median, value.* An ordered series of data has been arranged according to their magnitude. Once the values are placed in order, the median is the value of the number that has an equal number of values to its left and right. In the case of finding a median for an even number of values, the two center values of the ordered set of numbers are added together and the result is divided by 2. Figure 17.17 shows the calculation of several medians. Typically, the mean is used more often than the median, but the median is less affected by outliers than the mean.

Mode The *mode is the most frequently occurring number in a group of values.* In a set of numbers, a mode may or

---

Determining the Mean

Averages for each of the subgroups for the thicknesses of the clutch plate can be calculated.

1. Calculate the sum of each set of subgroup values:

   Subgroup 1:

   $\sum X_1 = 0.0625 + 0.0626 + 0.0624 + 0.0625 + 0.0627$
   $= 0.3127$

2. Calculate the subgroup average by dividing the sum by the number of samples in the subgroup (n = 5):

   Subgroup 1:

   $$\overline{X} = \frac{0.0625 + 0.0626 + 0.0624 + 0.0625 + 0.0627}{5}$$

   $$= \frac{0.3127}{5}$$

   $$= 0.0625$$

LEAN SIX SIGMA TOOLS at WORK

From earlier in the chapter, Table 17.4 gives a list of the sums and averages calculated for this example. Once the averages for each subgroup have been calculated, a grand average for all of the subgroups can be found by dividing the sum of the subgroup sums by the total number of items taken in all of the subgroups (150). A grand average is designated as $\overline{\overline{X}}$ (X double bar):

$$\overline{\overline{X}} = \frac{0.3127 + 0.3122 + 0.3121 + 0.3114 + \cdots + 0.3149}{150}$$

$$= \frac{9.3990}{150}$$

$$= 0.0627$$

Notice that an average of the averages is not taken. Taking an average of the averages will work only when the sample sizes are constant. Use the sums of each of the subgroups to perform the calculation.

Determining the Median

From the check sheet (Figure 17.7) the median of the clutch plate thickness data can be found. When the data are placed in an ordered series, the center or median number is found to be 0.0626. Each measurement must be taken into account when calculating a median. Do not use solely the cell midpoints of a frequency diagram or a histogram.

LEAN SIX SIGMA TOOLS at WORK

---

23 25 26 27 28 29 25 22 24 24 25 26 25

Unordered set of numbers

22 23 24 24 25 25 25 25 26 26 27 28 29

Ordered set of numbers

Median = 25

1 2 4 1 5 2 6 7

Unordered set of numbers

1 1 2 2 4 5 6 7

Ordered set of numbers

Median = (2 + 4) ÷ 2 = 3

FIGURE 17.17 Calculating Medians

---

may not occur (Figure 17.18). A set of numbers may also have two or more modes. If a set of numbers or measurements has one mode, it is said to be unimodal. If it has two numbers appearing with the same frequency, it is called bimodal. Distributions with more than two modes are referred to as multimodal. In a frequency distribution or a histogram, the cell with the highest frequency is the mode.

The Relationship Among the Mean, Median, and Mode As measures of central tendency, the mean, median, and mode can be compared with each other to determine where the data are located. Measures of central tendency describe the center position of the data. They show how the data tend to build up around a center value. When a distribution is symmetrical, the mean, mode, and median values are equal. For a skewed distribution, the values will be different (Figure 17.19). Comparing the mean (average), mode, and median determines whether a distribution is skewed and, if it is, in which direction.

Spread: Measures of Dispersion

The range and standard deviation are two measurements that enable the investigator to determine the spread of the data, that is, where the values fall in relation to each other and to the mean. Because these two describe where the data are dispersed on either side of a central value, they are often referred to as measures of dispersion. Used in conjunction with the mean, mode, and median, these values create a more complete picture of a distribution.

Range As was pointed out in the discussion of the histogram earlier in this chapter, the *range is the difference between the largest value in a series of values or sample and the smallest value in that same series.* A range value describes how far the data spread. All of the other values in a population or sample will fall between the highest and lowest values:

$$R = X_h - X_l$$

where

R = range
X_h = highest value in the series
X_l = lowest value in the series

---

23 25 26 27 28 29 25 22 24 24 25 26 25

Unordered set of numbers

22 23 24 24 25 25 25 25 26 26 27 28 29

Ordered set of numbers

Mode = 25

1 3 4 1 5 2 6 6 7

Unordered set of numbers

1 1 2 3 4 5 6 6 7

Ordered set of numbers

Bimodal = 1 and 6

100 101 103 104 106 107

No mode

658 659 659 659 670 670 670 671 672 672 672 674 674

Multimodal: 659, 670, 672

FIGURE 17.18 Calculating Modes

Determining the Mode

LEAN SIX SIGMA TOOLS at WORK

The mode can be found for the clutch plate thickness data. The check sheet (Figure 17.7) clearly shows that 0.0626 is the most frequently occurring number. It is tallied 30 times.

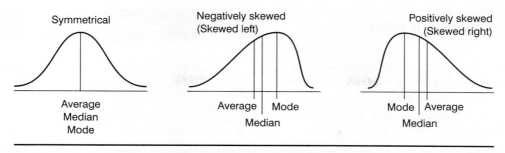

FIGURE 17.19 Comparison of Mean, Mode, and Median

Seeing the Relationship

LEAN SIX SIGMA TOOLS at WORK

Knowing the average, median, and mode of the clutch plate data provides information about the symmetry of the data. If the distribution is symmetrical, the average, mode, and median values will be equal. A skewed distribution will have different values. From previous examples, the values for the clutch plate are as follows:

Average = 0.0627 inch
Median = 0.0626 inch
Mode = 0.0626 inch

As seen in the frequency diagram (Figure 17.20), the mode marks the peak of the distribution. The average, slightly to the right of the mode and median, pulls the distribution to the right. This slight positive skew is due to the high values for clutch plate thickness that occur in later samples.

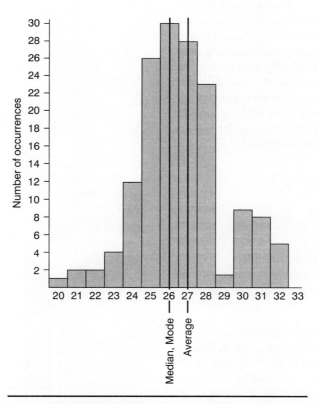

FIGURE 17.20 Comparison of Mean, Mode, and Median for the Clutch Plate

> ### Calculating Range Values
> **LEAN SIX SIGMA TOOLS at WORK**
>
> The flat round plate data comprises subgroups of sample size five (Table 17.4). For each sample, a range value can be calculated. For example:
>
> Subgroup 1   0.0625   0.0626   0.0624   0.0625   0.0627
> Range = $X_h - X_l$ = 0.0627 − 0.0624 = 0.0003
>
> Subgroup 2   0.0624   0.0623   0.0624   0.0626   0.0625
> Range = $X_h - X_l$ = 0.0626 − 0.0623 = 0.0003
>
> The other ranges are calculated in the same manner. These range values are used in the next chapter to study the variation present in the process over time.

Standard Deviation   The range shows where each end of the distribution is located, but it doesn't tell how the data are grouped within the distribution. In Figure 17.21, the three distributions have the same average and range, but all three are different. The *standard deviation shows the dispersion of the data within the distribution*. It describes how the individual values fall in relation to their means, the average distance from each data point to the mean. It describes the expected amount of variation present in a normally distributed set of data. The standard deviation, because it uses all of the measurements taken, provides more reliable information about the dispersion of the data. The range considers only the two extreme values in its calculation, giving no information concerning where the values may be grouped. Because it only considers the highest and lowest values, the range has the disadvantage of becoming a less accurate description of the data as the number of readings or sample values increases. The range is best used with small populations or small sample sizes of less than 10 values. However, because the range is easy to calculate, it is the most frequently used measure of dispersion. The standard deviation of the population is sometimes known as the *root mean square deviation*.

When the measurements have been taken from each and every item in the total population, the standard deviation is calculated through the use of the following formula:

$$\sigma = \sqrt{\frac{\sum_{i=1}^{n}(X_i - \mu)^2}{n}}$$

where

σ = standard deviation of the population
μ = mean value of the series of measurements
$X_i = X_1, X_2, \ldots, X_n$ = values of each reading
n = number of readings

A smaller standard deviation is desirable because it indicates greater similarity between data values—that is, the data are more precisely grouped, there is less variation in the process. In the case of products, a small standard deviation indicates that the products are nearly alike. Creating products or providing services that are similar to each other is optimal. When working with standard deviation values, remember that in a multiple step process, individual process step standard deviations cannot be combined for an overall standard deviation. To calculate overall process variation, add together the variances for each step and then take the square root.

When the measurements are taken from items sampled from the entire population, the previous formula is modified to reflect the fact that not every item in the population has

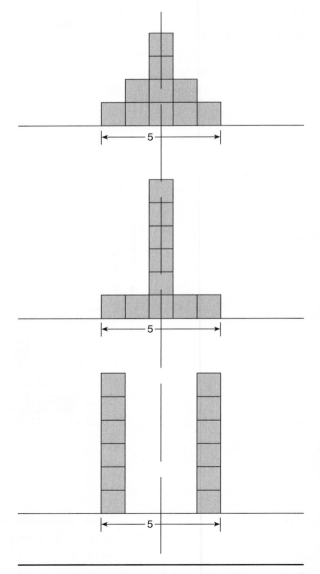

FIGURE 17.21   Different Distributions with Same Averages and Ranges

LEAN SIX SIGMA TOOLS at WORK

Determining the Standard Deviation of a Sample

In the case of subgroups comprising the clutch plate data, it is possible to calculate the sample standard deviation for each of the subgroups. For subgroup 1:

$$s_1 = \sqrt{\frac{\sum(X_i - \overline{X})^2}{n - 1}}$$

$$= \sqrt{\frac{(0.0624 - 0.0625)^2 + 2(0.0625 - 0.0625)^2 + (0.0626 - 0.0625)^2 + (0.0627 - 0.0625)^2}{5 - 1}}$$

$$= 0.0001$$

Standard deviations for the remaining subgroups can be calculated in the same manner.

---

been measured. This change is reflected in the denominator. The standard deviation of a sample is represented by the letter s:

$$s = \sqrt{\frac{\sum_{i=1}^{n}(X_i - \overline{X})^2}{n - 1}}$$

where

- s = standard deviation of the sample
- $\overline{X}$ = average value of the series of measurements
- $X_i = X_1, X_2, \ldots, X_n$ = values of each reading
- n = number of readings

Using the Mean, Mode, Median, Standard Deviation, and Range Together Measures of central tendency and measures of dispersion are critical when describing statistical data. As the following example shows, one without the other creates an incomplete picture of the values measured.

LEAN SIX SIGMA TOOLS at WORK

Seeing the Whole Picture I

Two engineers were keeping track of the rate of water pipe being laid by three different crews. Over the past 36 days, the amount of pipe laid per day was recorded and the frequency diagrams shown in Figure 17.22 were created. When they studied the data originally, the two engineers calculated only the mean, mode, and median for each crew.

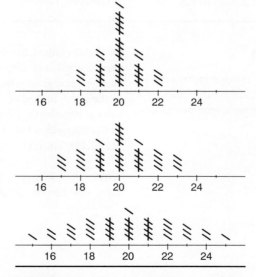

FIGURE 17.22 Frequency Diagrams of the Amount of Pipe Laid per Day in Feet

(Continued)

$$\text{Mean}_1 = 20 \quad \text{Median}_1 = 20 \quad \text{Mode}_1 = 20$$
$$\text{Mean}_2 = 20 \quad \text{Median}_2 = 20 \quad \text{Mode}_2 = 20$$
$$\text{Mean}_3 = 20 \quad \text{Median}_3 = 20 \quad \text{Mode}_3 = 20$$

From these numbers, the distributions appear the same. It was not until the range and standard deviation for each of the three pipe-laying crews' work were calculated that the differences became apparent:

$$\text{Range}_1 = 4$$
$$\text{Standard deviation}_1 = 1.03, \text{ rounded to } 1.0$$

$$\sigma = \sqrt{\frac{3(18-20)^2 + 7(19-20)^2 + 16(20-20)^2 + 7(21-20)^2 + 3(22-20)^2}{36}}$$

$$\text{Range}_2 = 6$$
$$\text{Standard deviation}_2 = 1.65, \text{ rounded to } 1.7$$

$$\sigma = \sqrt{\frac{3(17-20)^2 + 4(18-20)^2 + 6(19-20)^2 + 10(20-20)^2 + 6(21-20)^2 + 4(22-20)^2 + 3(23-20)^2}{36}}$$

$$\text{Range}_3 = 10$$
$$\text{Standard deviation}_3 = 2.42, \text{ rounded to } 2.4$$

$$\sigma = \sqrt{\frac{1(15-20)^2 + 2(16-20)^2 + 3(17-20)^2 + 4(18-20)^2 + 5(19-20)^2 + 6(20-20)^2 + 5(21-20)^2 + 4(22-20)^2 + 3(23-20)^2 + 2(24-20)^2 + 1(25-20)^2}{36}}$$

Once calculated, the ranges and standard deviations revealed that significant differences exist in the performance of the three crews. The first crew was much more consistent in the amount of pipe they laid per day. As the figures and calculations show, there is more day-to-day variation present in the amount of pipe the third crew lays.

Seeing the Whole Picture II

LEAN SIX SIGMA TOOLS at WORK

When we combine the analytical calculations with the graphical information from the previous Lean Six Sigma Tools at Work features, we see a more complete picture of the clutch plate data we are studying (Figure 17.11). The grand average, $\overline{\overline{X}} = 0.0627$ inches, median (0.0626), and mode (0.0626) confirm that the histogram is skewed slightly to the right. Because we know the grand average of the data, we also know that the distribution is not centered on the desired target value of 0.0625 inches. The frequency diagram gives us the critical information that there is only one plate with a thickness of 0.0629 inches. The range of our data is fairly broad; the frequency diagram shows an overall spread of the distribution of 0.0012 inches. In general, through their calculations and diagrams, the engineers have learned that they are making the plates too thick. They have also learned that the machining process is not producing plates of consistent thickness.

Other Measures of Dispersion

Skewness When a distribution lacks symmetry, it is considered **skewed**. A picture of the distribution is not necessary to determine skewness. Skewness can be measured by calculating the following value:

$$a_3 = \frac{\sum_{i=1}^{h} f_i(X_i - \overline{X})^3 / n}{s^3}$$

where

- a_3 = skewness
- X_i = individual data values under study
- $\overline{X}$ = average of individual values
- n = sample size
- s = standard deviation of sample
- f_i = frequency of occurrence

Once determined, the skewness figure is compared with zero. A skewness value of zero means that the distribution is symmetrical. A value greater than zero means that the data are skewed to the right; the tail of the distribution goes to the right. If the value is negative (less than zero), then the distribution is skewed to the left, with a tail of the distribution going to the left (Figure 17.23). The higher the value, the stronger the skewness.

Kurtosis Kurtosis, the peakedness of the data, is another value that can be calculated:

$$a_4 = \frac{\sum_{i=1}^{h} f_i(X_i - \overline{X})^4 / n}{s^4}$$

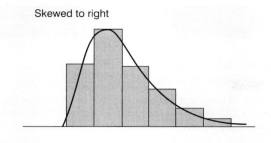

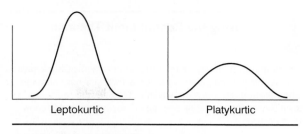

FIGURE 17.24 Kurtosis

more peaked than that on the right. Its kurtosis value would be larger.

CENTRAL LIMIT THEOREM

Much of statistical process control is based on the use of samples taken from a population of items. The central limit theorem enables conclusions to be drawn from the sample data and applied to a population. The **central limit theorem** states that a *group of sample averages tends to be normally distributed; as the sample size n increases, this tendency toward normality improves.* In other words, sampling distributions with sample size n = 30 will approximate a normal curve faster than a sampling distribution with a sample size n = 5. The population from which the samples are taken does not need to be normally distributed for the sample averages to tend to be normally distributed (Figure 17.25). In the field of quality, the central limit theorem supports the use of sampling to analyze the population. The mean of the sample averages will approximate the mean of the population. The variation associated with the sample averages will be less than that of the population. It is important to remember that it is the sample *averages* that tend toward normality, as the following Lean Six Sigma Tools at Work feature shows.

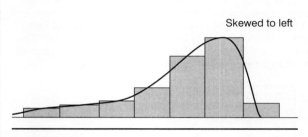

FIGURE 17.23 Skewness

where

a_4 = kurtosis
X_i = individual data values under study
$\overline{X}$ = average of individual values
n = sample size
s = standard deviation of sample

Once calculated, the kurtosis value must be compared with another distribution or with a standard in order to be interpreted. In Figure 17.24, the distribution on the left side is

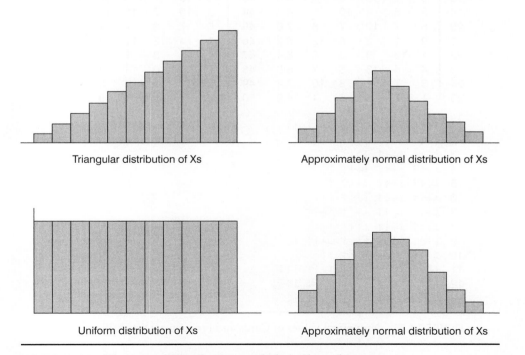

FIGURE 17.25 Nonnormal Distributions and Their Plots of Averages

Using the Central Limit Theorem

LEAN SIX SIGMA TOOLS at WORK

Chuck and Bob are trying to settle an argument. Chuck says that averages from a deck of cards will form a normal curve when plotted. Bob says they won't. They decide to try the following exercise involving averages. They are going to follow these rules:

1. They will randomly select five cards from a well-shuffled deck and write down the values (Figure 17.26). (An ace is worth 1 point, a jack 11, a queen 12, and a king 13.)
2. They will record the numerical values on a graph (Figure 17.26).
3. They will calculate the average for the five cards.
4. They will graph the results of step 3 on a graph separate from that used in step 2 (Figure 17.27).
5. They will then replace the five cards in the deck.
6. They will shuffle the deck.
7. They will repeat this process 50 times.

Figure 17.26 displays the results of steps 1 and 2. Because the deck was well shuffled and the selection of cards from the deck was random, each card had the same chance of being selected—1/52. The fairly uniform distribution of values in the frequency diagram in Figure 17.26 shows that each type of card was selected approximately the same number of times. The distribution would be even more uniform if a greater number of cards had been drawn.

Figure 17.27 graphs the results of step 4. Notice that as the number of averages recorded increases, the results look more and more like a normal curve. As predicted by the central limit theorem, the distribution of the sample averages in the final diagram in Figure 17.27 is approximately normal. This has occurred even though the original distribution was not normal.

1	11	13	10	12	7	10.6	26	8	10	1	9	10	7.6
2	1	7	12	9	3	6.4	27	9	13	2	2	2	5.6
3	9	5	12	1	11	7.6	28	12	4	12	3	13	8.8
4	11	5	7	9	12	8.8	29	12	4	7	6	9	7.6
5	7	12	13	7	4	8.6	30	1	12	3	12	11	7.8
6	11	9	5	1	13	7.8	31	12	3	10	11	6	8.4
7	1	4	13	12	13	8.6	32	3	5	10	2	7	5.4
8	13	3	2	6	12	7.2	33	9	1	2	3	11	5.2
9	2	4	1	10	13	6.0	34	6	8	6	13	9	8.4
10	4	5	12	1	9	6.2	35	2	12	5	10	4	6.6
11	2	5	7	7	11	6.4	36	6	4	8	9	12	7.8
12	6	9	8	2	12	7.4	37	9	13	3	10	1	7.2
13	2	3	6	11	11	6.6	38	2	1	13	7	5	5.6
14	2	6	9	11	13	8.2	39	10	11	5	12	13	10.2
15	6	8	8	9	1	6.4	40	13	2	8	2	11	7.2
16	3	4	12	1	6	5.2	41	2	10	5	4	11	6.4
17	8	1	8	6	10	6.6	42	10	4	12	7	11	8.8
18	5	7	6	8	8	6.8	43	13	13	7	1	10	8.8
19	2	5	4	10	1	4.4	44	9	10	7	11	11	9.6
20	5	7	12	7	8	7.8	45	6	7	8	7	4	6.4
21	9	1	3	6	12	6.2	46	1	4	12	11	13	8.2
22	1	13	9	3	6	6.4	47	9	11	8	1	11	8.0
23	4	5	13	5	7	6.8	48	8	13	10	13	4	9.6
24	3	7	9	8	10	7.4	49	12	11	11	2	3	7.8
25	1	7	6	6	1	4.2	50	2	12	5	11	9	7.8

```
 1  ||||  ||||  ||||  ||||  ||
 2  ||||  ||||  ||||  ||||
 3  ||||  ||||  ||||
 4  ||||  ||||  ||||  |
 5  ||||  ||||  ||||  |
 6  ||||  ||||  ||||  |||
 7  ||||  ||||  ||||  ||||  |
 8  ||||  ||||  ||||  |
 9  ||||  ||||  ||||  ||||  |
10  ||||  ||||  ||||  ||
11  ||||  ||||  ||||  ||||  |||
12  ||||  ||||  ||||  ||||  ||||
13  ||||  ||||  ||||  ||||  ||
```

FIGURE 17.26 Numerical Values of Cards and Frequency Distribution

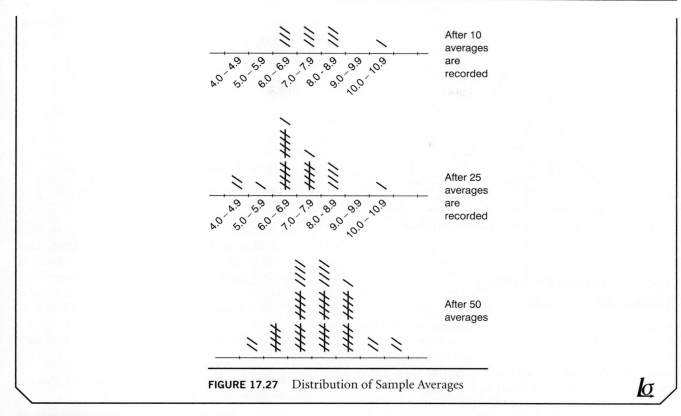

FIGURE 17.27 Distribution of Sample Averages

NORMAL FREQUENCY DISTRIBUTION

The normal frequency distribution, the familiar bell-shaped curve (Figure 17.28), is commonly called a *normal curve*. A **normal frequency distribution** is described by the normal density function:

$$f(x) = \frac{1}{\sigma\sqrt{2\pi}} e^{-(x-\mu)^2/2\sigma^2} \quad -\infty < x < \infty$$

where

$\pi \approx 3.14159$
$e \approx 2.71828$

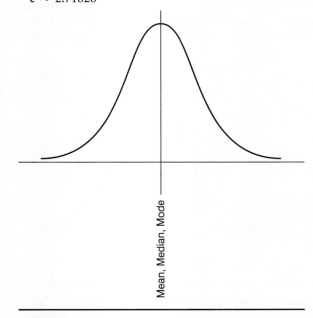

FIGURE 17.28 The Normal Curve

The normal frequency distribution has six distinct features:

1. A normal curve is symmetrical about μ, the central value.
2. The mean, mode, and median are all equal.
3. The curve is unimodal and bell-shaped.
4. Data values concentrate around the mean value of the distribution and decrease in frequency as the values get further away from the mean.
5. The area under the normal curve equals 1. One hundred percent of the data are found under the normal curve, 50 percent on the left-hand side, 50 percent on the right.
6. The normal distribution can be described in terms of its mean and standard deviation by observing that 99.7 percent of the measured values fall within ±3 standard deviations of the mean ($\mu \pm 3\sigma$), that 95.5 percent of the data fall within ±2 standard deviations of the mean ($\mu \pm 2\sigma$), and that 68.3 percent of the data fall within ±1 standard deviation ($\mu \pm 1\sigma$). Figure 17.29 demonstrates the percentage of measurements falling within each standard deviation.

These six features combine to create a peak in the center of the distribution, with the number of values decreasing as the measurements get farther away from the mean. As the data fall away toward the horizontal axis, the curve flattens. The tails of the normal distribution approach the horizontal axis and extend indefinitely, never reaching or crossing it.

Although not all symmetrical distributions are normal distributions, these six features are general indicators of a normal distribution. (There is a chi square test for normality. Refer to a statistics text for a complete description of the chi square test.)

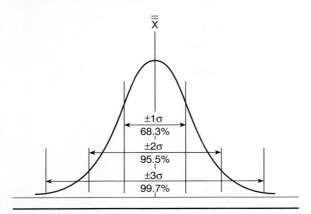

FIGURE 17.29 Percentage of Measurements Falling Within Each Standard Deviation

Standard Normal Probability Distribution: Z Tables

The area under the normal curve can be determined if the mean and the standard deviation are known. The mean, or in the case of samples, the average, locates the center of the normal distribution. The standard deviation defines the spread of the data about the center of the distribution.

The relationships discussed in features 5 and 6 of the normal frequency distribution make it possible to calculate the percentage of values that fall between any two readings. If 100 percent of the data are under the normal curve, then the amount of product above or below a particular value can be determined. These values may be dimensions such as the upper and lower specification limits set by the designer or they can be any value of interest. The formula for finding the area under the normal curve is as follows:

$$f(x) = \frac{1}{\sigma\sqrt{2\pi}} e^{-(x-\mu)^2/2\sigma^2} \quad -\infty < x < \infty$$

where

$\pi \approx 3.14159$
$e \approx 2.71828$

This formula can be simplified through the use of the standard normal probability distribution table (Appendix 1). This table uses the following formula:

$$f(Z) = \frac{1}{\sqrt{2\pi}} e^{-Z^2/2}$$

where

$Z = \dfrac{X_i - \overline{X}}{s}$ = standard normal value
X_i = individual X value of interest
$\overline{X}$ = average
s = standard deviation

This formula also works with population means and population standard deviations:

$$Z = \frac{X_i - \mu}{\sigma_{\overline{X}}} = \text{standard normal value}$$

where

X_i = individual X value of interest
μ = population mean
$\sigma_{\overline{X}}$ = population standard deviation = $\dfrac{\sigma}{\sqrt{n}}$

Z is used with the table in Appendix 1 to find the value of the area under the curve, which represents a percentage or proportion of the product or measurements produced. If Z has a positive value, then it is to the right of the center of the distribution and X_i is larger than $\overline{X}$. If the Z value is negative, then it is on the left side of the center and X_i is smaller than $\overline{X}$.

To find the area under the normal curve associated with a particular X_i, use the following procedure:

1. Use the information on normal curves to verify that the measurements are normally distributed.
2. Use the mean, standard deviation, and value of interest in the formula to calculate Z.
3. Find the Z value in the table in Appendix 1.
4. Use the table to convert the Z values to the area of interest.
5. Convert the area of interest value from the table to a percentage by multiplying by 100.

The table in Appendix 1 is a left-reading table, meaning that it will provide the area under the curve from negative infinity up to the value of interest (Figure 17.30). These values will have to be manipulated to find the area greater than the value of interest or between two values. Drawing a picture of the situation in question and shading the area of interest often helps clarify the Z calculations.

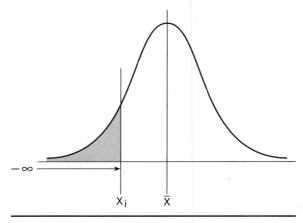

FIGURE 17.30 Normal Curve for Left-Reading Z Table

Using Standard Normal Probability Distribution

LEAN SIX SIGMA TOOLS at WORK

The engineers working with the clutch plate thickness data have determined that their data approximate a normal curve. They would like to determine what percentage of parts from the samples taken are below 0.0624 inch and above 0.0629 inch.

1. From the data in Table 17.3, they calculated an average of 0.0627 and a standard deviation of 0.00023. They used the Z tables to determine the percentage of parts under 0.0624 inch thick. In Figure 17.31 the area of interest is shaded.

$$Z = \frac{0.0624 - 0.0627}{0.00023} = -1.30$$

From Appendix 1: Area = 0.0968

or 9.68 percent of the parts are thinner than 0.0624 inch.

2. When determining the percentage of the parts that are 0.0629 inch thick or thicker, it is important to note that the table in Appendix 1 is a left-reading table. Because the engineers want to determine the percentage of parts thicker than 0.0629 (the area shaded in Figure 17.32), they will have to subtract the area up to 0.0629 from 1.00.

$$Z = \frac{0.0629 - 0.0627}{0.00023} = 0.87$$

Area = 0.8079

or 80.79 percent of the parts are *thinner* than 0.0629 inch. However, they want the percentage of parts that are *thicker* than 0.0629 inch. To find this area they must subtract the area from 1.0 (remember: 100 percent of the parts fall under the normal curve):

$$1.00 - 0.8079 = 0.1921$$

or 19.21 percent of the parts are thicker than 0.0629 inch.

3. The engineers also want to find the percentage of the parts between 0.0623 and 0.0626 inch thick: The area of interest is shaded in Figure 17.33. In this problem the engineers must calculate two areas of interest, one for those parts 0.0623 inches thick or thinner and the other for those parts 0.0626 inch thick or thinner. The area of interest for those parts 0.0623 inch and thinner will be subtracted from the area of interest for 0.0626 inch and thinner.

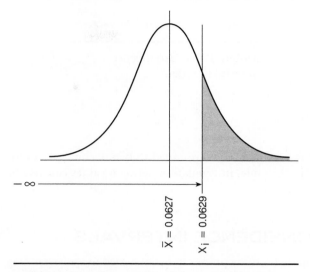

FIGURE 17.32 Area Under the Curve, $X_i = 0.0629$

First solve for parts 0.0623 inch or thinner:

$$Z_1 = \frac{0.0623 - 0.0627}{0.00023} = -1.74$$

Area$_1$ = 0.0409

or 4.09 percent of the parts are 0.0623 inches or thinner.

Then solve for parts 0.0626 inches or thinner:

$$Z_2 = \frac{0.0626 - 0.0627}{0.00023} = -0.44$$

Area$_2$ = 0.3300

or 33 percent of the parts are 0.0626 inches or thinner. Subtracting these two areas will determine the area in between:

$$0.3300 - 0.0409 = 0.2891$$

or 28.91 percent of the parts fall between 0.0623 and 0.0626 inches.

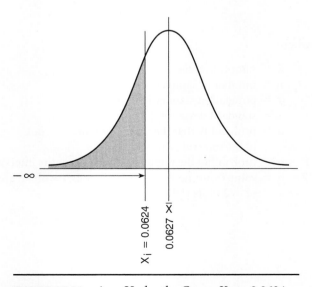

FIGURE 17.31 Area Under the Curve, $X_i = 0.0624$

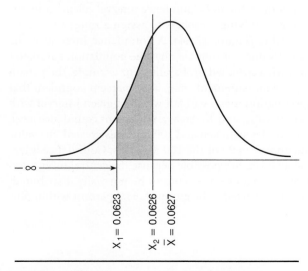

FIGURE 17.33 Area Under the Curve Between 0.0623 and 0.0626

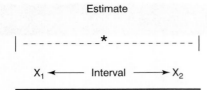

FIGURE 17.34 Confidence Interval Diagram

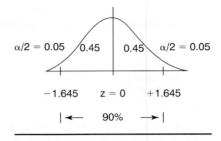

FIGURE 17.35 Normal Curve Showing 90 Percent Confidence Level

Values of Z should be rounded to two decimal places for use in the table. Interpolation between values can also be performed.

CONFIDENCE INTERVALS

People who study processes are interested in understanding process performance capability. Because it is often unfeasible to measure each item produced by a process or activity provided by a service, samples are taken. From these samples, averages are calculated. When samples in a subgroup are averaged and the standard deviation calculated, these values are called point estimates. As a result of the central limit theorem, as long as they are random and unbiased samples, these subgroup sample averages can serve as estimators of the population mean. Determining whether the sample average is a good approximation of the population mean depends on the spread of the sample data, the standard deviation of the distribution of the subgroup. The standard deviation, or standard error (SE), indicates the amount of error that will exist when the subgroup average is used to estimate the population mean. Because samples represent the population, it is possible to wonder whether the sample average is a good estimate of the population mean. In other words, how much confidence should be placed in the estimate? Confidence interval calculations enable those studying a process to assign a range to an estimated value (Figure 17.34). A confidence interval is the spread of values, or interval, that the population parameter fits in with a specified probability. For example, the person taking the measurements may be 90 percent confident that the population mean will fall within a given interval with end points at X_1 and X_2. So at a 90 percent confidence level, if 100 samples are taken and 100 intervals created, the value would appear in 90 of the 100 intervals. Confidence interval calculations are possible because the central limit theorem shows that averages tend to be normally distributed. Values are expected to be grouped, 68.3 percent within plus or minus one sigma, 95.5 percent within plus or minus two sigma, and 99.7 percent within plus or minus three sigma. The remaining 0.27 percent is split in the two tails. This probability is referred to as alpha (α) with $\alpha/2$ in each tail. Using the Z tables, if the degree of confidence is 90 percent, then the figure would appear as in Figure 17.35. The confidence level attached to the interval determines the size or width of the interval. Large sample sizes provide greater confidence, and therefore, a narrower interval, than small sample sizes. The degree of confidence is expressed as a percentage. Typically, the degrees of confidence chosen are 90 percent ($\alpha = 0.10$), 95 percent ($\alpha = 0.05$), and 99 percent ($\alpha = 0.01$).

Confidence interval testing is a technique that enables us to determine how well the subgroup average approximates the population mean. This straightforward calculation makes it possible to make statements such as: "there is a 95 percent probability that the sample average is a good estimator of the population mean" or "there is a 90 percent probability that the population mean is between X_1 and X_2." To determine X_1 and X_2, the end points for the confidence interval, use the formula:

$$\overline{X} \pm \frac{Z_{(\alpha/2)}(\sigma)}{\sqrt{n}}$$

where

$\overline{X}$ = sample average
n = number of samples
σ = population standard deviation
s = standard deviation (also $\sigma_{(n-1)}$)
α = probability that the population mean is not in the interval (alpha risk)
$1 - \alpha$ = probability the population mean is in the interval
$Z_{(\alpha/2)}$ = value from the Z table in Appendix 1 with an area of $\alpha/2$ to its right

LEAN SIX SIGMA TOOLS at WORK

Confidence Interval Calculation Using α

Manufacturing medical devices requires the ability to meet close tolerances. For a particularly critical machine setup, the manufacturing engineer conducted two runoffs for an injection molding machine. The first runoff, in which all of the molded parts were measured, the population mean was 0.800 mm with a population standard deviation, σ, of 0.007. From the second runoff, only 50 parts were sampled and their measurements taken. These randomly selected samples of parts have an average length of 0.822 mm and a standard deviation of 0.010. The manufacturing engineer would like to know, with 95 percent confidence, the interval values for the population mean for the second runoff.

$$0.822 \pm \frac{1.96(0.007)}{\sqrt{50}}$$

$$0.822 \pm 0.002$$

where

$\bar{X} = 0.822$
$n = 50$
$\sigma = 0.007$
$\alpha = 0.05$
$1 - 0.05 = 0.95$
$Z_{(\alpha/2)}$ = value from the Z table in Appendix 1 with an area of 0.025 to its right.

The interval is (0.820, 0.824). The engineer can be 95 percent confident that the population mean is between these two values (Figure 17.36).

The above formula, using the Z table, is considered a reasonable approximation if $n \geq 30$. For smaller sample sizes there is not an easy method to determine whether the population is normal. Under these circumstances, the t distribution is used.

$$\bar{X} \pm \frac{t_{(\alpha/2)}(s)}{\sqrt{n}}$$

where

$\bar{X}$ = sample average
n = number of samples
s = standard deviation (also $\sigma_{(n-1)}$)
α = probability that the population mean is not in the interval (alpha risk)
$1 - \alpha$ = probability the population mean is in the interval
$t_{(\alpha/2)}$ = value from the t table
df = degrees of freedom $(n - 1)$, the amount of data used by the measure of dispersion

The t value compensates for our lack of information about σ. The smaller the sample size, the more doubt exists, and the larger t must be. This t value is selected based on the degrees of freedom in the system, $n - 1$. Values for the t distribution appear in Appendix 3. Use Figure 17.37 as a guide when deciding to use either method.

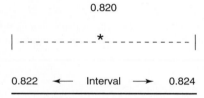

FIGURE 17.36 Lean Six Sigma Tools At Work Confidence Interval Diagram

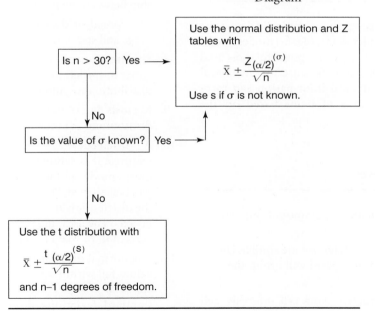

FIGURE 17.37 Confidence Interval Calculations: Choosing Between the Z and t Distribution

Confidence Interval Calculation Using the t Distribution at JF

LEAN SIX SIGMA TOOLS at WORK

Two machines have recently been installed at JF. To determine whether additional noise dampening devices are needed, their safety engineer has taken several noise exposure measurements at a water jet cutting work center and a conventional grinding machine. He would like to determine with 90 percent confidence that these samples represent the mean noise exposure expected to be experienced by workers throughout the plant. The values represent the percentage of allowable daily dose. A value of 50 percent during an 8-hour shift represents 85 dB.

For the water jet cutting work center, the ratios recorded were $X_1 = 0.45$, $X_2 = 0.47$, and $X_3 = 0.44$, resulting in an $\bar{X} = 0.45$ and $s = 0.015$. The water jet noise value is calculated below.

$$0.45 \pm \frac{2.920(0.15)}{\sqrt{3}}$$

where

$$\bar{X} = 0.45$$
$$n = 3$$
$$s = 0.015$$
$$\alpha = 0.10$$
$$1 - \alpha = 0.90$$
$$t_{(\alpha/2)} = 2.920$$
$$df = 3 - 1 = 2$$

The interval is (0.425, 0.475). The hygienist can be 90 percent confident that the population mean is between these two values for the water jet cutting work center. Because these values are below the legal permissible decibel levels, no additional noise dampening is required.

For the conventional grinding machine, the ratios recorded were $X_1 = 0.55$, $X_2 = 0.75$, and $X_3 = 0.57$, resulting in an $\bar{X} = 0.62$ and $s = 0.11$.

$$0.62 \pm \frac{2.920(0.11)}{\sqrt{3}}$$

where

$$\bar{X} = 0.62$$
$$n = 3$$
$$s = 0.11$$
$$\alpha = 0.10$$
$$1 - \alpha = 0.90$$
$$t_{(\alpha/2)} = 2.920$$
$$df = 2$$

The interval is (0.43, 0.81). The safety engineer can be 90 percent confident that the population mean is between these two values. Note that due to its large standard deviation, this is a much broader spread than the other interval. This broad interval includes values that are above the legal permissible decibel levels, and therefore, additional noise dampening is required.

SUMMARY

Frequency diagrams and histograms graphically depict the processes or occurrences under study. Means, modes, medians, standard deviations, and ranges are powerful tools used to describe processes statistically. Because of the central limit theorem, users of statistical information can form conclusions about populations of items based on the sample statistics.

TAKE AWAY TIPS

1. Accuracy and precision are of paramount importance in quality assurance.
2. Histograms and frequency diagrams are similar. Unlike a frequency diagram, a histogram will group the data into cells.
3. Histograms are constructed using cell intervals, cell midpoints, and cell boundaries.
4. The analysis of histograms and frequency diagrams is based on shape, location, and spread.
5. Shape refers to symmetry, skewness, and kurtosis.
6. The location or central tendency refers to the relationship between the mean (average), mode, and median.
7. The spread or dispersion of data is described by the range and standard deviation.
8. Skewness describes the tendency of data to be gathered either to the left or right side of a distribution. When a distribution is symmetrical, skewness equals zero.
9. Kurtosis describes the peakedness of data. Leptokurtic distributions are more peaked than platykurtic ones.
10. A normal curve can be identified by the following five features: It is symmetrical about a central value. The mean, mode, and median are all equal. It is unimodal and bell-shaped. Data cluster around the mean value of the distribution and then fall away toward the horizontal axis. The area under the normal curve equals 1; 100 percent of the data is found under the normal curve.
11. In a normal distribution 99.7 percent of the measured values fall within ±3 standard deviations of the mean ($\mu \pm 3\sigma$); 95.5 percent of the data fall within ±2 standard deviations of the mean ($\mu \pm 2\sigma$); and 68.3 percent of the data fall within ±1 standard deviation ($\mu \pm 1\sigma$).
12. The area under a normal curve can be calculated using the Z table and its associated formula.

FORMULAS

$$R = X_h - X_l$$

$$\mu \text{ or } \overline{X} = \frac{X_1 + X_2 + X_3 + \ldots + X_n}{n} = \frac{\sum X_i}{n}$$

$$\sigma = \sqrt{\frac{\sum (X_i - \mu)^2}{n}}$$

$$s = \sqrt{\frac{\sum (X_i - \overline{X})^2}{n - 1}}$$

$$a_3 = \frac{\sum f_i (X_i - \overline{X})^3 / n}{s^3}$$

$$a_4 = \frac{\sum f_i (X_i - \overline{X})^4 / n}{s^4}$$

$$f(x) = \frac{1}{\sigma \sqrt{2\pi}} e^{-\frac{(x-\mu)^2}{2\sigma^2}} \quad -\infty < X < \infty$$

where

$$\pi \approx 3.14159$$
$$e \approx 2.71828$$

$$f(Z) = \frac{1}{\sqrt{2\pi}} e^{-\frac{Z^2}{2}}$$

$$Z = \frac{X_i - \overline{X}}{s}$$

$$Z = \frac{X_i - \mu}{\sigma}$$

CHAPTER QUESTIONS

1. Describe the concepts of a sample and a population.
2. Describe a situation that is accurate, one that is precise, and one that is both. A picture may help you with your description.
3. Make a frequency distribution of the following data. Is this distribution bimodal? Multimodal? Skewed to the left? Skewed to the right? Normal?

   225, 226, 227, 226, 227, 228, 228, 229, 222, 223, 224, 226, 227, 228, 225, 221, 227, 229, 230

4. NB Plastics uses injection molds to produce plastic parts that range in size from a marble to a book. Parts are pulled off the press by one operator and passed on to another member of the team to be finished or cleaned up. This often involves trimming loose material, drilling holes, and painting. After a batch of parts has completed its cycle through the finishing process, a sample of five parts is chosen at random and certain dimensions are measured to ensure that each part is within certain tolerances. This information (in mm) is recorded for each of the five pieces and evaluated. Create a frequency diagram. Are the two operators trimming off the same amount of material? How do you know?

   Part Name: Mount
   Critical Dimension: 0.654 ± 0.005
   Tolerance: ±0.001
   Method of Checking: Caliper

Date	Time	Press	Oper	Samp 1	Samp 2	Samp 3	Samp 4	Samp 5
9/20/92	0100	#1	Jack	0.6550	0.6545	0.6540	0.6540	0.6545
9/20/92	0300	#1	Jack	0.6540	0.6540	0.6545	0.6545	0.6545
9/20/92	0500	#1	Jack	0.6540	0.6540	0.6540	0.6540	0.6535
9/20/92	0700	#1	Jack	0.6540	0.6540	0.6540	0.6540	0.6540
9/21/92	1100	#1	Mary	0.6595	0.6580	0.6580	0.6595	0.6595
9/21/92	1300	#1	Mary	0.6580	0.6580	0.6585	0.6590	0.6575
9/21/92	1500	#1	Mary	0.6580	0.6580	0.6580	0.6585	0.6590
9/22/92	0900	#1	Mary	0.6575	0.6570	0.6580	0.6585	0.6580

5. Create a histogram with the data in Question 4. Describe the distribution's shape, spread, and location.
6. Make a histogram of the following sample data:

   225, 226, 227, 226, 227, 228, 228, 229, 222, 223, 224, 226, 227, 228, 225, 221, 227, 229, 230, 225, 226, 227, 229, 228, 224, 223, 222, 225, 226, 227, 224, 223, 222, 228, 229, 225, 226

7. A manufacturer of CDs has a design specification for the width of the CD of 120 ± 0.3 mm. Create a histogram using the following data. Describe the distribution's shape, spread, and location.

Measurement	Tally
119.4	///
119.5	////
119.6	||||/
119.7	||||/ //
119.8	||||/ ||||/
119.9	||||/ ///
120.0	||||/ //
120.1	||||/
120.2	///

8. What is meant by the following expression: the central tendency of the data?

9. What is meant by the following expression: measures of dispersion?

10. Find the mean, mode, and median of the following numbers: 34, 35, 36, 34, 32, 34, 45, 46, 45, 43, 44, 43, 34, 30, 48, 38, 38, 40, 34.

11. Using the following sample data, calculate the mean, mode, and median:

1.116	1.122	1.125
1.123	1.122	1.123
1.133	1.125	1.118
1.117	1.121	1.123
1.124	1.136	1.122
1.119	1.127	1.122
1.129	1.125	1.119
1.121	1.124	
1.128	1.122	

12. For the data from Question 4, determine the mean, mode, median, standard deviation, and range. Use these values to describe the distribution. Compare this mathematical description with the description you created for the histogram problem.

13. For the CD data of Question 7, determine the mean, mode, median, standard deviation, and range. Use these values to describe the distribution. Compare this mathematical description with the description you created in Question 7.

14. If the average wait time is 12 minutes with a standard deviation of 3 minutes, determine the percentage of patrons who wait less than 15 minutes for their main course to be brought to their tables.

15. The thickness of a part is to have an upper specification of 0.925 and a lower specification of 0.870 mm. The average of the process is currently 0.917 with a standard deviation of 0.005. Determine the percentage of product above 0.93 mm.

16. The Rockwell hardness of specimens of an alloy shipped by your supplier varies according to a normal distribution with mean 70 and standard deviation 3. Specimens are acceptable for machining only if their hardness is greater than 65. What percentage of specimens will be acceptable?

17. If the mean value of the weight of a particular brand of dog food is 20.6 lb and the standard deviation is 1.3, assume a normal distribution and calculate the amount of product produced that falls below the lower specification value of 19.7 lb.

18. For the CD data from Question 7, determine what percentage of the CDs produced are above and below the specifications of 120 ± 0.3 mm.

19. NB Manufacturing has ordered the construction of a new machine to replace an older machine in a machining cell. Now that the machine has been built, a runoff is to be performed. The diameters on the test pieces were checked for runout. From the 32 parts sampled, the average was 0.0015 with a standard deviation of 0.0008. The engineers would like to know, with 90 percent confidence, the interval values for the population mean.

20. An automotive manufacturer has selected 10 car seats to study the Rockwell hardness of the seat recliner mechanism. A sample of 8 has an average of 44.795 and a standard deviation of 0.402. At a 95 percent confidence level, what is the interval for the population mean?

VARIABLE CONTROL CHARTS

When suppliers experienced difficulties inserting diodes, resistors, and capacitors in the circuit boards without shattering them, the Six Sigma team used control charts to study variation in the supplier's manufacturing process. Control charts pointed to the problem: the original hole size specification for the circuit board was too small.

Paraphrased from Six Sigma Report in Business Magazine, October 1997

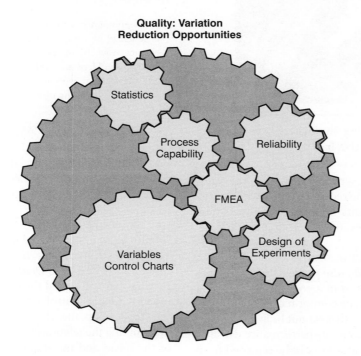

LEARNING OPPORTUNITIES

1. To understand the concept of variation
2. To understand the difference between assignable causes and chance causes
3. To learn how to construct control charts for variables, either $\bar{X}$ and R charts or $\bar{X}$ and s charts
4. To recognize when a process is under control and when it is not
5. To understand the importance of the R and s charts when interpreting variable control charts
6. To know how to revise a control chart in which assignable causes have been identified and corrected

VARIABLE CONTROL CHARTS
Dr. Walter Shewhart

Like most of us, Dr. Walter Shewhart (1891–1967) believed that we could make great decisions if we had perfect knowledge of the situation. However, life rarely provides perfect knowledge and who has time to wait for it anyway? Work needs to be done and decisions need to be made, so Dr. Shewhart (Figure 18.1) developed statistical methods in the form of control charts that can be used to improve the quality of the processes that provide goods and services.

While working at Bell Laboratories in the 1920s and 1930s, Dr. Shewhart was the first to encourage the use of statistics to identify, monitor, and eventually remove the sources of variation found in repetitive processes. His work combined two aspects of quality: the subjective aspect, what the customer wants; and the objective side, the physical

FIGURE 18.1 Dr. Walter Shewhart
(**Source:** Alcatel-Lucent USA Inc.)

properties of the goods or services, including the value received for the price paid. He recognized that when translating customer requirements to actual products and services, statistical measures of key characteristics are important to ensure quality.

Dr. Shewhart identified two sources of variation in a process. **Controlled variation,** also termed **chance or common causes,** is variation present in a process due to the very nature of the process. Chance, or common causes, are small random changes in the process that cannot be avoided. They consistently affect the process and its performance day after day, every day. This type of variation can be removed from the process only by changing the process. Lean Six Sigma organizations realize that removing chance causes from a system usually involves management intervention. For example, consider a person who has driven the same route to work dozens of times and determined that it takes about 20 minutes to get from home to work, regardless of minor changes in weather or traffic conditions. If this is the case, then the only way the person can improve on this time is to change the process by finding a new route.

Uncontrolled variation, also known as **special** or **assignable causes,** comes from sources external to the process. This type of variation is not normally part of the process. Assignable causes are variations in the process that can be identified and isolated as the specific cause of a change in the behavior of the process. This type of variation arises because of special circumstances. Sources of variation can be found in the process itself, the materials used, the operator's actions, or the environment. Examples of factors that can contribute to process variation include tool wear, machine vibration, and work-holding devices. Changes in material thickness, composition, or hardness are sources of variation. Operator actions affecting variation include overadjusting the machine, making an error during the inspection activity, changing the machine settings, or failing to properly align the part before machining. Environmental factors affecting variation include heat, light, radiation, and humidity. For instance, a commuter would experience uncontrolled variation if a major traffic accident stopped traffic or a blizzard made traveling nearly impossible. Uncontrolled variation prevents the process from performing to the best of its ability.

It was Dr. Shewhart who put forth the fundamental principle that once a process is under control, exhibiting only controlled variation, future process performance can be predicted, within limits, on the basis of past performance. He wrote:

> A phenomenon will be said to be controlled when, through the use of past experience, we can predict, at least within limits, how the phenomenon may be expected to vary in the future. Here it is understood that prediction within limits means that we can state, at least approximately, the probability that the observed phenomenon will fall within the given limits.*

Though he was a physicist, Dr. Shewhart studied process control through the use of charting techniques. Based on his understanding of variation and the belief that assignable causes of variation could be found and eliminated, Dr. Shewhart developed the formulas and table of constants used to create the most widely used statistical control charts in quality: the $\overline{X}$ and R charts. These charts (Figure 18.2) first appeared in a May 16, 1924, internal Bell Telephone Laboratories report. Later in his 1931 text, *Economic Control of Quality of Manufactured Product,* Dr. Shewhart presented the foundation principles on which modern quality control is based.

To develop the charts, Dr. Shewhart first set about determining the relationship between the standard deviation of the mean and the standard deviation of the individual observations. He demonstrated the relationship by using numbered, metal-lined, disk-shaped tags. From a bowl borrowed from his wife's kitchen, he drew these tags at random to confirm the standard deviation of subgroup sample means is the standard deviation of individual samples divided by the square root of the subgroup size.

$$s_{\overline{x}} = \frac{s}{\sqrt{n}}$$

where

$s_{\overline{x}}$ = standard deviation of the mean (standard error)
s = standard deviation of individual observations
n = number of observations in each subgroup mean

The control charts, as designed by Dr. Shewhart, have three purposes: to define standards for the process, to aid in problem-solving efforts to attain the standards, and to

---

*Walter Shewhart, *Economic Control of Quality of Manufactured Product.* New York: Van Nostrand Reinhold, 1931, p. 6.

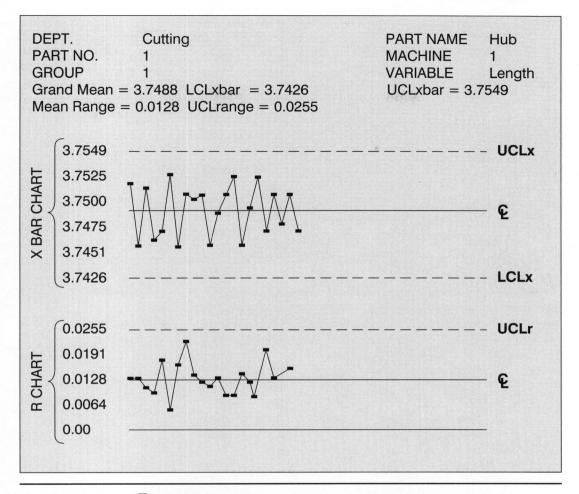

FIGURE 18.2 Typical $\overline{X}$ and R Chart

judge whether the standards have been met. Although Dr. Shewhart concentrated his efforts on manufacturing processes, his ideas and charts are applicable to any process found in nonmanufacturing environments.

Statistical process control charts are more than a tool. They provide a framework for monitoring the behavior of a process and provide a feedback loop that enables organizations to achieve dramatic process improvements. Since their introduction in 1931, Dr. Shewhart's control charts have served to advance process improvement efforts in nearly every type of industry despite differing opinions about their appropriateness, applicability, limits derivations, sampling frequency, and use. It is a tribute to the ruggedness of Shewhart's invention that they remain the preeminent statistical process control tool.

The only shortcoming in a histogram analysis is its failure to show process performance over time. Look at the data in Table 18.1. When these averages are graphed in a histogram, the result closely resembles a normal curve (Figure 18.3a). Graphing the averages by subgroup number, according to when they were produced, gives a different impression of the data (Figure 18.3b). From the chart, it appears that the thickness of the clutch plate is increasing as production continues. This was not evident during the creation of the histogram or

an analysis of the average, range, and standard deviation. A control chart enhances the analysis of the process by showing how that process is performing over time. Using statistical control charting creates a feedback loop enabling lean Six Sigma organizations to improve their processes, products, and services.

Control charts serve two basic functions:

1. As decision-making tools. They provide an economic basis for making a decision as to whether to investigate for potential problems, to adjust the process, or to leave the process alone.

    a. Control charts provide information for timely decisions concerning recently produced items. If an out-of-control condition is shown by the control chart, then a decision can be made about sorting or reworking the most recent production.

    b. Control chart information is used to determine the process capability, or the level of quality the process is capable of producing. Samples of completed product can be statistically compared with the process specifications. This comparison provides information concerning the process's ability to meet the specifications set by the product designer.

TABLE 18.1 Clutch Plate Thickness: Sums and Averages

						$\sum X_i$	$\bar{X}$
Subgroup 1	0.0625	0.0626	0.0624	0.0625	0.0627	0.3127	0.0625
Subgroup 2	0.0624	0.0623	0.0624	0.0626	0.0625	0.3122	0.0624
Subgroup 3	0.0622	0.0625	0.0623	0.0625	0.0626	0.3121	0.0624
Subgroup 4	0.0624	0.0623	0.0620	0.0623	0.0624	0.3114	0.0623
Subgroup 5	0.0621	0.0621	0.0622	0.0625	0.0624	0.3113	0.0623
Subgroup 6	0.0628	0.0626	0.0625	0.0626	0.0627	0.3132	0.0626
Subgroup 7	0.0624	0.0627	0.0625	0.0624	0.0626	0.3126	0.0625
Subgroup 8	0.0624	0.0625	0.0625	0.0626	0.0626	0.3126	0.0625
Subgroup 9	0.0627	0.0628	0.0626	0.0625	0.0627	0.3133	0.0627
Subgroup 10	0.0625	0.0626	0.0628	0.0626	0.0627	0.3132	0.0626
Subgroup 11	0.0625	0.0624	0.0626	0.0626	0.0626	0.3127	0.0625
Subgroup 12	0.0630	0.0628	0.0627	0.0625	0.0627	0.3134	0.0627
Subgroup 13	0.0627	0.0626	0.0628	0.0627	0.0626	0.3137	0.0627
Subgroup 14	0.0626	0.0626	0.0625	0.0626	0.0627	0.3130	0.0626
Subgroup 15	0.0628	0.0627	0.0626	0.0625	0.0626	0.3132	0.0626
Subgroup 16	0.0625	0.0626	0.0625	0.0628	0.0627	0.3131	0.0626
Subgroup 17	0.0624	0.0626	0.0624	0.0625	0.0627	0.3126	0.0625
Subgroup 18	0.0628	0.0627	0.0628	0.0626	0.0630	0.3139	0.0627
Subgroup 19	0.0627	0.0626	0.0628	0.0625	0.0627	0.3133	0.0627
Subgroup 20	0.0626	0.0625	0.0626	0.0625	0.0627	0.3129	0.0626
Subgroup 21	0.0627	0.0626	0.0628	0.0625	0.0627	0.3133	0.0627
Subgroup 22	0.0625	0.0626	0.0628	0.0625	0.0627	0.3131	0.0626
Subgroup 23	0.0628	0.0626	0.0627	0.0630	0.0627	0.3138	0.0628
Subgroup 24	0.0625	0.0631	0.0630	0.0628	0.0627	0.3141	0.0628
Subgroup 25	0.0627	0.0630	0.0631	0.0628	0.0627	0.3143	0.0629
Subgroup 26	0.0630	0.0628	0.0620	0.0628	0.0627	0.3142	0.0628
Subgroup 27	0.0630	0.0628	0.0631	0.0628	0.0627	0.3144	0.0629
Subgroup 28	0.0632	0.0632	0.0628	0.0631	0.0630	0.3153	0.0631
Subgroup 29	0.0630	0.0628	0.0631	0.0632	0.0631	0.3152	0.0630
Subgroup 30	0.0632	0.0631	0.0630	0.0628	0.0628	0.3149	0.0630
						9.3981	

2. As problem-solving tools. They point out where improvement is needed.
    a. Control chart information can be used to help locate and investigate the causes of the unacceptable or marginal quality. By observing the patterns on the chart the investigator can determine what adjustments need to be made.
    b. During daily production runs, the operator can monitor machine production and determine when to make the necessary adjustments to the process or when to leave the process alone to ensure quality production.

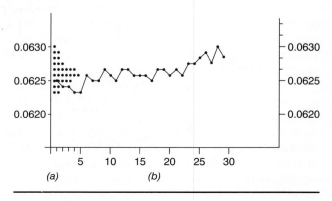

FIGURE 18.3 Chart with Histogram

LEAN SIX SIGMA TOOLS at WORK

Understanding Variation

An industrial engineering department is seeking to decrease the amount of time it takes to perform a printer assembly operation. An analysis of the methods used by the operators performing the assembly has revealed that one of the operators completes the assembly in 75 percent of the time it takes another operator performing the same assembly operation. Further investigation determines that the two operators use parts produced on different machines. The histograms in Figure 18.4 are based on measurements of the parts from the two different processes. In Figure 18.4b the spread of the process is considerably smaller, enabling the faster operator to assemble the parts much more quickly and with less effort. The slower operator must try a part in the assembly, discard the part if it doesn't fit, and try another part. Repeating the operation when the parts do not fit has made the operator much less efficient. This operator's apparent lack of speed is actually caused by the process and is not the fault of the operator. Had managers not investigated variation in the process, they might have made incorrect judgments about the operator's performance. Instead, they realize that management intervention will be necessary to improve production at the previous operation and produce parts with less variation.

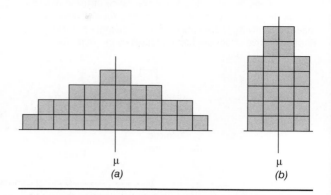

FIGURE 18.4 Histograms of Part Measurements from Two Machines

Variation

Variation, where no two items or services are exactly the same, exists in all processes. Although it may take a very precise measuring instrument or a very astute consumer to notice the variation, any process in nature will exhibit variation. Understanding variation and its causes results in better decisions.

Several types of variation are tracked with statistical methods. These include:

1. **Within-piece variation**, or the variation within a single item or surface. For example, a single square yard of fabric may be examined to see if the color varies from one location to another.

2. **Piece-to-piece variation**, or the variation that occurs among pieces produced at approximately the same time. For example, in a production run filling gallon jugs with milk, when each of the milk jugs is checked after the filling station, the fill level from jug to jug will be slightly different.

3. **Time-to-time variation**, or the variation in the product produced at different times of the day—for example, the comparison of a part that has been stamped at the beginning of a production run with the part stamped at the end of a production run.

Control Charts for Variables

Control charts enable those studying a process to analyze the variation present. To create a control chart, samples, arranged into subgroups, are taken during the process. The averages of the subgroup values are plotted on the control chart. The **centerline** (₵) of this chart shows where the process average is centered, the central tendency of the data. The **upper control limit (UCL)** and **lower control limit (LCL)**, calculated based on $\pm 3\sigma$, describe the spread of the process (Figure 18.2). In other words, we can expect future production to fall between these $\pm 3\sigma$ limits 99.7 percent of the time, providing the process does not change and is under control.

Instead of waiting until an entire production run is complete or until the product reaches the end of the assembly line, management can have critical part dimensions checked and charted throughout the process. If a part or group of parts has been made incorrectly, production can be stopped, adjusted, or otherwise modified to produce parts correctly.

Variables are the measurable characteristics of a product or service. Examples of variables include the height, weight, or length of a part. One of the most commonly used variable chart combinations in statistical process control is the combination of the $\overline{X}$ and R charts. Typical $\overline{X}$ and R charts are shown in Figure 18.2. $\overline{X}$ and R charts are used together to determine the distribution of the subgroup averages of sample measurements taken from a process. The importance of using these two charts in conjunction with each other will become apparent shortly.

$\overline{X}$ and R Charts

The $\overline{X}$ *chart is used to monitor the variation of the subgroup averages that are calculated from the individual sampled data.* Averages rather than individual observations are used on control charts because average values will indicate a change in the amount of variation much faster than will individual values. Control limits on this chart are used to evaluate the variation from one subgroup to another.

LEAN SIX SIGMA TOOLS at WORK

Defining the Problem

An assembly area has been experiencing serious delays in the construction of computer printers. As a Six Sigma Black Belt, you have been asked to determine the cause of these delays and fix the problems as soon as possible. You convened a meeting involving those closest to the assembly problems. Representatives from production, supervision, manufacturing, engineering, industrial engineering, quality assurance, and maintenance created a cause-and-effect diagram showing the potential causes for the assembly difficulties (Figure 18.5). Discussions during the meeting revealed that the shaft which holds the roller in place could be the major cause of assembly problems.

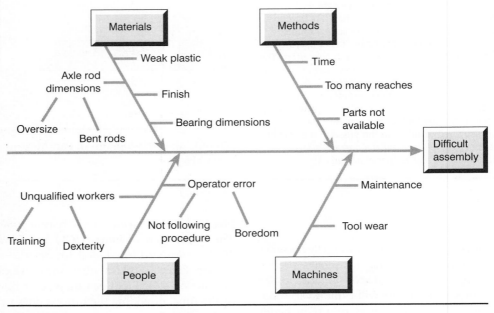

FIGURE 18.5 Printer Assembly: Cause-and-Effect Diagram

The following steps and examples explain the construction of an $\overline{X}$ chart.

1. Define the Problem In any situation it is necessary to determine the goal of monitoring a particular quality characteristic or group of characteristics. Control charts can be placed on a process to help determine where the true source of the problem is located.

2. Select the Quality Characteristic to Be Measured Variable control charts are based on measurements. The characteristics selected for measurement should be ones that affect product or service performance. Characteristic choice depends on whether the process is being monitored for within-piece variation, piece-to-piece variation, or variation over time. Product or service characteristics such as length, height, viscosity, color, temperature, and velocity are typically used in manufacturing settings. Delivery times, checkout times, and service times are examples of characteristics chosen in a service industry.

3. Choose a Rational Subgroup Size to Be Sampled Subgroups, and the samples composing them, must be homogeneous. A ***homogeneous subgroup*** *will have been produced under the same conditions, by the same machine, the same operator, the same mold, and so on.* Homogeneous lots can also be designated by equal time intervals. Samples should be taken in an unbiased, random fashion. They

LEAN SIX SIGMA TOOLS at WORK

Identifying the Quality Characteristic

As the lean Six Sigma project meeting described in the earlier Lean Six Sigma Tools at Work feature continues, further investigation reveals that the length of the shaft is hindering assembly operations. The characteristic to measure has been identified as piece-to-piece variation in the length of the shafts. To begin to study the situation, measurements of the lengths of the shafts will be sampled.

should be representative of the entire population. The letter n is used to designate the number of samples taken within a subgroup. When constructing $\overline{X}$ and R charts, keep the subgroup sample size constant for each subgroup taken.

Decisions concerning the specific size of the subgroup—n, or the number of samples—require judgment. Sampling should occur frequently enough to detect changes in the process. Ask how often is the system expected to change. Examine the process and identify the factors causing change in the process. To be effective, sampling must occur as often as the system's most frequently changing factor. Once the number and frequency of sampling have been selected, they should not be changed unless the system itself has changed.

Realistically, sampling frequency must balance the value of the data obtained with the costs of taking the samples. Sampling is usually more frequent when control charts are first used to monitor the process. As process improvements are made and the process stabilizes, the frequency of sampling and subgroup size can be decreased.

When gathering sample data, it is important to have the following information in order to properly analyze the data:

1. *Who* will be collecting the data?
2. *What* aspect of the process is to be measured?
3. *Where* or at what point in the process will the sample be taken?
4. *When* or how frequently will the process be sampled?
5. *Why* is this particular sample being taken?
6. *How* will the data be collected?
7. *How many* samples will be taken (subgroup size)?

Some other guidelines to be followed include:

- The larger the subgroup size, the more sensitive the chart becomes to small variations in the process average. This will provide a better picture of the process because it allows the investigator to detect changes in the process quickly.
- Although a larger subgroup size makes for a more sensitive chart, it also increases inspection costs.
- Destructive testing may make large subgroup sizes unfeasible. For example, it would not make sense for a fireworks manufacturer to test each and every one of its products.
- Subgroup sizes smaller than four do not create a representative distribution of subgroup averages. Subgroup averages are nearly normal for subgroups of four or more even when sampled from a nonnormal population.
- When the subgroup size exceeds 10, the standard deviation (s) chart, rather than the range (R) chart, should be used. For large subgroup sizes, the s chart gives a better representation of the true dispersion or true differences between the individuals sampled than does the R chart.

4. Collect the Data To create a control chart, an amount of data sufficient to accurately reflect the statistical control of the process must be gathered. A minimum of 20 subgroups of sample size n = 4 is suggested. Each time a subgroup of sample size n is taken, an average is calculated for the subgroup. To do this, the individual values are recorded, summed, and then divided by the number of samples in the subgroup. This average, $\overline{X}_i$, is then plotted on the control chart.

5. Determine the Trial Centerline for the $\overline{X}$ Chart The centerline of the control chart is the process average. It would be the mean, μ, if the average of the population measurements for the entire process were known. Because the value of the population mean μ cannot be determined unless all of the parts being produced are measured, in its place the grand average of the subgroup averages, $\overline{\overline{X}}$ (X double bar, covered previously in Chapter 17), is used. The grand average, or $\overline{\overline{X}}$, is calculated by summing all the subgroup averages and then dividing by the number of subgroups. This value is plotted as the centerline of the $\overline{X}$ chart:

$$\overline{\overline{X}} = \frac{\sum_{i=1}^{m} \overline{X}_i}{m}$$

where

$\overline{\overline{X}}$ = average of the subgroup averages
$\overline{X}_i$ = average of the ith subgroup
m = number of subgroups

6. Determine the Trial Control Limits for the $\overline{X}$ Chart Control limits are established at ±3 standard deviations from the centerline for the process using the following formulas:

$$UCL_{\overline{X}} = \overline{\overline{X}} + 3\sigma_{\overline{x}}$$
$$LCL_{\overline{X}} = \overline{\overline{X}} - 3\sigma_{\overline{x}}$$

where

UCL = upper control limit of the $\overline{X}$ chart
LCL = lower control limit of the $\overline{X}$ chart
$\sigma_{\overline{x}}$ = population standard deviation of the subgroup averages

---

Selecting Subgroup Sample Size

LEAN SIX SIGMA TOOLS at WORK

The production from the machine making the shafts is consistent at 150 per hour. Because the process is currently exhibiting problems, your team has decided to take a sample of five measurements every 10 minutes from the production. The values for the day's production run are shown in Figure 18.6.

(Continued)

DEPT.	Roller		PART NAME	Shaft	
PART NO.	1		MACHINE	1	
GROUP	1		VARIABLE	length	
Subgroup	1	2	3	4	5
Time	07:30	07:40	07:50	08:00	08:10
Date	07/02/10	07/02/10	07/02/10	07/02/10	07/02/10
1	11.95	12.03	12.01	11.97	12.00
2	12.00	12.02	12.00	11.98	12.01
3	12.03	11.96	11.97	12.00	12.02
4	11.98	12.00	11.98	12.03	12.03
5	12.01	11.98	12.00	11.99	12.02
X̄	11.99	12.00	11.99	11.99	12.02
Range	0.08	0.07	0.04	0.06	0.03
Subgroup	6	7	8	9	10
Time	08:20	08:30	08:40	08:50	09:00
Date	07/02/95	07/02/95	07/02/95	07/02/95	07/02/95
1	11.98	12.00	12.00	12.00	12.02
2	11.98	12.01	12.01	12.02	12.00
3	12.00	12.03	12.04	11.96	11.97
4	12.01	12.00	12.00	12.00	12.05
5	11.99	11.98	12.02	11.98	12.00
X̄	11.99	12.00	12.01	11.99	12.01
Range	0.03	0.05	0.04	0.06	0.08
Subgroup	11	12	13	14	15
Time	09:10	09:20	09:30	09:40	09:50
Date	07/02/95	07/02/95	07/02/95	07/02/95	07/02/95
1	11.98	11.92	11.93	11.99	12.00
2	11.97	11.95	11.95	11.93	11.98
3	11.96	11.92	11.98	11.94	11.99
4	11.95	11.94	11.94	11.95	11.95
5	12.00	11.96	11.96	11.96	11.93
X̄	11.97	11.94	11.95	11.95	11.97
Range	0.05	0.04	0.05	0.06	0.07
Subgroup	16	17	18	19	20
Time	10:00	10:10	10:20	10:30	10:40
Date	07/02/95	07/02/95	07/02/95	07/02/95	07/02/95
1	12.00	12.02	12.00	11.97	11.99
2	11.98	11.98	12.01	12.03	12.01
3	11.99	11.97	12.02	12.00	12.02
4	11.96	11.98	12.01	12.01	12.00
5	11.97	11.99	11.99	11.99	12.01
X̄	11.98	11.99	12.01	12.00	12.01
Range	0.04	0.05	0.03	0.06	0.03
Subgroup	21				
Time	10:50				
Date	07/02/95				
1	12.00				
2	11.98				
3	11.99				
4	11.99				
5	12.02				
X̄	12.00				
Range	0.04				

$R = 0.05 = 12.02 - 11.97$

$$\frac{12.00 + 11.98 + 11.99 + 11.96 + 11.97}{5} = 11.98$$

FIGURE 18.6 Values for a Day's Production

Collecting Data

A sample of size n = 5 is taken at 10-minute intervals from the process making shafts. As shown in Figure 18.6, a total of 21 subgroups of sample size n = 5 have been taken. Each time a subgroup sample is taken, the individual values are recorded [Figure 18.6, (1)], summed, and then divided by the number of samples taken to get the average for the subgroup [Figure 18.6, (2)]. This subgroup average is then plotted on the control chart [Figure 18.7, (1)].

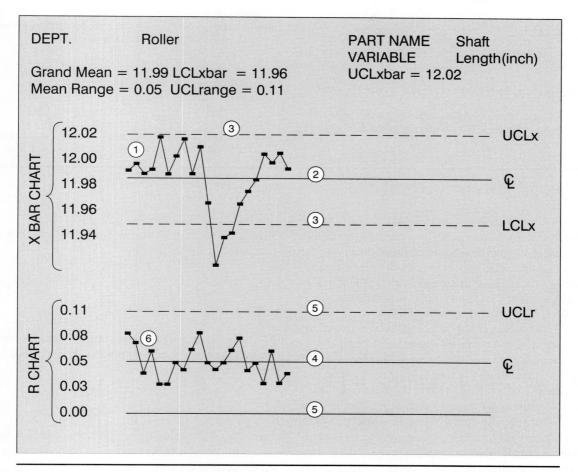

FIGURE 18.7 $\overline{X}$ and R Control Charts for Roller Shaft Length

The population standard deviation σ is needed to calculate the upper and lower control limits. Because control charts are based on sample data, Dr. Shewhart developed a good approximation of $3\sigma_{\overline{x}}$ using the product of an A_2 factor multiplied by $\overline{R}$, the average of the ranges. The $A_2\overline{R}$ combination uses the sample data for its calculation. $\overline{R}$ is calculated by summing the values of the individual subgroup ranges and dividing by the number of subgroups m:

$$\overline{R} = \frac{\sum_{i=1}^{m} R_i}{m}$$

where

$\overline{R}$ = average of the ranges
R_i = individual range values for the sample
m = number of subgroups

A_2, the factor that allows the approximation $A_2\overline{R} \approx 3\sigma_{\overline{x}}$ to be true, is selected based on the subgroup sample size n. See Appendix 2 for the A_2 factors.

Upon replacement, the formulas for the upper and lower control limits become:

$$UCL_{\overline{X}} = \overline{\overline{X}} + A_2\overline{R}$$
$$LCL_{\overline{X}} = \overline{\overline{X}} - A_2\overline{R}$$

After calculating the control limits, we place the centerline $\overline{\overline{X}}$ and the upper and lower control limits (UCL and LCL, respectively) on the chart. The upper and lower control limits are shown by dashed lines. The grand average, or $\overline{\overline{X}}$, is shown by a solid line. The control limits on the $\overline{X}$ chart will be symmetrical about the central line.

7. Determine the Trial Control Limits for the R Chart When an $\overline{X}$ chart is used to evaluate the variation

LEAN SIX SIGMA TOOLS at WORK

Calculating the X̄ Chart Centerline and Control Limits

Construction of an $\bar{X}$ chart begins with the calculation of the centerline $\bar{\bar{X}}$. Using the 21 subgroups of sample size n = 5 provided in Figure 18.6, we calculate $\bar{\bar{X}}$ by summing all the subgroup averages based on the individual samples taken and then dividing by the number of subgroups, m:

$$\bar{\bar{X}} = \frac{11.99 + 12.00 + 11.99 + \cdots + 12.00}{21}$$

$$= \frac{251.77}{21} = 11.99$$

This value is plotted as the centerline of the $\bar{X}$ chart [Figure 18.7, (2)].

$\bar{R}$ is calculated by summing the values of the individual subgroup ranges (Figure 18.6) and dividing by the number of subgroups, m:

$$\bar{R} = \frac{0.08 + 0.07 + 0.04 + \cdots + 0.04}{21}$$

$$= \frac{1.06}{21} = 0.05$$

The A_2 factor for a sample size of five is selected from the table in Appendix 2. The values for the upper and lower control limits of the $\bar{X}$ chart are calculated as follows:

$$UCL_{\bar{X}} = \bar{\bar{X}} + A_2\bar{R}$$
$$= 11.99 + 0.577(0.05) = 12.02$$
$$LCL_{\bar{X}} = \bar{\bar{X}} - A_2\bar{R}$$
$$= 11.99 - 0.577(0.05) = 11.96$$

Once calculated, the upper and lower control limits (UCL and LCL, respectively) are placed on the chart [Figure 18.7, (3)].

---

in quality from subgroup to subgroup, the range chart is a method of determining the amount of variation among the individual samples. The importance of the range chart is often overlooked. Without the range chart, or the standard deviation chart to be discussed later, it would not be possible to fully understand process capability. Where the $\bar{X}$ chart shows the average of the individual subgroups, giving the viewer an understanding of where the process is centered, the range chart shows the spread or dispersion of the individual samples within the subgroup. If the product displays a wide spread or a large range, then the individuals being produced are not similar to each other. The optimal situation from a quality perspective is when the parts are grouped closely around the process average. This situation will yield a small value for both the range and the standard deviation, meaning that the measurements are very similar to each other. $\bar{R}$, the centerline of the R chart, is calculated as:

$$\bar{R} = \frac{\sum_{i=1}^{m} R_i}{m}$$

To create the upper and lower control limits for the $\bar{R}$ chart, the average of the subgroup ranges ($\bar{R}$) multiplied by the D_3 and D_4 factors is used:

$$UCL_R = D_4\bar{R}$$
$$LCL_R = D_3\bar{R}$$

Along with the value of A_2, the values of D_3 and D_4 are found in the table in Appendix 2. These values are selected on the basis of the subgroup sample size n.

LEAN SIX SIGMA TOOLS at WORK

Calculating the R Chart Centerline and Control Limits

Constructing an R chart is similar to creating an $\bar{X}$ chart. To begin the process, individual range values are calculated for each of the subgroups by subtracting the highest value in the subgroup from the lowest value [Figure 18.6, (3)]. Once calculated, these individual range values (R_i) are plotted on the R chart [Figure 18.7, (6)].

To determine the centerline of the R chart, individual range (R_i) values are summed and divided by the total number of subgroups to give $\bar{R}$ [Figure 18.7, (4)].

$$\bar{R} = \frac{0.08 + 0.07 + 0.04 + \cdots + 0.04}{21}$$

$$= \frac{1.06}{21} = 0.05$$

With n = 5, the values of D_3 and D_4 are found in the table in Appendix 2. The control limits for the R chart are calculated as follows:

$$UCL_R = D_4\bar{R}$$
$$= 2.114(0.05) = 0.11$$
$$LCL_R = D_3\bar{R}$$
$$= 0(0.05) = 0$$

The control limits are placed on the R chart [Figure 18.7, (5)].

The control limits, when displayed on the R chart, should theoretically be symmetrical about the centerline ($\bar{R}$). However, because range values cannot be negative, a value of zero is given for the lower control limit with sample sizes of six or less. This results in an R chart that is asymmetrical. As with the $\bar{X}$ chart, control limits for the R chart are shown with a dashed line. The centerline is shown with a solid line.

CONTROL CHART INTERPRETATION

8. Examine the Process: Control Chart Interpretation

Correct interpretation of control charts is essential to managing a process. Understanding the sources and potential causes of variation is critical to good management decisions. Managers must be able to determine whether the variation present in a process is indicating a trend that must be dealt with or is merely random variation natural to the process. Misinterpretation can lead to a variety of losses, including the following:

- Blaming people for problems that they cannot control
- Spending time and money looking for problems that do not exist
- Spending time and money on process adjustments or new equipment that are not necessary
- Taking action where no action is warranted
- Asking for worker-related improvements where process or equipment improvements need to be made first

Once the performance of a process is predictable, there is a sound basis for making plans and decisions concerning the process, the system, and its output. Costs to manufacture the product or provide the service become predictable.

A *process is considered to be in a state of process control, or* **under control**, *when the performance of the process falls within the statistically calculated control limits and exhibits only chance, or common, causes*. When a process is under control, it is considered stable and the amount of future variation is predictable. A stable process does not necessarily meet the specifications set by the designer or exhibit minimal variation; a stable process merely has a predictable amount of variation.

Lean Six Sigma practitioners appreciate the benefits of a stable process with predictable variation. When the process performance is predictable, there is a rational basis for planning. It is fairly straightforward to determine costs associated with a stable process. Quality levels from time period to time period are predictable. When changes, additions, or improvements are made to a stable process, the effects of the change can be determined quickly and reliably.

When an assignable cause is present, the process is considered unstable, out of control, or beyond the expected normal variation. In an unstable process the variation is unpredictable, meaning that the magnitude of the variation could change from one time period to another. Lean Six Sigma practitioners need to determine whether the variation that exists in a process is common or assignable. To treat an assignable cause as a chance cause could result in a disruption to a system or a process that is operating correctly except for the assignable cause. To treat chance causes as assignable causes is an ineffective use of resources because the variation is inherent in the process.

When a system is subject to only chance causes of variation, 99.7 percent of the parts produced will fall within $\pm 3\sigma$. This means that if 1,000 subgroups are sampled, 997 of the subgroups will have values within the upper and lower control limits. Based on the normal curve, a control chart can be divided into three zones (Figure 18.8). Zone A is ± 1 standard deviation from the centerline and should contain approximately 68.3 percent of the calculated sample averages or ranges. Zone B is ± 2 standard deviations from the centerline and should contain 27.2 percent (95.5 percent − 68.3 percent) of the points. Zone C is ± 3 standard deviations from the centerline and should contain only approximately 4.2 percent of the points (99.7 percent − 95.5 percent). With these zones as a guide, a control chart exhibits a state of control when:

1. Two-thirds of the points are near the center value
2. A few of the points are on or near the center value
3. The points appear to float back and forth across the centerline
4. The points are balanced (in roughly equal numbers) on both sides of the centerline
5. There are no points beyond the control limits
6. There are no patterns or trends on the chart

While analyzing $\bar{X}$ and R charts, take a moment to study the scale of the range chart. The spread of the upper and lower control limits will reveal whether a significant amount of variation is present in the process. This clue to the amount of variation present may be overlooked if the R chart is checked only for patterns or out-of-control points.

Identifying Patterns. A process that is not under control or is unstable displays patterns of variation. Patterns signal the need to investigate the process and determine whether an assignable cause can be found for

FIGURE 18.8 Zones on a Control Chart

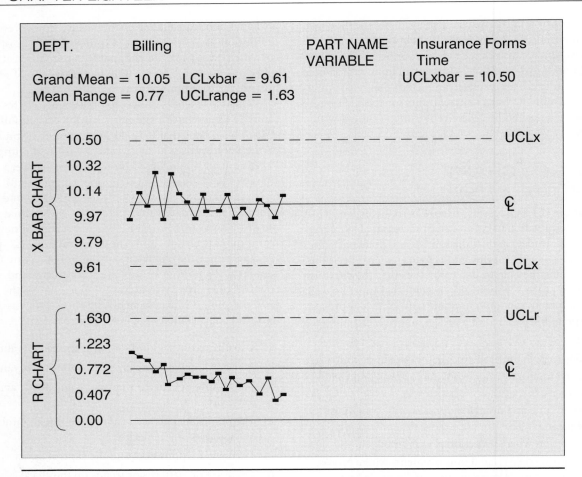

FIGURE 18.9 Control Chart with Grand Mean = 10.05

the variation. Figures 18.9 through 18.15 display a variety of out-of-control conditions and give some reasons why those conditions may exist.

Trends or Steady Changes in Level. A trend is a steady, progressive change in the location where the data are centered on the chart. Figure 18.9 displays a downward trend on the R chart. Note that the points were found in the upper half of the control chart at the beginning of the process and on the lower half of the chart at the end. The key to identifying a trend or steady change in level is to recognize that the points are slowly and steadily working their way from one level of the chart to another.

A trend may appear on the $\overline{X}$ chart because of tool or die wear, a gradual deterioration of the equipment, a buildup of chips, a slowly loosening work-holding device, a breakdown of the chemicals used in the process, or some other gradual change. R chart trends could be due to changes in worker skills, shifting work-holding devices, or wear out. Improvements would lead to less variation; increases in variation would reflect a decrease in skill or a change in the quality of the incoming material.

An oscillating trend would also need to be investigated (Figure 18.10). In this type of trend the points oscillate up and down for approximately 14 points or more. This could be due to a lack of homogeneity, perhaps a mixing of the output from two machines making the same product.

Change, Jump, or Shift in Level. Figure 18.11 displays what is meant by a change, jump, or shift in level. Note that the process begins at one level (Figure 18.11a) and jumps quickly to another level (Figure 18.11b) as the process continues to operate. This change, jump, or shift in level is fairly abrupt, unlike a trend described above. A change, jump, or shift can occur either on the $\overline{X}$ or R chart or on both charts. Causes for sudden shifts in level tend to reflect some new and fairly significant difference in the process. When investigating a sudden shift or jump in level, look for significant changes that can be pinpointed to a specific moment in time. For the $\overline{X}$ chart, causes include new machines, dies, or tooling; the minor failure of a machine part; new or inexperienced workers; new batches of raw material; new production methods; or changes to the process settings. For the R chart, potential sources of jumps or shifts in level causing a change in the process variability or spread include a new or inexperienced operator, a sudden increase in the play

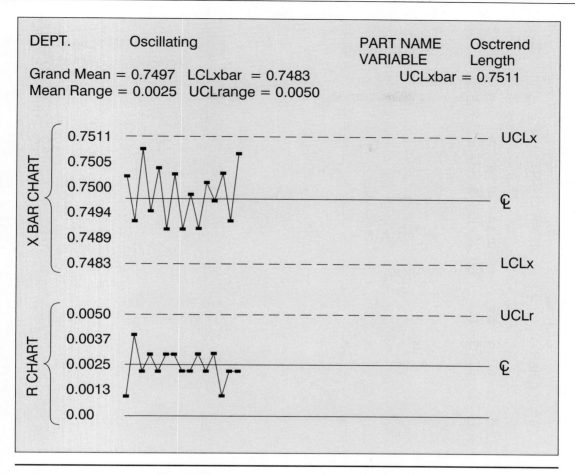

FIGURE 18.10 An Oscillating Trend

associated with gears or work-holding devices, or greater variation in incoming material.

Runs. A process can be considered out of control when there are unnatural runs present in the process. Imagine tossing a coin. If two heads occur in a row, the onlooker would probably agree that this occurred by chance. Even though the probability of the coin landing with heads showing is 50-50, no one expects coin tosses to alternate between heads and tails. If, however, an onlooker saw someone toss six heads in a row, that onlooker would probably be suspicious that this set of events is not due to chance. The same principle applies to control charts. Although the points on a control chart do not necessarily alternate above and below the centerline in a chart that is under control, the points are normally balanced above and below the centerline. A cluster of seven points in a row above or below the centerline would be improbable and would likely have an assignable cause. The same could be said for situations where 10 out of 11 points or 12 out of 14 points are located on one side or the other of the centerline (Figure 18.12). A run may also be considered a trend if it displays increasing or decreasing values.

Runs on the $\overline{X}$ chart can be caused by temperature changes; tool or die wear; gradual deterioration of the process; or deterioration of the chemicals, oils, or cooling fluids used in the process. Runs on the R chart (Figure 18.13) signal a change in the process variation. Causes for these R chart runs could be a change in operator skill, either an improvement or a decrement, or a gradual improvement in the homogeneity of the process because of changes in the incoming material or changes to the process itself.

Recurring Cycles. Recurring cycles are caused by systematic changes related to the process. When investigating what appears to be cycles (Figure 18.14) on the chart, it is important to look for causes that will change, vary, or cycle over time. For the $\overline{X}$ chart, potential causes are tool or machine wear conditions, an accumulation and then removal of chips or other waste material around the tooling, maintenance schedules, periodic rotation of operators, worker fatigue, periodic replacement of cooling fluid or cutting oil, or changes in the process environment such as temperature or humidity. Cycles on an R chart are not as common; an R chart displays the variation or spread of the process, which usually does not cycle. Potential causes are related to lubrication cycles and operator fatigue.

Cycles can be difficult to locate because the entire cycle may not be present on a single chart. The frequency of inspection could potentially cause a cycle to be overlooked. For example, if the cycle occurs every 15 minutes and

216 CHAPTER EIGHTEEN

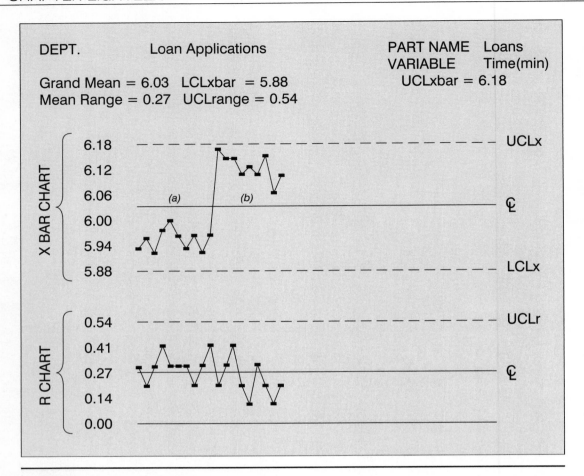

FIGURE 18.11 Change in Level

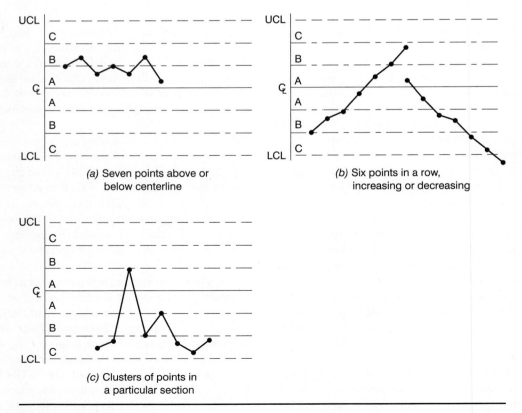

FIGURE 18.12 Runs

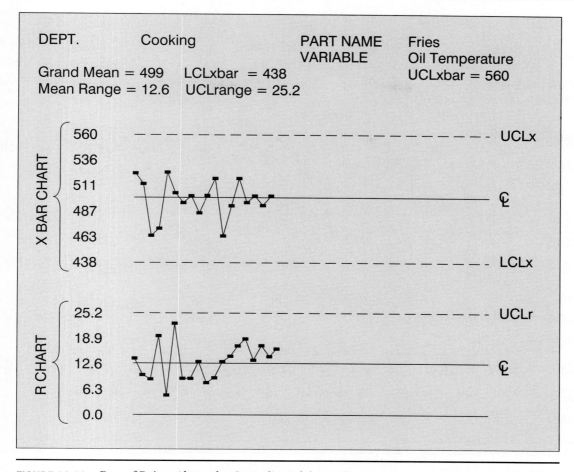

FIGURE 18.13 Run of Points Above the Centerline of the R Chart

samples are taken only every 30 minutes, then it is possible for the cycle to be overlooked.

Two Populations. When a control chart is under control, approximately 68 percent of the sample averages will fall within $\pm 1\sigma$ of the centerline. When a large number of the sample averages appear near or outside the control limits, two populations of samples might exist. "Two populations" refers to the existence of two (or more) sources of data.

On an $\overline{X}$ chart, the different sources of production might be due to the output of two or more machines being combined before sampling takes place. It might also occur because the work of two different operators is combined or two different sources of raw materials are brought together in the process. A two-population situation means that the items being sampled are not homogeneous (Figure 18.15). Maintaining the homogeneity of the items being sampled is critical for creating and using control charts.

This type of pattern on an R chart signals that different workers are using the same chart or that the variation is due to the fact that raw materials are coming from different suppliers.

9. Revise the Charts There are two circumstances under which the control chart is revised and new limits calculated. Existing calculations can be revised if a chart exhibits good control and any changes made to improve the process are permanent. When the new operating conditions become routine and no out-of-control signals have been seen, the chart may be revised. The revisions provide a better estimate of the population standard deviation, representing the spread of all of the individual parts in the process. With this value, a better understanding of the entire process can be gained.

Control limits are also revised if patterns exist, provided that the patterns have been identified and eliminated. Once the causes have been determined, investigated, and corrected in such a way that they will not affect the process in the future, the control chart can be revised. The new limits will reflect the changes and improvements made to the process. In both cases the new limits are used to judge the process behavior in the future.

The following four steps are taken to revise the charts.

A. Interpret the Original Charts. The R chart reflects the stability of the process and should be analyzed first. A lack of control on the R chart shows that the process is not producing parts that are very similar to each other. The process is not precise. If the R chart exhibits process control, study the $\overline{X}$ chart. Determine whether cycles, trends, runs, two populations, mistakes, or other examples of lack of control exist. If both the $\overline{X}$ and R charts are exhibiting good control, proceed to step D. If the charts display out-of-control conditions, then continue to step B.

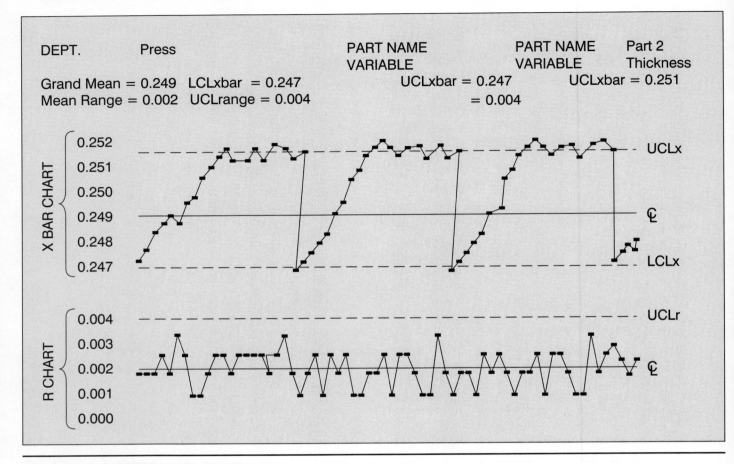

FIGURE 18.14 Cycle in Part Thickness

B. Isolate the Cause. If either the $\overline{X}$ or R chart is not exhibiting good statistical control, find the cause of the problem. Problems shown on the control chart may be removed only if the causes of those problems have been isolated and steps have been taken to eliminate them.

C. Take Corrective Action. Take the necessary steps to correct the causes associated with the problems exhibited on the chart. Once the causes of variation have been removed from the process, these points may be removed from the control chart and the calculations revised.

D. Revise the Chart. To determine the new limits against which the process will be judged in the future, it is necessary to remove any undesirable points, the causes of which have been determined and corrected, from the charts. The criteria for removing points are based on finding the cause behind the out-of-control condition. If no cause can be found and corrected, then the points *cannot* be removed from the chart. Groups of points, runs, trends, and other patterns can be removed in the same manner as removing individual points. In the case of charts that are exhibiting good statistical control, the points removed will equal zero and the calculations will continue from there.

Two methods can be used to discard the data. When it is necessary to remove a subgroup from the calculations, it can be removed from only the out-of-control chart or it can be removed from both charts. In the first case, when an $\overline{X}$ value must be removed from the $\overline{X}$ control chart, its corresponding R value is *not* removed from the R chart, and vice versa. In this text, the points are removed from *both* charts. This second approach has been chosen because the values on both charts are interrelated. The R chart values describe the spread of the data on the $\overline{X}$ chart. Removing data from one chart or the other negates this relationship.

The formulas for revising both the $\overline{X}$ and R charts are as follows:

$$\overline{\overline{X}}_{new} = \frac{\Sigma \overline{X} - \overline{X}_d}{m - m_d}$$

$$\overline{R}_{new} = \frac{\Sigma R - R_d}{m - m_d}$$

where

$\overline{X}_d$ = discarded subgroup averages
m_d = number of discarded subgroups
R_d = discarded subgroup ranges

The newly calculated values of $\overline{X}$ and $\overline{R}$ are used to establish updated values for the centerline and control limits on the chart. These new limits reflect that improvements have been made to the process and future production

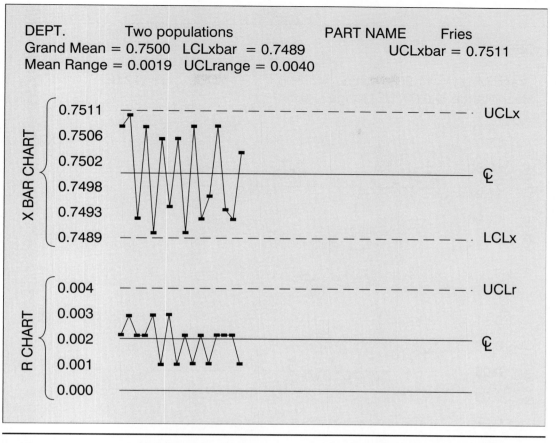

FIGURE 18.15 Two Populations

should be capable of meeting these new limits. The formulas for the revised limits are:

$$\overline{X}_{new} = \overline{X}_0 \qquad \overline{R}_{new} = R_0$$
$$\sigma_0 = R_0/d_2$$
$$UCL_{\overline{X}} = \overline{X}_0 + A\sigma_0$$
$$LCL_{\overline{X}} = \overline{X}_0 - A\sigma_0$$
$$UCL_R = D_2\sigma_0$$
$$LCL_R = D_1\sigma_0$$

where d_2, A, D_1, and D_2 are factors from the table in Appendix 2.

Achieve the Purpose Users of control charts are endeavoring to decrease the variation inherent in a process over time. Once established, control charts enable the user to understand where the process is currently centered and what the distribution of that process is. To know this information and not utilize it to improve the process defeats the purpose of creating control charts.

As the process improves, the average should come closer to the center of the specifications. The spread of the data, as shown by the range or the standard deviation, should decrease, and the parts produced or services provided should become more similar to each other.

Examining the Control Charts

Returning to the computer printer roller shaft example, an examination of the $\overline{X}$ and R charts begins by investigating the R chart, which displays the variation present in the process. Evidence of excessive variation would indicate that the process is not producing consistent product. The R chart (Figure 18.16) exhibits good control. The points are evenly spaced on both sides of the centerline and there are no points beyond the control limits. There are no unusual patterns or trends in the data. Given these observations, it can be said that the process is producing parts of similar dimensions.

LEAN SIX SIGMA TOOLS at WORK

Next the $\overline{X}$ chart is examined. An inspection of the $\overline{X}$ chart reveals an unusual pattern occurring at points 12, 13, and 14. These measurements are all below the lower control limit. When compared with other samples throughout the day's production, the parts produced during the time when samples 12, 13, and 14 were taken were much shorter than parts produced during other times in the production run. A glance at the R chart reveals that the range of the individual measurements taken in the samples is small, meaning that the parts produced during samples 12, 13, and 14 are all similar in size. An investigation into the cause of the production of undersized parts needs to take place.

(*Continued*)

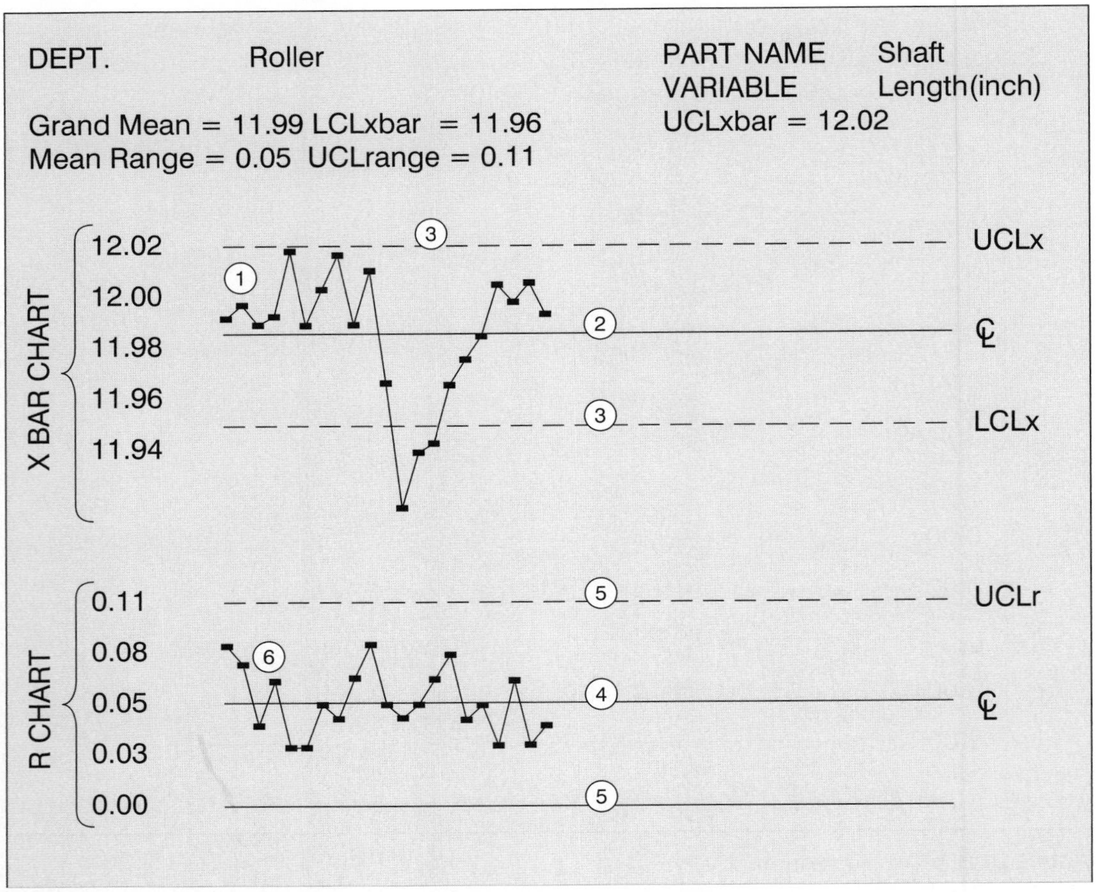
DEPT. Roller
PART NAME Shaft
VARIABLE Length(inch)
Grand Mean = 11.99 LCLxbar = 11.96
UCLxbar = 12.02
Mean Range = 0.05 UCLrange = 0.11
X BAR CHART
12.02
12.00
11.98
11.96
11.94
UCLx
LCLx
R CHART
0.11
0.08
0.05
0.03
0.00
UCLr
1
2
3
4
5
6

FIGURE 18.16 $\overline{X}$ and R Control Charts for Roller Shaft Length

A Further Examination of the Control Charts

LEAN SIX SIGMA TOOLS at WORK

An investigation into the differences in shaft lengths has been conducted. The $\overline{X}$ and R charts aid the investigators by allowing them to isolate when the differences were first noticed. Because the R chart (Figure 18.16) exhibits good control, the investigators are able to concentrate their attention on possible causes for a consistent change in shaft length for those three subgroups. Their investigation reveals that the machine settings had been bumped during the loading of the machine. For the time being, operators are being asked to check the control panel settings after loading the machine. To take care of the problem for the long term, manufacturing engineers are looking into possible design changes to protect the controls against accidental manipulation.

Revising the Control Limits

LEAN SIX SIGMA TOOLS at WORK

Because a cause for the undersized parts has been determined for the values for 12, 13, and 14, they can be removed from the calculations for the $\overline{X}$ and R chart. The points removed from calculations remain on the chart; but they are crossed out. The new or revised limits will be used to monitor future production. The new limits will extend from the old limits, as shown in Figure 18.17.

Revising the calculations is performed as follows:

$$\overline{\overline{X}}_{new} = \overline{X}_0 = \frac{\sum_{i=1}^{m} \overline{X} - \overline{X}_d}{m - m_d}$$

$$= \frac{251.77 - 11.94 - 11.95 - 11.95}{21 - 3}$$

$$= 12.00$$

$$\bar{R}_{new} = R_0 = \frac{\sum_{i=1}^{m} R - R_d}{m - m_d}$$

$$= \frac{1.06 - 0.04 - 0.05 - 0.06}{21 - 3}$$

$$= 0.05$$

Calculating the σ_0 for the process, when n = 5,

$$\sigma_0 = \frac{R_0}{d_2} = \frac{0.05}{2.326} = 0.02$$

$$UCL_{\bar{X}} = 12.00 + 1.342(0.02) = 12.03$$
$$LCL_{\bar{X}} = 12.00 - 1.342(0.02) = 11.97$$
$$UCL_R = 4.918(0.02) = 0.10$$
$$LCL_R = 0(0.02) = 0$$

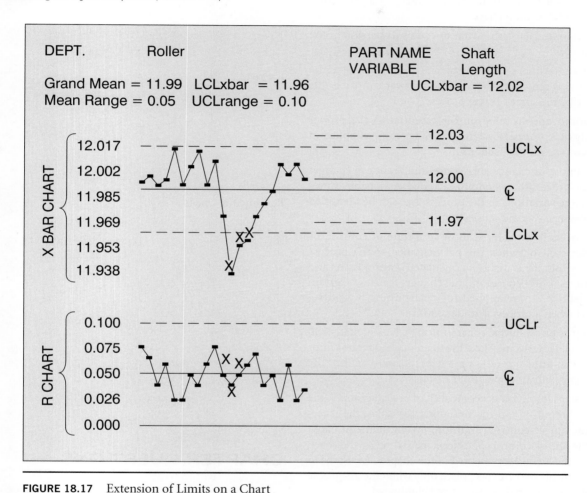

FIGURE 18.17 Extension of Limits on a Chart

SUMMARY

Control charts are easy to construct and use in studying a process, whether that process is in a manufacturing or service environment. Control charts indicate areas for improvement. Once the root cause has been identified, changes can be proposed and tested, and improvements can be monitored through the use of control charts. Through the use of control charts, similar gains can be realized in the manufacturing sector. Users of control charts report savings in scrap, including material and labor; lower rework costs; reduced inspection; higher product quality; more consistent part characteristics; greater operator confidence; lower troubleshooting costs; reduced completion times; faster deliveries; and others.

TAKE AWAY TIPS

1. Control charts enhance the analysis of a process by showing how that process performs over time. Control charts allow for early detection of process changes.

2. Control charts serve two basic functions: They provide an economic basis for making a decision as to whether to investigate for potential problems, adjust the process, or leave the process alone; and they assist in the identification of problems in the process.

3. Variation, differences between items, exists in all processes. Variation can be within-piece, piece-to-piece, and time-to-time.

4. The $\bar{X}$ chart is used to monitor the variation in the average values of the measurements of groups of samples. Averages rather than individual observations are used on control charts because average values will indicate a change in the amount of variation much faster than individual values will.

5. The $\bar{X}$ chart, showing the central tendency of the data, is always used in conjunction with either a range or a standard deviation chart.

6. The R and s charts show the spread or dispersion of the data.

7. The centerline of a control chart shows where the process is centered. The upper and lower control limits describe the spread of the process.

8. A homogeneous subgroup is essential to the proper study of a process. Certain guidelines can be applied in choosing a rational subgroup.

9. Common, or chance, causes are small random changes in the process that cannot be avoided. Assignable causes are large variations in the process that can be identified as having a specific cause.

10. A process is considered to be in a state of control, or under control, when the performance of the process falls within the statistically calculated control limits and exhibits only common, or chance, causes. Certain guidelines can be applied for determining by control chart when a process is under control.

11. Patterns on a control chart indicate a lack of statistical control. Patterns may take the form of changes or jumps in level, runs, trends, or cycles or may reflect the existence of two populations or mistakes.

12. The steps for revising a control chart are (a) examine the chart for out-of-control conditions; (b) isolate the causes of the out-of-control condition; (c) eliminate the cause of the out-of-control condition; and (d) revise the chart, using the formulas presented in the chapter. Revisions to the control chart can take place only when the assignable causes have been determined and eliminated.

FORMULAS

AVERAGE AND RANGE CHARTS

$\bar{X}$ chart:

$$\bar{\bar{X}} = \frac{\sum_{i=1}^{m} \bar{X}_i}{m}$$

$$UCL_{\bar{X}} = \bar{\bar{X}} + A_2\bar{R}$$
$$LCL_{\bar{X}} = \bar{\bar{X}} - A_2\bar{R}$$

R chart:

$$\bar{R} = \frac{\sum_{i=1}^{m} R_i}{m}$$

$$UCL_R = D_4\bar{R}$$
$$LCL_R = D_3\bar{R}$$

Revising the charts:

$$\bar{\bar{X}} = \bar{\bar{X}}_{new} = \frac{\sum_{i=1}^{m} \bar{X} - \bar{X}_d}{m - m_d}$$

$$UCL_{\bar{X}} = \bar{\bar{X}}_0 + A\sigma_0$$
$$LCL_{\bar{X}} = \bar{\bar{X}}_0 - A\sigma_0$$
$$\sigma_0 = R_0/d_2$$

$$R_{new} = \frac{\sum_{i=1}^{m} R - R_d}{m - m_d}$$

$$UCL_R = D_2\sigma_0$$
$$LCL_R = D_1\sigma_0$$

CHAPTER QUESTIONS

1. Describe the difference between chance and assignable causes.

2. A large bank establishes $\bar{X}$ and R charts for the time required to process applications for its charge cards. A sample of five applications is taken each day. The first four weeks (20 days) of data give

$$\bar{\bar{X}} = 16 \text{ min} \quad \bar{s} = 3 \text{ min} \quad \bar{R} = 7 \text{ min}$$

Based on the values given, calculate the centerline and control limits for the $\bar{X}$ and R charts.

3. The data below are $\bar{X}$ and R values for 25 samples of size n = 4 taken from a process filling bags of fertilizer. The measurements are made on the fill weight of the bags in pounds.

Subgroup Number	$\bar{X}$	Range
1	50.3	0.73
2	49.6	0.75
3	50.8	0.79
4	50.9	0.74
5	49.8	0.72
6	50.5	0.73
7	50.2	0.71
8	49.9	0.70
9	50.0	0.65
10	50.1	0.67
11	50.2	0.65
12	50.5	0.67
13	50.4	0.68
14	50.8	0.70
15	50.0	0.65
16	49.9	0.66
17	50.4	0.67
18	50.5	0.68
19	50.7	0.70
20	50.2	0.65
21	49.9	0.60
22	50.1	0.64
23	49.5	0.60
24	50.0	0.62
25	50.3	0.60

Set up the $\bar{X}$ and R charts on this process. Interpret the charts. Does the process seem to be in control? If necessary, assume assignable causes and revise the trial control limits. If the average fill of the bags is to be 50.0 pounds, how does this process compare?

4. The data below are $\bar{X}$ and R values for 12 samples of size $n = 5$. They were taken from a process producing bearings. The measurements are made on the inside diameter of the bearing. The data have been coded from 0.50; in other words, a measurement of 0.50345 has been recorded as 345. Range values are coded from 0.000; that is, 0.00013 is recorded as 13.

Subgroup Number	$\bar{X}$	Range
1	345	13
2	347	14
3	350	12
4	346	11
5	350	15
6	345	16
7	349	14
8	348	13
9	348	12
10	354	15
11	352	13
12	355	16

Set up the $\bar{X}$ and R charts on this process. Does the process seem to be in control? Why or why not? If necessary, assume assignable causes and revise the trial control limits.

5. Describe how both an $\bar{X}$ and R or s chart would look if they were under normal statistical control.

6. Why is the use and interpretation of an R or s chart so critical when examining an $\bar{X}$ chart?

7. Create an $\bar{X}$ and R chart for the clutch plate information in Table 18.1 on page 000. You will need to calculate the range values for each subgroup. Calculate the control limits and centerline for each chart. Graph the data with the calculated values. Beginning with the R chart, how does the process look?

8. RM Manufacturing makes thermometers for use in the medical field. These thermometers, which read in degrees Celsius, are able to measure temperatures to a level of precision of two decimal places. Each hour, RM Manufacturing tests eight randomly selected thermometers in a solution that is known to be at a temperature of 3°C. Use the following data to create and interpret an $\bar{X}$ and R chart. Based on the desired thermometer reading of 3°, interpret the results of your plotted averages and ranges.

Subgroup	Average Temperature	Range
1	3.06	0.10
2	3.03	0.09
3	3.10	0.12
4	3.05	0.07
5	2.98	0.08
6	3.00	0.10
7	3.01	0.15
8	3.04	0.09
9	3.00	0.09
10	3.03	0.14
11	2.96	0.07
12	2.99	0.11
13	3.01	0.09
14	2.98	0.13
15	3.02	0.08

CHAPTER NINETEEN

PROCESS CAPABILITY

She had the virtuous sense of one who has made an intuitive leap from the most fragile collection of evidence.

From Bodies of Water by J. S. Borthwick

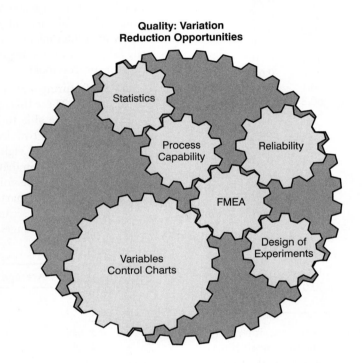

 LEARNING OPPORTUNITIES

1. To gain an understanding of the relationship between individual values and their averages
2. To understand the difference between specification limits and control limits
3. To learn to calculate and interpret the process capability indices: C_p, C_r, and C_{pk}

Consider the quote that begins this chapter. Unlike the person in the quote, lean Six Sigma practitioners know that they cannot make quick judgments without the data to support their conclusions. They need evidence. They need the answer to the question: How do we know the process will generate products or services that meet the customer's specifications? *Process capability refers to the ability of a process to produce products or provide services capable of meeting the specifications set by the customer or designer.* As discussed in previous chapters, variation affects a process and may prevent the process from producing products or services that meet customer specifications. Reducing process variability and creating consistent quality both increase the viability of predictions of future process performance (Figure 19.1).

Manufacturers of products and providers of services can use process capability concepts to assist in decisions concerning product or process specifications, appropriate production methods, equipment to be used, and time commitments.

Process Capability 225

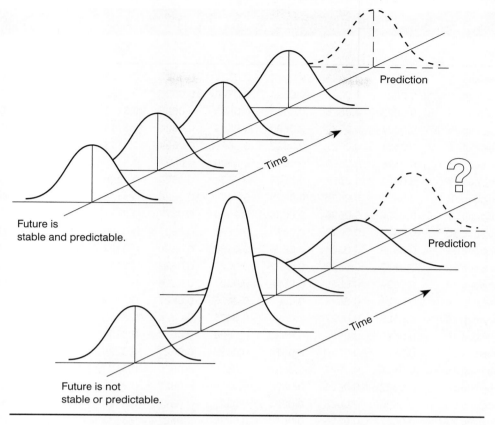

FIGURE 19.1 Future Predictions

INDIVIDUAL VALUES COMPARED WITH AVERAGES

Process capability is based on the performance of individual products or services against specifications. Information from samples helps determine the behavior of individuals in a process.

A relationship exists between subgroup sample averages and individual values. Table 19.1 repeats the tally of individual and average values of clutch plate thickness data from Table 18.1. With this actual production line data, two frequency diagrams have been created in Figure 19.2. One frequency diagram is constructed of individual values; the other is made up of subgroup averages. Both distributions are approximately normal. The important difference to note is that individual values spread much more widely than their averages. When the two diagrams are compared, the averages are grouped closer to the center value than are the individual values, as described by the central limit theorem. Average values smooth out the highs and lows associated with individuals. This comparison will be important to keep in mind when comparing the behavior of averages and control limits with that of individual values and specification limits (Figure 19.3).

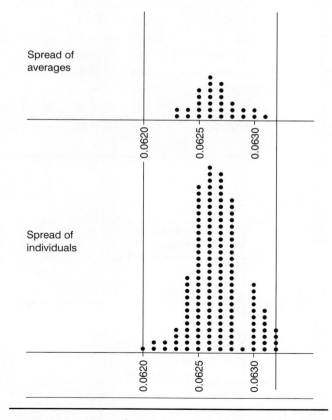

FIGURE 19.2 Normal Curves for Individuals and Averages

226 CHAPTER NINETEEN

TABLE 19.1 Clutch Plate Thickness: Sums and Averages

						$\sum X_i$		$\bar{R}$
Subgroup 1	0.0625	0.0626	0.0624	0.0625	0.0627	0.3127	0.0625	0.0003
Subgroup 2	0.0624	0.0623	0.0624	0.0626	0.0625	0.3122	0.0624	0.0003
Subgroup 3	0.0622	0.0625	0.0623	0.0625	0.0626	0.3121	0.0624	0.0004
Subgroup 4	0.0624	0.0623	0.0620	0.0623	0.0624	0.3114	0.0623	0.0004
Subgroup 5	0.0621	0.0621	0.0622	0.0625	0.0624	0.3113	0.0623	0.0004
Subgroup 6	0.0628	0.0626	0.0625	0.0626	0.0627	0.3132	0.0626	0.0003
Subgroup 7	0.0624	0.0627	0.0625	0.0624	0.0626	0.3126	0.0625	0.0003
Subgroup 8	0.0624	0.0625	0.0625	0.0626	0.0626	0.3126	0.0625	0.0002
Subgroup 9	0.0627	0.0628	0.0626	0.0625	0.0627	0.3133	0.0627	0.0003
Subgroup 10	0.0625	0.0626	0.0628	0.0626	0.0627	0.3132	0.0626	0.0003
Subgroup 11	0.0625	0.0624	0.0626	0.0626	0.0626	0.3127	0.0625	0.0002
Subgroup 12	0.0630	0.0628	0.0627	0.0625	0.0627	0.3134	0.0627	0.0005
Subgroup 13	0.0627	0.0626	0.0628	0.0627	0.0626	0.3137	0.0627	0.0002
Subgroup 14	0.0626	0.0626	0.0625	0.0626	0.0627	0.3130	0.0626	0.0002
Subgroup 15	0.0628	0.0627	0.0626	0.0625	0.0626	0.3132	0.0626	0.0003
Subgroup 16	0.0625	0.0626	0.0625	0.0628	0.0627	0.3131	0.0626	0.0003
Subgroup 17	0.0624	0.0626	0.0624	0.0625	0.0627	0.3126	0.0625	0.0003
Subgroup 18	0.0628	0.0627	0.0628	0.0626	0.0630	0.3139	0.0627	0.0004
Subgroup 19	0.0627	0.0626	0.0628	0.0625	0.0627	0.3133	0.0627	0.0003
Subgroup 20	0.0626	0.0625	0.0626	0.0625	0.0627	0.3129	0.0626	0.0002
Subgroup 21	0.0627	0.0626	0.0628	0.0625	0.0627	0.3133	0.0627	0.0003
Subgroup 22	0.0625	0.0626	0.0628	0.0625	0.0627	0.3131	0.0626	0.0003
Subgroup 23	0.0628	0.0626	0.0627	0.0630	0.0627	0.3138	0.0628	0.0004
Subgroup 24	0.0625	0.0631	0.0630	0.0628	0.0627	0.3141	0.0628	0.0006
Subgroup 25	0.0627	0.0630	0.0631	0.0628	0.0627	0.3143	0.0629	0.0004
Subgroup 26	0.0630	0.0628	0.0629	0.0628	0.0627	0.3142	0.0628	0.0003
Subgroup 27	0.0630	0.0628	0.0631	0.0628	0.0627	0.3144	0.0629	0.0004
Subgroup 28	0.0632	0.0632	0.0628	0.0631	0.0630	0.3153	0.0631	0.0004
Subgroup 29	0.0630	0.0628	0.0631	0.0632	0.0631	0.3152	0.0630	0.0004
Subgroup 30	0.0632	0.0631	0.0630	0.0628	0.0628	0.3149	0.0630	0.0004
						9.3981		

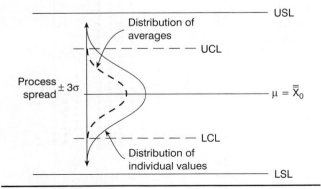

FIGURE 19.3 Comparison of the Spread of Individual Values with Averages

ESTIMATION OF POPULATION SIGMA FROM SAMPLE DATA

The larger the sample size, n, the more representative the sample average, $\bar{X}$, is of the mean of the population, μ. In other words, $\bar{X}$ becomes a more reliable estimate of μ as the sample size is increased. The ability of $\bar{X}$ to approximate μ is measured by the expression $\sigma/\sqrt{n}$, the standard error of the mean. It is possible to estimate the spread of the population of individuals using the sample data. This formula shows the relationship between the population standard deviation (σ) and the standard deviation of the subgroup averages ($\sigma_{\bar{X}}$):

$$\sigma_{\bar{X}} = \frac{\sigma}{\sqrt{n}}$$

where

$\sigma_{\bar{X}}$ = standard deviation of subgroup averages
σ = population standard deviation
n = number of observations in each subgroup

The population standard deviation, σ, is determined by measuring the individuals. This necessitates measuring every value. To avoid complicated calculations, if the process can be assumed to be normal, the population standard deviation can be estimated from either the standard deviation associated with the sample standard deviation (s) or the range (R):

$$\hat{\sigma} = \frac{\bar{s}}{c_4} \quad \text{or} \quad \hat{\sigma} = \frac{\bar{R}}{d_2}$$

where

$\hat{\sigma}$ = estimate of population standard deviation
$\bar{s}$ = sample standard deviation calculated from subgroup samples
$\bar{R}$ = average range of subgroups
c_4 as found in Appendix 2
d_2 as found in Appendix 2

Because of the estimators (c_4 and d_2), these two formulas will yield similar but not identical values for $\hat{\sigma}$. Dr. Shewhart confirmed that the standard deviation of subgroup sample means is the standard deviation of individual samples divided by the square root of the subgroup size. He did this by drawing, at random from a large bowl, numbered, metal-lined, disk-shaped tags. He used this information to determine the estimators c_4 and d_2.

CONTROL LIMITS VERSUS SPECIFICATION LIMITS

A process is in control only when its process centering and the amount of variation present in the process remains constant over time. If both are constant, then the behavior of the process will be predictable. As discussed in Chapter 18, a process under control exhibits the following characteristics:

1. Two-thirds of the points are near the center value.
2. A few of the points are close to the center value.
3. The points float back and forth across the centerline.
4. The points are balanced (in roughly equal numbers) on both sides of the centerline.
5. There are no points beyond the control limits.
6. There are no patterns or trends on the chart.

It is important to note that a process in statistical control will not necessarily meet specifications as established by the customer. There is a difference between a process conforming to specifications and a process performing within statistical control.

Established during the design process or from customer requests, specifications communicate what the customers expect, want, or need from the process. Specifications can be considered the voice of the customer.

Control limits are the voice of the process. The centerline on the $\bar{X}$ chart represents process centering. The R and s chart limits represent the amount of variation present in the process. Control limits are a prediction of the variation that the process will exhibit in the near future. The difference between specifications and control limits is that specifications relay wishes and control limits tell of reality.

Occasionally, creators of control charts inappropriately place specification limits on control charts. Processes are unaware of specifications; they perform to the best of their capabilities. Unfortunately, specification limits on control charts encourage users to adjust the process on the basis of them instead of on the true limits of the process, the control limits. The resulting miscued changes can potentially disrupt the process and increase process variation.

As Figure 19.2 shows, the spread of individual values is wider than the spread of the averages. For this reason, control limits cannot be compared directly with specification limits. An $\bar{X}$ chart does not reflect how widely the individual values composing the plotted averages spread. This is one reason why an R or s chart is always used in conjunction with the $\bar{X}$ chart. The spread of the individual data can be seen only by observing what is happening on the R or s chart. If the values on the R or s chart are large, then the variation associated with the average is large. Figure 19.4 shows a control chart created using the concepts from Chapter 18 and the values in Table 19.1. The $\bar{X}$s are circled. Individual values shown as Xs are also plotted on this chart. Note where the individual values fall in relation to the control limits established for the process. The individuals spread more widely than the averages and follow the pattern established by the R chart. Studying the R chart in conjunction with the $\bar{X}$ chart can significantly increase the understanding of how the process is performing.

For explanatory purposes, both control limits and specification limits appear on the charts in Figure 19.5, a practice not to be followed in industry. The variation in Figure 19.5, top chart, exceeds control limits marking the expected process variation but not the specification limits. Although the process is out of control, for the time being the customer's needs are being met. The process in Figure 19.5, bottom chart, is under control and within the control limits, but the specification limits do not correspond with the control limits. This situation reveals that the process is performing to the best of its abilities, but not well enough to meet the specifications set by the customer.

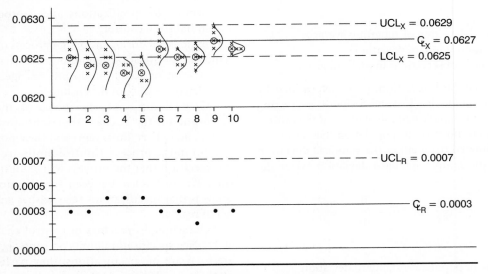

FIGURE 19.4 $\bar{X}$ and R Chart Showing Averages and Individuals

THE 6σ SPREAD VERSUS SPECIFICATION LIMITS

The spread of the individuals in a process, 6σ, is the measure used to compare the realities of production with the desires of the customers. The process standard deviation is based on either s or R from control chart data:

$$\hat{\sigma} = \frac{\bar{s}}{c_4} \quad \text{or} \quad \hat{\sigma} = \frac{\bar{R}}{d_2}$$

where

$\hat{\sigma}$ = estimate of population standard deviation
$\bar{s}$ = sample standard deviation calculated from process
$\bar{R}$ = average range of subgroups calculated from process
c_4 as found in Appendix 2
d_2 as found in Appendix 2

Remember, because of the estimators (c_4 and d_2), these two formulas will yield similar but not identical values for $\hat{\sigma}$.

Specification limits, the allowable spread of the individuals, are compared with the 6σ spread of the process to determine how capable the process is of meeting the specifications. Three different situations can exist when specifications and 6σ are compared: (1) the 6σ process spread can be less than the spread of the specification limits; (2) the 6σ process spread can be equal to the spread of the specification limits; (3) the 6σ process spread can be greater than the spread of the specification limits.

Case I: $6\sigma < \text{USL} - \text{LSL}$ This is the most desirable case. Figure 19.7 illustrates this relationship. The control limits have been placed on the diagram, as well as the spread of the process averages (dotted line). The 6σ spread of the process individuals is shown by the solid line. As expected, the spread of the individual values is greater than the spread of the averages; however, the values are still within the specification limits. The 6σ spread of the individuals is less than the spread of the specifications. This allows for more room for process shifts while staying within the specifications. Notice that even if the process drifts out of control (Figure 19.7b), the change must be dramatic before the parts are considered out of specification.

Case II: $6\sigma = \text{USL} - \text{LSL}$ In this situation, 6σ is equal to the tolerance (Figure 19.8a). As long as the process remains in control and centered, with no change in process variation, the parts produced will be within specification. However, a shift in the process mean (Figure 19.8b) will

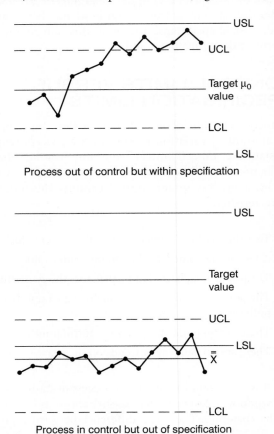

FIGURE 19.5 For Explanation Purposes Only, These Charts Show Control Limits versus Specification Limits

Using the X̄ and R Charts to Assess the Process

LEAN SIX SIGMA TOOLS at WORK

A lean Six Sigma team has been studying the results of a tensile strength test. To better understand the process performance and the spread of the individual values that compose the averages, the team members have overlaid the R chart pattern on the X̄ chart. To do this easily, they divided the X̄ chart into three sections (Figure 19.6) chosen on the basis of how the data on the X̄ chart appear to be grouped. X̄ values in section A are centered at the mean and are very similar. In section B, the X̄ values are above the mean and more spread out. Section C values have a slight downward trend.

Studying the sections on the R chart reveals that the spread of the data is changing (Figure 19.6). The values in section A have an average amount of variation, denoted by the normal curve corresponding to section A. Variation increases in section B, resulting in a much broader spread on the normal curve (B). The significant decrease in variation in section C is shown by the narrow, peaked distribution. The R chart describes the spread of the individuals on the X̄ chart.

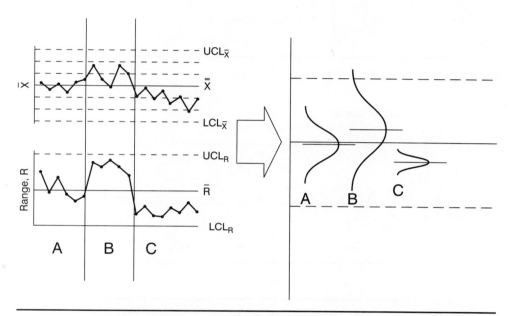

FIGURE 19.6 Overlaying the R Chart Pattern on the X̄ Chart

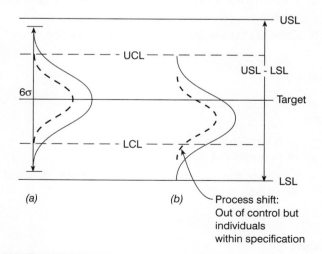

FIGURE 19.7 Case I: $6\sigma < USL - LSL$

Case III: $6\sigma > USL - LSL$ Any time that the 6σ spread is greater than the tolerance spread, an undesirable situation exists (Figure 19.9a). Even though the process is exhibiting only natural patterns of variation, it is incapable of meeting the specifications set by the customer. To correct this problem, management intervention will be necessary in order to change the process to decrease the variation or to recenter the process if necessary. The capability of the process cannot be improved without changing the existing process. To achieve a substantial reduction in the standard deviation or spread of the data, management will have to authorize the utilization of different materials, the overhaul of the machine, the purchase of a new machine, the retraining of the operator, or other significant changes to the process. Other, less desirable approaches to dealing with this problem are to perform 100 percent inspection on the product, increase the specification limits, or shift the process average so that all of the nonconforming products occur at one end of the distribution (Figure 19.9b). In certain cases, shifting the process average can eliminate scrap and increase the amount of rework, thus saving scrap costs by increasing rework costs.

result in the production of parts that are out of specification. An increase in the variation present in the process also creates an out-of-specification situation (Figure 19.8c).

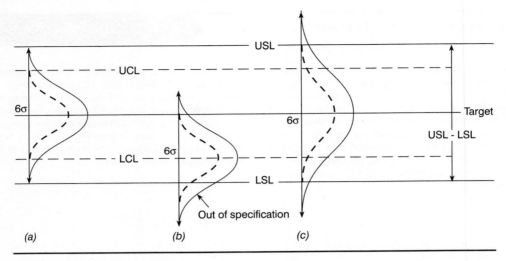

FIGURE 19.8 Case II: $6\sigma = $ USL $-$ LSL

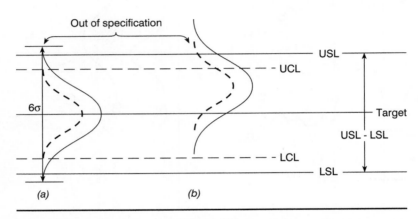

FIGURE 19.9 Case III: $6\sigma >$ USL $-$ LSL

CALCULATING PROCESS CAPABILITY INDICES

Process capability indices are mathematical ratios that quantify the ability of a process to produce products within the specifications. The capability indices compare the spread of the individuals created by the process with the specification limits set by the customer or designer. The 6σ spread of the individuals can be calculated for a new process that has not produced a significant number of parts or for a process currently in operation. In either case, a true 6σ value cannot be determined until the process is under control, as described by the $\overline{X}$ and R charts or $\overline{X}$ and s charts. If the process is not stable, the calculated values may not be representative of the true process capability.

Calculating $6\hat{\sigma}$

Assuming the process is under statistical control, we use one of the two following methods to calculate $6\hat{\sigma}$ of a new process:

1. Take at least 20 subgroups of sample size 4 for a total of 80 measurements.

2. Calculate the sample standard deviation, s_i, for each subgroup.

3. Calculate the average sample standard deviation, $\overline{s}$:

$$\overline{s} = \frac{\sum_{i=1}^{m} s_i}{m}$$

where

s_i = standard deviation for each subgroup
m = number of subgroups

4. Calculate the estimate of the population standard deviation:

$$\hat{\sigma} = \frac{\overline{s}}{c_4}$$

where

c_4 is obtained from Appendix 2.

5. Multiply the population standard deviation by 6.

A second method of calculating $6\hat{\sigma}$ is to use the data from a control chart. Once again, it is assumed that the process is under statistical control, with no unusual patterns of variation.

LEAN SIX SIGMA TOOLS at WORK

Calculating $6\hat{\sigma}$

The lean Six Sigma team members monitoring a process making roller shafts for printers wish to calculate $6\hat{\sigma}$ using the data in Table 19.2. They have 21 subgroups of sample size 5 for a total of 105 measurements, more than the 80 recommended.

They calculate the sample standard deviation, s_i, for each subgroup (Table 19.2). From these values they calculate the average sample standard deviation, $\bar{s}$:

$$\bar{s} = \frac{\sum_{i=1}^{m} s_i}{m} = \frac{0.031 + 0.029 + \cdots + 0.015}{21}$$

$$= \frac{0.414}{21} = 0.02$$

The next step involves calculating the estimate of the population standard deviation:

$$\hat{\sigma} = \frac{\bar{s}}{c_4} = \frac{0.02}{0.9400} = 0.021$$

The value of c_4 is obtained from Appendix 2 and is based on a sample size of 5.

As the final step they multiply the population standard deviation by 6:

$$6\hat{\sigma} = 6(0.021) = 0.126$$

This value can be compared with the spread of the specifications to determine how the individual products produced by the process compare with the specifications set by the designer.

TABLE 19.2 $\bar{X}$ and s Values of Roller Shafts

Subgroup Number	X_1					$\bar{X}$	s
1	11.950	12.000	12.030	11.980	12.010	11.994	0.031
2	12.030	12.020	11.960	12.000	11.980	11.998	0.029
3	12.010	12.000	11.970	11.980	12.000	11.992	0.016
4	11.970	11.980	12.000	12.030	11.990	11.994	0.023
5	12.000	12.010	12.020	12.030	12.020	12.016	0.011
6	11.980	11.980	12.000	12.010	11.990	11.992	0.013
7	12.000	12.010	12.030	12.000	11.980	12.004	0.018
8	12.000	12.010	12.040	12.000	12.020	12.014	0.017
9	12.000	12.020	11.960	12.000	11.980	11.992	0.023
10	12.020	12.000	11.970	12.050	12.000	12.008	0.030
11	11.980	11.970	11.960	11.950	12.000	11.972	0.019
12	11.920	11.950	11.920	11.940	11.960	11.938	0.018
13	11.980	11.930	11.940	11.950	11.960	11.952	0.019
14	11.990	11.930	11.940	11.950	11.960	11.954	0.023
15	12.000	11.980	11.990	11.950	11.930	11.970	0.029
16	12.000	11.980	11.970	11.960	11.990	11.980	0.016
17	12.020	11.980	11.970	11.980	11.990	11.988	0.019
18	12.000	12.010	12.020	12.010	11.990	12.006	0.011
19	11.970	12.030	12.000	12.010	11.990	12.000	0.022
20	11.990	12.010	12.020	12.000	12.010	12.006	0.011
21	12.000	11.980	11.990	11.990	12.020	11.996	0.015

1. Take the past 20 subgroups, sample size of 4 or more.
2. Calculate the range, R, for each subgroup.
3. Calculate the average range, $\bar{R}$:

$$\bar{R} = \frac{\sum_{i=1}^{m} R_i}{m}$$

where

R_i = individual range values for the subgroups
m = number of subgroups

4. Calculate the estimate of the population standard deviation, $\hat{\sigma}$:

$$\hat{\sigma} = \frac{\bar{R}}{d_2}$$

where

d_2 is obtained from Appendix 2.

5. Multiply the population standard deviation by 6.

Using more than 20 subgroups will improve the accuracy of the calculations.

Calculating $6\hat{\sigma}$

LEAN SIX SIGMA TOOLS at WORK

The lean Six Sigma team members used the data in Table 19.1 to calculate $6\hat{\sigma}$ for the clutch plate. Thirty subgroups of sample size 5 and their ranges are used to calculate the average range, $\overline{R}$:

$$\overline{R} = \frac{\sum_{i=1}^{m} R_i}{m} = \frac{0.0003 + 0.0003 + \cdots + 0.0004}{21}$$
$$= 0.0003$$

Next, the engineers calculate the estimate of the population standard deviation, $\hat{\sigma}$:

$$\hat{\sigma} = \frac{\overline{R}}{d_2} = \frac{0.0003}{2.326} = 0.0001$$

Using a sample size of 5, they take the value for d_2 from Appendix 2.

To determine $6\hat{\sigma}$, they multiply the population standard deviation by 6:

$$6\hat{\sigma} = 6(0.0001) = 0.0006$$

They now compare this value with the specification limits to determine how well the process is performing.

The Capability Index

Once calculated, the σ values can be used to determine several indices related to process capability. The **capability index C_p** is the ratio of tolerance (USL − LSL) and $6\hat{\sigma}$:

$$C_p = \frac{\text{USL} - \text{LSL}}{6\hat{\sigma}}$$

where

C_p = capability index
USL − LSL = upper specification limit − lower specification limit, or tolerance

The capability index is interpreted as follows: If the capability index is larger than 1.00, a Case I situation exists (Figure 19.7a). This is desirable. The greater this value, the better. If the capability index is equal to 1.00, then a Case II situation exists (Figure 19.8a). This is not optimal, but it is feasible. If the capability index is less than 1.00, then a Case III situation exists (Figure 19.9a). Values of less than 1 are undesirable and reflect the process's inability to meet the specifications.

The Capability Ratio

Another indicator of process capability is called the **capability ratio**. This ratio is similar to the capability index, though it reverses the numerator and the denominator. It is defined as follows:

$$C_r = \frac{6\hat{\sigma}}{\text{USL} - \text{LSL}}$$

A capability ratio less than 1 is the most desirable situation. The larger the ratio, the less capable the process is of meeting specifications. Be aware that it is easy to confuse the two indices. The most commonly used index is the capability index.

Finding the Capability Index I

LEAN SIX SIGMA TOOLS at WORK

The clutch plate from the earlier Lean Six Sigma Tools at Work feature has specification limits of 0.0625 ± 0.0003. The upper specification limit is 0.0628 and the lower specification limit is 0.0622. To calculate C_p:

$$C_p = \frac{\text{USL} - \text{LSL}}{6\hat{\sigma}} = \frac{0.0628 - 0.0622}{0.0006}$$
$$= 1.0$$

A value of 1.0 means that the process is just capable of meeting the demands placed on it by the customer's specifications. To be on the safe side, changes will need to occur to improve the process performance.

Finding the Capability Index II

LEAN SIX SIGMA TOOLS at WORK

The lean Six Sigma team members monitoring the roller shaft process want to calculate its C_p. They use specification limits of USL = 12.05, LSL = 11.95:

$$C_r = \frac{\text{USL} - \text{LSL}}{6\hat{\sigma}} = \frac{12.05 - 11.95}{0.126}$$
$$= 0.794$$

This process is not capable of meeting the demands placed on it. Improvements will need to take place to meet the customer's expectations.

C_{pk}

The centering of the process is shown by C_{pk}. A process operating in the center of the specifications set by the designer is usually more desirable than one that is consistently producing parts to the high or low side of the specification limits. In Figure 19.10, all three distributions have the same C_p index value of 1.3. Though each of these processes has the same capability index, they represent three different scenarios. In the first situation the process is centered as well as capable. In the second, a further upward shift in the process would result in an out-of-specification situation. The reverse holds true in the third situation. C_p and C_r do not take into account the centering of the process. *The ratio that reflects how the process is performing in terms of a nominal, center, or target value is C_{pk}.* C_{pk} can be calculated using the following formula:

$$C_{pk} = \frac{Z(\min)}{3}$$

where Z(min) is the smaller of

$$Z(USL) = \frac{USL - \overline{X}}{\hat{\sigma}}$$

$$\text{or } Z(LSL) = \frac{\overline{X} - LSL}{\hat{\sigma}}$$

When $C_{pk} = C_p$ the process is centered. Figure 19.11 illustrates C_p and C_{pk} values for a process that is centered

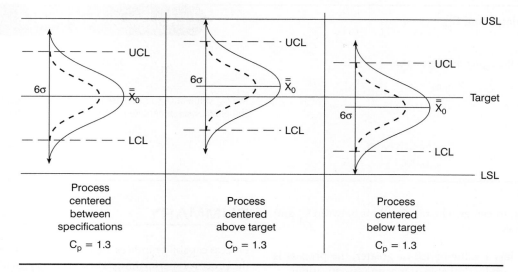

FIGURE 19.10 Shifts in Process Centering

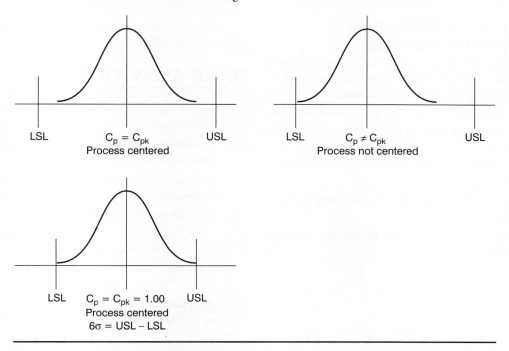

FIGURE 19.11 Process Centering: C_p versus C_{pk}

Finding C_{pk}

LEAN SIX SIGMA TOOLS at WORK

Determine the C_{pk} for the roller shaft values. The average, $\bar{X}$, is equal to 11.990.

$$C_{pk} = \frac{Z(min)}{3}$$

where

$$Z(min) = \text{smaller of } \frac{(USL - \bar{X})}{\hat{\sigma}} \text{ or } \frac{(\bar{X} - LSL)}{\hat{\sigma}}$$

$$Z(USL) = \frac{(12.050 - 11.990)}{0.021} = 2.857$$

$$Z(LSL) = \frac{(11.990 - 11.950)}{0.021} = 1.905$$

$$C_{pk} = \frac{1.905}{3} = 0.635$$

A C_{pk} value of less than 1 means that the process is not capable. Because the C_p value (0.794) and the C_{pk} value (0.635) are not equal, the process is not centered between the specification limits.

Calculating C_{pk}

LEAN SIX SIGMA TOOLS at WORK

To determine the C_{pk} for the clutch plate:

$$C_{pk} = \frac{Z(min)}{3} = \frac{1}{3} = 0.3333$$

where

$$Z(USL) = \frac{0.0628 - 0.0627}{0.0001} = 1$$

$$Z(LSL) = \frac{0.0627 - 0.0622}{0.0001} = 5$$

The C_{pk} value of 0.3333 is less than 1 and not equal to $C_p = 1$. The process is not centered between the specification limits.

and one that is off center. The relationships between C_p and C_{pk} are as follows:

1. When C_p has a value of 1.0 or greater, the process is producing product capable of meeting specifications.
2. The C_p value does not reflect process centering.
3. When the process is centered, $C_p = C_{pk}$.
4. C_{pk} is always less than or equal to C_p.
5. When C_p is greater than or equal to 1.0 and C_{pk} has a value of 1.00 or more, it indicates the process is producing product that conforms to specifications.
6. When C_{pk} has a value less than 1.00, it indicates the process is producing product that does not conform to specifications.
7. A C_p value of less than 1.00 indicates that the process is not capable.
8. A C_{pk} value of zero indicates the process average is equal to one of the specification limits.
9. A negative C_{pk} value indicates that the average is outside the specification limits.

SUMMARY

Process capability indices are used to judge how consistently the process is performing. These indices provide a great deal of information concerning process centering and the ability of the process to meet specifications. Process capability indices can guide the improvement process toward uniformity about a target value.

TAKE AWAY TIPS

1. Process capability refers to the ability of a process to meet the specifications set by the customer or designer.
2. Individuals in a process spread more widely around a center value than do the averages.
3. Specification limits are set by the designer or customer. Control limits are determined by the current process.
4. $6\hat{\sigma}$ is the spread of the process or process capability.
5. C_p, the capability index, is the ratio of the tolerance (USL − LSL) and the process capability ($6\hat{\sigma}$).
6. C_r, the capability ratio, is the ratio of the process capability (6σ) and the tolerance (USL − LSL).
7. C_{pk} is the ratio that reflects how the process is performing in relation to a nominal, center, or target value.

FORMULAS

$$\hat{\sigma} = \frac{\bar{s}}{c_4}$$

$$\hat{\sigma} = \frac{\bar{R}}{d_2}$$

CAPABILITY INDICES

$$C_p = \frac{USL - LSL}{6\hat{\sigma}}$$

$$C_r = \frac{6\hat{\sigma}}{USL - LSL}$$

$$C_{pk} = \frac{Z(min)}{3}$$

where Z(min) is the smaller of $Z(USL) = (USL - \bar{\bar{X}})/\hat{\sigma}$ or $Z(LSL) = (\bar{\bar{X}} - LSL)/\hat{\sigma}$.

CHAPTER QUESTIONS

1. What do control limits represent? What do specification limits represent? Describe the three cases that compare specification limits to control limits.

2. Why can a process be in control but not be capable of meeting specifications?

3. A hospital is using $\bar{X}$ and R charts to record the time it takes to process patient account information. A sample of five applications is taken each day. The first four weeks' (20 days') data give the following values:

$$\bar{\bar{X}} = 16 \text{ min} \quad \bar{R} = 7 \text{ min}$$

If the upper and lower specifications are 21 minutes and 13 minutes, respectively, calculate $6\hat{\sigma}$, C_p, and C_{pk}. Interpret the indices.

4. For the data in Question 3 of Chapter 18, calculate $6\hat{\sigma}$, C_p, and C_{pk}. Interpret the indices. The specification limits are 50 ± 0.5.

5. Stress tests are used to study the heart muscle after a person has had a heart attack. Timely information from these stress tests can help doctors prevent future heart attacks. The team investigating the turnaround time of stress tests has managed to reduce the amount of time it takes for a doctor to receive the results of a stress test from 68 to 32 hours on average. The team had a goal of reducing test turnaround times to between 30 and 36 hours. Given that the new average test turnaround time is 32 hours, with a standard deviation of 1, and n = 9, calculate and interpret C_p and C_{pk}.

6. Hotels use statistical information and control charts to track their performance on a variety of indicators. Recently a hotel manager has been asked whether his team is capable of maintaining scores between 8 and 10 (on a scale of 1 to 10) for "overall cleanliness of room." The most recent data has a mean of 8.624, a standard deviation of 1.446, and n = 10. Calculate and interpret C_p and C_{pk}.

7. The Tasty Morsels Chocolate Company tracks the amount of chocolate found in its chocolate bars. The target is 26 grams and the upper and lower specifications are 29 and 23 grams, respectively. If their most recent $\bar{X}$ has a centerline of 25 and the R chart has a centerline of 2, and n = 4, is the process capable? Calculate and interpret C_p and C_{pk}.

8. From the information in Question 7 of Chapter 18, calculate $6\hat{\sigma}$, C_p, and C_{pk}. Interpret the indices. The specification limits are 3 ± 0.05.

9. For the data in Question 10 of Chapter 18, use $\bar{s}/c_4$ to calculate $6\hat{\sigma}$, C_p, and C_{pk}. Interpret the indices. Specifications: 400 ± 150 particulates per million.

10. A quality analyst is checking the process capability associated with the production of struts, specifically the amount of torque used to tighten a fastener. Twenty-five samples of size 4 have been taken. These were used to create $\bar{X}$ and R charts. The values for these charts are as follows. The upper and lower control limits for the $\bar{X}$ chart are 74.80 Nm and 72.37 Nm, respectively. $\bar{\bar{X}}$ is 73.58 Nm. $\bar{R}$ is 1.66. The specification limits are 80 Nm ± 10. Calculate $6\hat{\sigma}$, C_p, and C_{pk}. Interpret the values.

RELIABILITY

Quality is a customer determination which is based on the customer's actual experience with the product or service.

Dr. Armand Feigenbaum

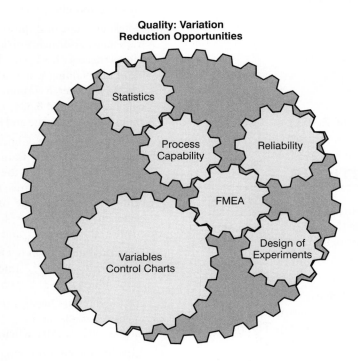

 LEARNING OPPORTUNITIES

1. To understand the importance of system reliability
2. To understand how performance during the life of a product, process, or system is affected by its design and configuration
3. To be able to compute the reliability of systems, including systems in series, parallel, and hybrid combinations
4. To know what to look for in a comprehensive reliability program

RELIABILITY CONCEPTS

Like Dr. Feigenbaum, lean Six Sigma organizations know that their customers are interested in quality over the long run, their actual experience with the product or service when they use it. **Reliability**, *or quality over the long term, is the ability of a product to perform its intended function over a period of time and under prescribed environmental conditions.* The reliability of a system, subsystem, component, or part is dependent on many factors, including the quality of research performed at its conception, the original design and any subsequent design changes, the complexity of the design, the manufacturing processes, the handling received during shipping, the environment surrounding its use, the end user, and numerous other factors. The causes of unreliability are many. Improper design, less-than-specified construction materials, faulty manufacturing or assembly, inappropriate testing leading to unrealistic conclusions, and damage during shipment are all factors that may contribute to a lack of product reliability. Once the product reaches the user, improper start-ups, abuse, lack of maintenance, or misapplication can seriously affect the reliability of an item. Reliability studies are looking for the answers to such questions as: Will it work? How long will it last?

Recognizing that reliability, or quality over the long run, plays a key role in the consumer's perception of quality, lean

Six Sigma companies understand the importance of having a sound reliability program. Reliability testing enables companies to better comprehend how their products will perform under normal usage as well as extreme or unexpected situations. Reliability programs provide information about product performance by systematically studying the product. To ensure product quality, reliability tests subject a product to a variety of conditions besides expected operational parameters. These conditions may include excessive use, vibration, damp, heat, cold, humidity, dust, corrosive materials, and other environmental or user stresses. By providing information about product performance, a sound reliability program can have a significant effect on a company's financial statements. Early product testing can prevent poorly designed products from reaching the marketplace. Later testing can improve on products already in use.

RELIABILITY PROGRAMS

Lean Six Sigma organizations have reliability programs that endeavor to incorporate reliability concepts into product or service design. Reliability issues surface in nearly every facet of product or service design, development, creation, use, and service. Reliability programs include the areas of design, testing, manufacture, raw material and component purchases, production, packaging, shipping, marketing, field service, and maintenance. A sound reliability program developed and implemented to support an entire system will consider the following aspects:

1. **The entire system.** What composes the system? What goals does the system meet? What are the components of the system? How are they interrelated? How do the interrelationships influence system reliability? What are the system reliability requirements set by the customer?

   Critical to ensuring system reliability is a complete understanding of purpose of the product. The consumer's reliability needs and expectations must be a part of product reliability. Also important is an understanding of the life of a product. How is it shipped? How is it transported and stored? What type of environment will it face during usage? Comprehensive knowledge of a product and its life is the foundation of a strong reliability program.

2. **The humans in the system.** What are their limitations? What are their capabilities? What knowledge do they have of the product? How will this knowledge affect their use of the product? How might they misuse the product?

   Product reliability can be increased if greater emphasis is placed on proper training and education of the users of the product. Appropriate steps must also be taken to ensure that the sales department does not promise more than the product can deliver. Setting the expectations of the users is as important as showing them how to use a product.

3. **Maintenance of the system.** Can the components or subassemblies be maintained separately from the system? Are the system components accessible? Is the system designed for replacement components or subsystems? Are those components or subsystems available? Can those components or subsystems be misapplied or misused? Will maintenance be performed under difficult circumstances?

   If and when a component or a system fails, those using the system or component will be interested in the length of time it takes to return the failed system, subsystem, or component to full operational status. Changes to product, process, or system design based on maintainability considerations are sure to include an investigation of factors such as ease of repair or replacement of components, costs, and time. Well-designed policies, practices, and procedures for effective preventive maintenance before failures occur and timely corrective maintenance after a failure occurs increase overall system reliability.

4. **Simplicity of design.** Will a simple design be more effective than a complicated one? Will a reduction in the number of elements increase the system reliability? Will the addition of elements increase the system reliability?

   Simple, straightforward designs will increase system reliability. Designs of this sort are less likely to break down and are more easily manufactured. In some instances, the effort to impress with the current technological wizardry decreases the product's reliability. The greater the number of components, the greater the complexity and the easier it is for some aspect of the product, process, or system to fail.

5. **Redundant and fail-safe features.** Can the addition of redundant components or fail-safe features prevent overall system failure? At what cost?

   Incremental increases in reliability should be balanced against costs. Products can be overdesigned for their purposes. There is a diminishing return on investment for this approach to achieving a reliable product. Is it preferable to avoid specially designed parts and components for the sake of easy availability? Will interchangeability decrease maintenance complexity?

6. **Manufacturing methods and purchasing requirements.** Have the chosen manufacturing methods enhanced system quality and reliability? Have purchasing policies been designed to support quality and reliability?

   Purchasing and manufacturing must be made aware of how critical it is to purchase and use the materials deemed appropriate for the life cycle of this product.

7. **Maintenance of complete product or system performance records.** Can this information be used to increase future system reliability?

   System reliability information from tests or from actual experience with the product or system should be gathered in a manner that will be meaningful when used for decisions concerning product or system design changes, development of new products or systems, product or system manufacture, operation, maintenance, or support.

8. Communication. Have clear channels of communication been established among all those involved in the design, manufacture, shipment, maintenance, and use of the product or system?

Although perhaps not a complete list, the above areas for consideration form the basis of a well-structured reliability program. When the program provides answers to the questions, system reliability is enhanced.

Disasters That Could Have Been Averted

LEAN SIX SIGMA TOOLS at WORK

Sometimes it is difficult to comprehend that a complete reliability program improves overall system reliability. Consider some examples from real life events where critical aspects we discussed were ignored:

Aspect 1. The Entire System. In 1989, United Airlines Flight 232 crashed in Sioux City, Iowa, despite heroic efforts on the part of the flight crew. The DC-10's tail engine exploded in flight and destroyed the plane's hydraulic lines. Although the plane could fly on its two wing-mounted engines, the loss of hydraulic power rendered the plane's main control systems (flaps and gear) inoperable. A study of the plane's design revealed that, although backup hydraulic systems existed, all three hydraulic lines, including backups, went through one channel, which rested on top of the third engine in the tail. When the engine failed, debris from the engine severed all the hydraulic lines at the same time. In his March 28, 1997, *USA Today* article, "Safety Hearings End Today on Cracks in Jet Engines," Robert Davis stated, "The nation's airlines could be forced to spend more time and money using new techniques for inspecting engines." Those familiar with reliability issues know that the problems are more complicated than an airline's maintenance practices.

Aspect 1. The Entire System. During the summer of 1996, power outages plagued Western states, leaving offices, businesses, and homes without lights, air conditioning, elevators, computers, cash registers, faxes, electronic keys, ventilation, and a host of other services. Traffic chaos resulted when traffic signals failed to function. During one outage, 15 states were without power. This particular power failure occurred when circuit breakers at a transmission grid were tripped. The interrupted flow of power caused generators further down the line to overload. Automatic systems, activated in the event of an overload, cut power to millions of customers. Six weeks later, a second power grid failure resulted when sagging transmission lines triggered a domino effect similar to the previous outage. This outage affected 10 states and at least 5.6 million people for nearly 24 hours. When millions of people rely on highly complex and integrated systems such as power transmission and satellites, these systems need careful consideration from a reliability point of view.

Not much has changed since 1996. Power outages continue to be prevalent. One massive power outage on August 14, 2003, left 50 million people in the United States and Canada without power for days. In the summer of 2004, Athens, Greece, experienced a series of power outages that threatened to disrupt the Olympics. Studies of power outages reveal the shortcomings that contribute to these massive failures. Blackouts can be caused by unreliable or faulty equipment; voluntary reliability standards; overloaded transmission lines; untrained, unprepared, or overtaxed repair personal; and poor planning.

Aspect 2. The Humans in the System. Often we think of reliability as an equipment-based system; however, reliable systems are needed in other, more human-oriented systems. As reported by Jeffrey Zaslow in the February 25, 2004, *Wall Street Journal* article "It's 9 P.M., Do You Know Where You Parked Your Car?" guest services managers at all major parking lots have systems in place designed to reunite patrons with their vehicles. Although some guest services managers operate handheld computers running a license plate information location program that finds cars based on license plate information scanned from parked cars, most still rely on the guest's memory.

Aspect 3. Maintenance of the System. According to a 2004 Purdue University study of 1,300 aviation incidents and accidents, maintenance errors, such as incomplete or incorrect tasks, were the primary contributing factors in 14 percent of aviation incidents and 8 percent of accidents. These figures, compiled from 1984–2002 records, are for all types of planes, from propeller to jumbo jet. The leading cause: Failure to Follow Maintenance Procedures. Though weather, turbulence, and pilot error still contribute the largest share of accidents and incidents, these maintenance figures reveal that a problem exists.

Aspect 4. Simplicity of Design. The design of the rocket joints on the space shuttle Challenger was based on the highly successful Titan III rocket. The joints of the Titan contained a single O-ring. When adapting the design for the space shuttle, designers felt that adding a second O-ring would make the design even more reliable. Investigation of the design after the explosion revealed that the additional O-ring had contributed to the disaster. Designers have redesigned the rocket to include a *third* O-ring, but will it increase reliability?

Aspect 5. Redundant and Fail-Safe Features. On December 20, 1997, air traffic over the central United States was disrupted by a multiple level power failure. Air traffic control systems maintain three backup systems. The combined reliability of such systems is considered to be greater than 99.5 percent. On December 20, however, one system was shut down for scheduled maintenance, the commercial power failed, and a technician mistakenly pulled a circuit card from the remaining power system. With no other backups, the air traffic control information system, including computers, software, and displays, was inoperable for about five hours. The unexpected failure of the redundant and backup systems needs to be carefully considered as part of overall reliability.

PRODUCT LIFE CYCLE CURVE

The life cycle of a product is commonly broken down into three phases: early failure, chance failure, and wear-out (Figure 20.1). The early failure, or infant mortality, phase is characterized by failures occurring very quickly after the product has been produced or put into use by the consumer. The curve during this phase is exponential, with the number of failures decreasing the longer the product is in use. Failures at this stage have a variety of causes. Some early failures are due to inappropriate or inadequate materials, marginal components, incorrect installation, or poor manufacturing techniques. Incomplete testing may not have revealed design weaknesses that become apparent only as the consumer uses the system. Inadequate quality checks could have allowed substandard products to leave the manufacturing area. The manufacturing processes or tooling may not have been capable of producing to the specifications needed by the designer. The consumer also plays a role in product reliability. Once the product reaches the consumer, steps must be taken to ensure that the consumer understands the appropriate environment for product use. Improper usage will affect the reliability of a product.

During the chance failure portion of a product's useful life, failures occur randomly. This may be due to inadequate or insufficient design margins. Manufacturing or material problems have the potential to cause intermittent failures. At this stage the consumer can also affect product reliability. Misapplication or misuse of the product by the consumer

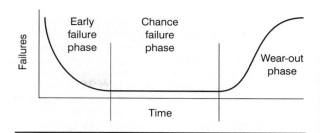

FIGURE 20.1 Life Cycle Curve

can lead to product failure. Overstressing the product is a common cause of random failures.

As the product ages, it approaches the final stage of its life cycle, the wear-out phase. During this phase, failures increase in number until few, if any, of the product are left. Wear-out failures are due to a variety of causes, some related to actual product function, some cosmetic. A system's reliability, useful operation, or desirability may decrease if it becomes scratched, dented, chipped, or otherwise damaged. Age and the associated wear, discolorations, and brittleness may lead to material failure. Normal wear could decrease reliability through misalignments, loose fittings, and interference between components. Combined stresses placed on the product (such as the zipper or latches on a suitcase) during its lifetime of use are a source of decreased reliability. Neglect or inadequate preventive maintenance lessens product reliability and shortens product life.

Aircraft Landing Gear Life Cycle Management

LEAN SIX SIGMA TOOLS at WORK

Ever since Charles E. Taylor, the father of aviation maintenance, began working with the Wright Brothers to repair parts and components on such early models as the Wright B Flyer, aircraft maintenance programs have existed. Airlines recognize that the effectiveness of their organization depends on aircraft operational readiness, the overall reliability of their planes. The wisdom of participating in a preventive maintenance program is often justified by tracking reliability measures such as mean time to failure, mean time to repair, and availability rates.

Knowing that the reliability of landing gear components can materially affect aircraft operational readiness, leaders at CLP Corporation reviewed their two approaches to maintaining the reliability of their two separate fleets of aircraft. The objective of their review was to determine whether life cycle management of landing gear increased aircraft reliability and cost effectiveness.

To better manage maintenance, repair, and replacement costs, aircraft owners study their fleets, monitoring the reliability of key components. Aircraft landing gear reliability is often judged based on the number of takeoff-and-landing cycles completed. Planes, as well as their individual components, follow the life cycle curve presented in Figure 20.1. Fortunately, robust designs of key components and routine maintenance and repair often keep planes flying long after their predicted number of takeoff-and-landing cycles or flights has passed. In this example, the expected life of landing gear for both larger and smaller jets is 8,000 takeoff-and-landing cycles.

CLP's two jet fleets are managed as separate business entities. The smaller jets make short hops around the country, whereas the larger jets, with their greater fuel capacity, tend to be used for cross-country or cross-continent, longer duration flights. A study of the maintenance records revealed that CLP's smaller, 150-passenger jets experienced a greater number of landing gear reliability issues than the larger, 250-passenger jets. In short, the smaller jets are not as reliable as the larger jets.

At first glance, it may appear that the greater number of landing gear mishaps may be due to the smaller jets' greater number of landings and takeoffs. Further investigation uncovered that, as they aged, the smaller jets experienced lower landing gear reliability than the larger jets. Landing gear related mishaps accounted for 9 percent of total aircraft mishaps from 1995 through 2004. Of 13 landing gear related mishaps, 11 occurred on smaller jets and two occurred on larger jets. True to the life cycle curve, as the jets have aged, failures have increased.

Although mishaps cannot always be prevented through maintenance, aircraft that had a program of landing gear maintenance experienced reduced mishap rates and had greater reliability. A fully defined life cycle management process for the larger jets' landing gear from the time of acquisition through production and deployment has been in place for several years. Because the smaller jets did not have this process in place, they were less reliable, needing maintenance, repair, and replacement at a greater rate and at higher costs than the larger jets.

The smaller jets have historically been maintained on an as-needed, on-location program rather than returning to a single

(Continued)

maintenance facility. The larger jets, with their maintenance process, have all but emergency work taken care of at a single maintenance facility. One significant benefit of having repairs performed at a single location is the increase in worker knowledge of the aircraft of and its components. Another critical factor in reducing mean time to repair is that one location provides easy access to replacement parts. As the smaller jets aged, flight line maintenance becomes less effective, reliability levels decrease, and maintenance, repair, and replacement costs increase.

The larger jet management team maintained thorough records related to maintenance, repair, and replacement costs both before and after implementing the preventive maintenance program for the larger jets. In Table 20.1 these differences have been quantified using reliability concepts. The mean time between failure (MTBF) rates were calculated based on the number of failures observed by the maintenance staff and the number of takeoff-and-landing cycles the jets had completed. Aircraft availability calculations include the mean time to repair, that is, the average time it takes to return the jet to service after repairing a landing gear system component. The maintenance costs are the average total costs to repair or replace the failed landing gear.

The reliability calculations related to MTBF and availability show the significant difference between the costs associated with planned maintenance and unplanned maintenance, as well as the difference between availabilities of aircraft. Based on operational maintenance records, CLP Corporation recognized that reduced reliability related to landing gear, coupled with aging aircraft concerns, necessitated a change in its maintenance processes for its smaller jets. By comparison, the larger jets, with their maintenance process approach, have highly maintainable landing gear with an extended life and increased aircraft reliability. The landing gear maintenance process is only one of their many preventive maintenance programs. Overall, the larger jets' preventive maintenance program is projected to save the corporation over $100 million in a 10-year period.

Having a preventive maintenance process increases product reliability resulting in the following:

Increased effectiveness (availability for deployment)

Lower risk of failure

Increased component life

Reduced remote location maintenance workload

Greater maintenance knowledge at single facility

Increased ability to project life cycle costs associated with maintenance and component replacement

Lower risk of grounding an aircraft or an entire fleet

Fewer incidents needing crisis management

Increased morale

TABLE 20.1 Cost Analysis for Planned Maintenance Using Mean Time Between Failures

	MTBF with Planned Maintenance	Cost with Planned Maintenance	Aircraft Availability	MTBF w/o Planned Maintenance	Cost w/o Planned Maintenance	Aircraft Availability
Left Main Landing Gear*	8,300	$380,140	99%	3,917	$506,680	78%
Right Main Landing Gear*	8,186	$421,127	99%	3,043	$568,438	74%
Nose Landing Gear*	7,400	$407,392	98%	5,138	$473,413	68%

*Expected life: 8,000 takeoff-and-landing cycles.

MEASURES OF RELIABILITY

Overall system reliability depends on the individual reliabilities associated with the parts, components, and subassemblies. Reliability tests exist to aid in determining whether distinct patterns of failure exist during the product's or system's life cycle. Reliability tests determine what failed, how it failed, and the number of hours, cycles, actuations, or stresses it was able to bear before failure. Once these data are known, lean Six Sigma practitioners can make decisions concerning product reliability expectations, corrective action steps, maintenance procedures, and costs of repair or replacement. Several different types of tests exist to judge the reliability of a product, including failure-terminated, time-terminated, and sequential tests. The name of each of these tests says a good deal about the type of the test. *Failure-terminated tests are ended when a predetermined number of failures occur within the sample being tested.* The decision concerning whether the product is acceptable is based on the number of products that have failed during the test. *A time-terminated test is concluded when an established number of hours is reached.* For this test, product is accepted on the basis of how many products failed before reaching the time limit. *A sequential test relies on the accumulated results of the tests.*

Failure Rate, Mean Life, and Availability

When system performance is time dependent, such as the length of time a system is expected to operate, then reliability is measured in terms of mean life, failure rates, availability, mean time between failures, and specific mission reliability. As a system is used, data concerning failures become available. This information can be utilized to estimate the mean life and failure rate of the system. Failure rate, λ, the probability of a failure during a stated period of time, cycle, or number of impacts, can be calculated as follows:

$$\lambda_{estimated} = \frac{\text{number of failures observed}}{\text{sum of test multi or cycles}}$$

Calculating Failure Rate and Average Life

LEAN SIX SIGMA TOOLS at WORK

Twenty windshield wiper motors are being tested using a time-terminated test. The test is concluded when a total of 200 hours of continuous operation have been completed. During this test, the number of windshield wipers that fail before reaching the time limit of 200 hours is counted. If three wipers failed after 125, 152, and 189 hours, calculate the failure rate λ and the average life θ:

$$\lambda_{estimated} = \frac{\text{number of failures observed}}{\text{sum of test times of cycles}}$$
$$= \frac{3}{125 + 152 + 189 + (17)200} = 0.0008$$

From this, θ, the average life, can be calculated:

$$\theta_{estimated} = \frac{1}{\lambda} = \frac{1}{0.0008} = 1250 \text{ hours}$$

or

$$\theta_{estimated} = \frac{\text{sum of test times or cycles}}{\text{number of failures observed}}$$
$$= \frac{125 + 152 + 189 + (17)200}{3} = 1289 \text{ hours}$$

The average life of the windshield wiper motor is 1,289 hours.

Here, the difference between $\theta_{estimated}$ based on $1/\lambda$ and $\theta_{estimated}$ calculated directly is due to rounding.

From this, θ, the average life, can be calculated:

$$\theta_{estimated} = \frac{1}{\lambda}$$

or

$$\theta_{estimated} = \frac{\text{sum of test times or cycles}}{\text{number of failures observed}}$$

The average life θ is also known as the mean time between failures or the mean time to failure. Mean time between failures (MTBF), how much time has elapsed between failures, is used when speaking of repairable systems. Mean time to failure is used for nonrepairable systems.

Mean time between failures (MTBF) and mean time to failure (MTTF) describe reliability as a function of time. Here the amount of time that the system is actually operating is of great concern. For example, without their radar screen, air traffic controllers are sightless and therefore out of operation. To be considered reliable, the radar must be functional for a significant amount of the expected operating time. Because many systems need preventive or corrective maintenance, a system's reliability can be judged in terms of the amount of time it is available for use:

$$\text{Availability} = \frac{\text{mean time to failure (MTTF)}}{\text{MTTF} + \text{mean time to repair}}$$

MTBF values can be used in place of MTTF.

Calculating System Reliability

When system performance is dependent on the number of cycles completed successfully, such as the number of times a coin-operated washing machine accepts the coins and begins operation, then reliability is measured in terms of the probability of successful operation. *Reliability is the probability that a product will not fail during a particular time period.* Like probability, reliability takes on numerical values between 0.0 and 1.0. A reliability value of 0.78 is interpreted as 78 out of 100 parts will function as expected during a particular time period and 22 will not. If n is the total number of units being tested and s represents those units performing satisfactorily, then reliability R is given by the following formula:

$$R = \frac{s}{n}$$

System reliability is determined by considering the reliability of the components and parts of the system.

Determining Availability

LEAN SIX SIGMA TOOLS at WORK

Windshield wiper motors are readily available and easy to install. Calculate the availability of the windshield wipers on a bus driven eight hours a day, if the mean time between failures or average life θ is 1,250 hours. When the windshield wiper motor must be replaced, the bus is out of service for a total of 24 hours.

$$\text{Availability} = \frac{\text{mean time between failure (MTBF)}}{\text{MTBF} + \text{mean time to repair}}$$
$$= \frac{1250}{1250 + 24} = 0.98$$

The bus is available 98 percent of the time.

Reliability in Series A *system in series* exists if proper system functioning depends on whether all the components in the system are functioning. Failure of any one component will cause system failure. Figure 20.2 portrays an example of a system in series. In the diagram, the system will function only if all four components are functioning. If one component fails, the entire system will cease functioning. The reliability of a system in series is dependent on the individual component reliabilities. To calculate series system reliability, the individual independent component reliabilities are multiplied together:

$$R_s = r_1 \cdot r_2 \cdot r_3 \cdot \cdots \cdot r_n$$

where

R_s = reliability of series system
r_i = reliability of component
n = number of components in system

The reliability of a system decreases as more components are added in series. This means that the series system

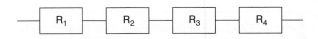

FIGURE 20.2 System in Series

reliability will never be greater than the reliability of the least reliable component. Sustained performance over the life of a product or process can be enhanced in several ways. Parallel systems and redundant or standby component configurations can be used to increase the overall system reliability.

Reliability in Parallel A *parallel system* is a system that is able to function if at least one of its components is functioning. Figure 20.4 displays a system in parallel. The system will function provided at least one component has not failed. In a parallel system, all of the components that are in parallel with each other must fail in order to have a system failure. This is the opposite of a series system in which, if one component

---

LEAN SIX SIGMA TOOLS at WORK

Calculating Reliability in a Series System

The reliability values for the flashlight components pictured in Figure 20.3 are listed below. If the components work in series, what is the reliability of the system?

Battery	0.75	Screw cap	0.97
Lightbulb	0.85	Body	0.99
Switch	0.98	Lens	0.99
Spring	0.99		

$$R_s = r_1 \cdot r_2 \cdot r_3 \cdot \cdots \cdot r_n$$
$$= 0.75 \cdot 0.98 \cdot 0.85 \cdot 0.99 \cdot 0.97 \cdot 0.99 \cdot 0.99$$
$$= 0.59$$

This flashlight has a reliability of only 0.59. Perhaps the owner of this flashlight ought to keep two on hand!

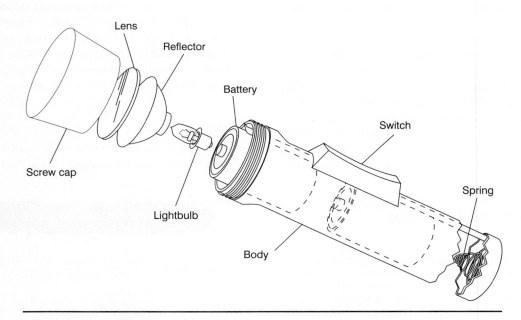

FIGURE 20.3 Flashlight

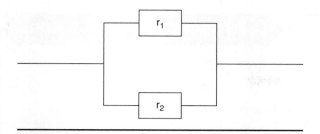

FIGURE 20.4 Parallel System

fails, the entire system fails. Because the duplicated or paralleled component takes over functioning for the failed part, the reliability of this type of system is calculated on the basis of the sum of the probabilities of the favorable outcomes: the probability that no components fail and the combinations of the successful operation of one component but not the other(s). The reliability of a parallel system is given by

$$R_p = 1 - (1 - r_1)(1 - r_2)(1 - r_3) \cdots (1 - r_n)$$

where

R_p = reliability of parallel system
r_i = reliability of component
n = number of components in system

System reliability increases as components are added to a parallel system. Reliability in a parallel system will be no less than the reliability of the most reliable component. Parallel redundancy is often used to increase the reliability of a critical system; however, at some point there is a diminishing rate of return where the added costs outweigh the increased reliability. Parallel components are just one type of redundancy used to increase system reliability.

Reliability in Redundant Systems and Backup Components Backup or spare components, used only if a primary component fails, increase overall system reliability. The likelihood of needing to access the spare or backup component in relation to the reliability of the primary component is shown mathematically as follows:

$$R_b = r_1 + r_b(1 - r_1)$$

where

R_b = reliability of backup system
r_1 = reliability of primary component
r_b = reliability of backup component
r_1 = chance of having to use backup

Reliability in Systems Designers often use the advantages of parallel and redundant systems on critical components of an overall system. As the Lean Six Sigma Tools at Work feature on p. 244 shows, calculating the reliability of such a system involves breaking down the overall system into groups of series, parallel, and redundant components.

Determining Reliability in a Parallel System

LEAN SIX SIGMA TOOLS at WORK

On a twin-engine aircraft, two alternators support a single electric system. If one were to fail, the other would allow the electrical system to continue to function. As shown in Figure 20.5, these alternators each have a reliability of 0.9200. Calculate their parallel reliability:

$$R_p = 1 - (1 - r_1)(1 - r_2)$$
$$= 1 - (1 - 0.9200)(1 - 0.9200) = 0.9936$$

Even though the alternators' reliability values individually are 0.9200, when combined they have a system reliability of 0.9936.

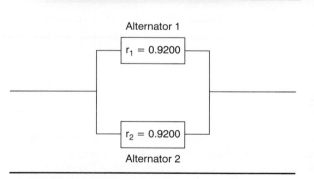

FIGURE 20.5 Systems in Parallel: Alternators

Calculating Reliability in a Redundant System

LEAN SIX SIGMA TOOLS at WORK

A local hospital uses a generator to provide backup power in case of a complete electrical power failure. This generator is used to fuel the equipment in key areas of the hospital such as the operating rooms and intensive care units. In the past few years, the power supply system has been very reliable (0.9800). There has been only one incident where the hospital lost power. The generator is tested frequently to ensure that it is capable of operating at a moment's notice. It too is very reliable (0.9600). Calculate the reliability of this system:

$$R_b = r_1 + r_b(1 - r_1)$$
$$= 0.9800 + 0.9600(1 - 0.9800) = 0.9992$$

The overall reliability of this redundant system is 0.9992.

Calculating the Reliability of a Combination System

LEAN SIX SIGMA TOOLS at WORK

In an endeavor to discourage thievery, a local firm has installed the alarm system shown in Figure 20.6. This system contains series, parallel, and backup components. Calculate the reliability of the system.

To calculate the overall system reliability, begin by determining the overall reliability of the components in parallel. The combined reliability of the five sensors is as follows:

$$R_p = 1 - (1 - r_1)(1 - r_2)(1 - r_3) \cdots (1 - r_n)$$
$$= 1 - (1 - 0.99)(1 - 0.99)(1 - 0.99)(1 - 0.99)(1 - 0.99)$$
$$= 1.0$$

The combined reliability of the power source and battery is as follows:

$$R_b = r_1 + r_b(1 - r_1)$$
$$= 0.92 + 0.88(1 - 0.92)$$
$$= 0.99$$

The overall reliability of the system is as follows:

$$R_s = r_p \cdot r_{cpu} \cdot r_{keypad} \cdot r_{siren} \cdot r_b$$
$$= 1.0 \cdot 0.99 \cdot 0.90 \cdot 0.95 \cdot 0.99$$
$$= 0.84$$

The overall reliability of the system is 0.84.

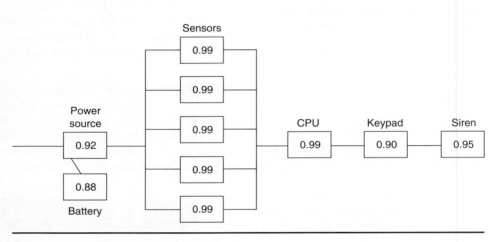

FIGURE 20.6 Reliability Diagram: Alarm System

SUMMARY

Lean Six Sigma organizations study product or system reliability to predict its useful life. Reliability tests are conducted to determine whether the predicted reliability has been achieved. Reliability programs are designed to improve product or system reliability through improved product design, manufacturing processes, maintenance, and servicing. Reliability engineers work to incorporate the aspects of reliability into products and services from their conception to useful life.

TAKE AWAY TIPS

1. Reliability refers to quality over the long term. The system's intended function, expected life, and environmental conditions all play a role in determining system reliability.

2. The three phases of a product's life cycle are early failure, chance failure, and wear-out.

3. Reliability tests aid in determining whether distinct patterns of failure exist.

4. Failure rates can be determined by dividing the number of failures observed by the sum of their test times.

5. θ, or the average life, is the inverse of the failure rate.

6. A system's availability can be calculated by determining the mean time to failure and dividing that value by the total of the mean time to failure plus the mean time to repair.

7. Reliability is the probability that failure will not occur during a particular time period.

8. For a system in series, failure of any one component will cause system failure. The reliability of a series system will never be greater than that of its least reliable component.

9. For a system in parallel, all of the components in parallel must fail to have system failure. The reliability in a

parallel system will be no less than the reliability of the most reliable component.
10. Overall system reliability can be increased through the use of parallel, backup, or redundant components.
11. Reliability programs are enacted to incorporate reliability concepts into system designs. A well thought out reliability program will include the eight considerations presented in this chapter.

FORMULAS

$$\lambda_{estimated} = \frac{\text{number of failures observed}}{\text{sum of test times or cycles}}$$

$$\theta_{estimated} = \frac{1}{\lambda}$$

or

$$\theta_{estimated} = \frac{\text{sum of test times or cycles}}{\text{number of failures observed}}$$

$$\text{Availability} = \frac{\text{mean time to failure (MTTF)}}{\text{MTTF + mean time to repair}}$$

(MTBF values can be used in place of MTTF.)

$$R = \frac{s}{n}$$
$$R_s = r_1 \cdot r_2 \cdot r_3 \cdot \cdots \cdot r_n$$
$$R_p = 1 - (1 - r_1)(1 - r_2)(1 - r_3) \cdots (1 - r_n)$$
$$R_b = r_1 + r_b(1 - r_1)$$

CHAPTER QUESTIONS

1. Define reliability in your own words. Describe the key elements and why they are important.
2. Describe the three phases of the life cycle curve. Draw the curve and label it in detail (the axes, phases, type of product failure, etc.).
3. Determine the failure rate λ for the following: You have tested circuit boards for failures during a 500-hour continuous use test. Four of the 25 boards failed. The first board failed in 80 hours, the second failed in 150 hours, the third failed in 350 hours, the fourth in 465 hours. The other boards completed the 500-hour test satisfactorily. What is the mean life of the product?
4. Determine the failure rate for a 90-hour test of 12 items where 2 items fail at 45 and 72 hours, respectively. What is the mean life of the product?
5. A power station has installed 10 new generators to provide electricity for a local metropolitan area. In the past year (8,760 hours), two of those generators have failed, one at 2,460 hours and one at 5,962 hours. It took five days, working 24 hours a day, to repair *each* generator. Using one year as the test period, what is the mean time between failures for these generators? Given the repair information, what is the availability of all 10 generators?
6. Why is a parallel system more reliable than a system in series?
7. Given the system below, what is the system reliability?

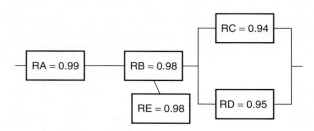

8. Given the diagram of the system below, what is the system reliability?

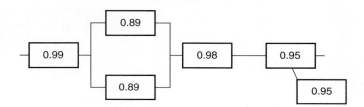

9. What is the reliability of the system below?

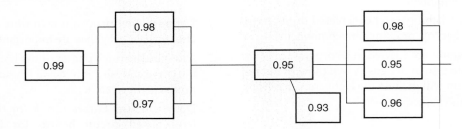

10. Airbags are concealed in the center of the steering wheel and in dashboards. When the front end of the vehicle strikes another object, the airbag inflates. The airbag will fully inflate upon impact in one-tenth of a second, providing a barrier to protect the driver. More advanced airbag systems have smart sensors that sense key elements related to the accident and determine the amount of force the detonator needs to emit. The diagram below shows the key elements that must operate in order for the airbag to inflate. Calculate the reliability of the system.

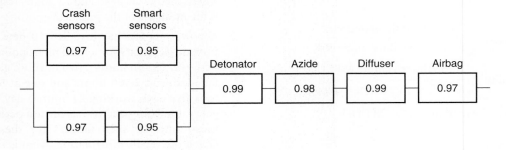

11. Often we think of reliability as an equipment-based system; however, reliable systems are needed in other, more human-oriented systems. As reported by Jeffrey Zaslow in the February 25, 2004, *Wall Street Journal* article "It's 9 P.M., Do You Know Where You Parked Your Car?" guest services managers at all major parking lots have systems in place designed to reunite patrons with their vehicles. The reliability of these systems is dependent on the reliability of the components, the location finding aids. In many major parking lots, two methods are used at once: the guest's memory as well as a license plate information location computer program. First, though, guest services needs to recognize that a guest is lost. Find the reliability of the car locating system shown below.

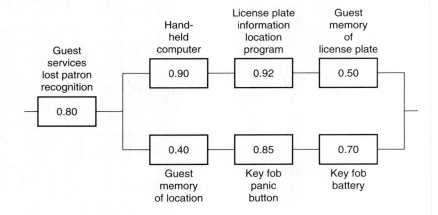

ONE

DESIGN OF EXPERIMENTS

Analyzing historical data is a reactive approach to process improvement. To proactively study processes, lean Six Sigma practitioners use design of experiments to determine which process changes will work best.

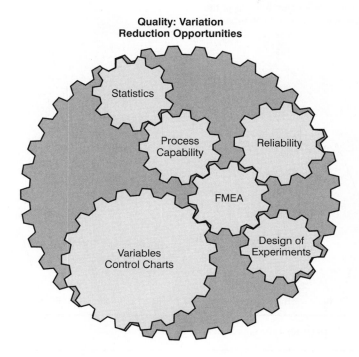

 LEARNING OPPORTUNITIES

1. To introduce the concept of design of experiments
2. To introduce the different styles of experiments
3. To introduce the terminology associated with design of experiments
4. To introduce the structure of a designed experiment

DESIGN OF EXPERIMENTS

Lean Six Sigma organizations seek to optimize their processes, products, and services. A technique called design of experiments enables users to study the changes they plan to make to a process or product and determine whether these changes result in optimization. ***Design of experiments (DOE) is a method of experimenting with the complex interactions among** parameters in a process or product with the objective of optimizing the process or product.* To design an experiment means creating a situation in which an organized investigation into all the different factors that can affect process or product parameters occurs. The design of the experiment provides a layout of the different factors and the values at which those factors are to be tested.

A complete study of designing experiments is beyond the scope of this text. The coverage of experiment design in this text is designed to provide the reader with an introduction to the terminology, concepts, and setups associated with experiment design. Readers interested in an in-depth study of experiment design should seek one of the many excellent texts in the area.

TRIAL AND ERROR EXPERIMENTS

A trial and error experiment involves making an educated guess about what should be done to effect change in a process or system. Trial and error experiments lack direction and focus.

247

A Trial and Error Experiment

LEAN SIX SIGMA TOOLS at WORK

Researchers are studying the effects of vehicle speed, tire pressure, oil type, and gas type on gas mileage.

Speed	55 mph, 65 mph
Tire pressure	28 psi, 35 psi
Oil	30 weight, 40 weight
Gas	Regular, Premium

As seen in Figure 21.1, many different permutations of the variables exist. As more variables are factored into an experiment, the complexity increases dramatically. If an experimenter were to randomly select speed, tire pressure, oil, and gas settings, he or she may have a difficult time determining which settings provide the best gas mileage. Because random selection is not an organized approach, it is not the best use of time and materials. Essentially, it is guesswork. Design of experiments is a method to arrive at the optimum settings more effectively.

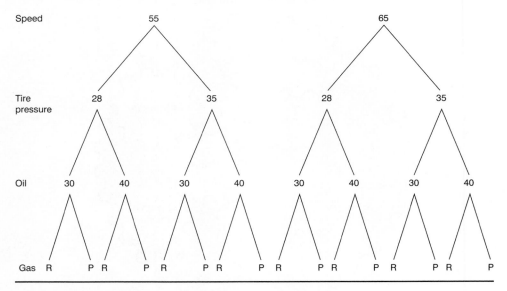

FIGURE 21.1 Experiment on Gas Mileage

They are hindered by the effectiveness of the guesswork of those designing the experiments. In other words, a good solution to the problem may be found by using this method, but in all likelihood this hit-or-miss approach will yield nothing useful.

A properly designed experiment seeks to determine the following:

- Which factors significantly affect the system under study
- How the magnitude of the factors affect the system
- The optimal level for each of the selected factors
- How to manipulate the factors to control the response

The experiment design must determine the following:

- The number of factors to include in the experiment
- The levels at which each factor will be tested
- The response variable
- The number of trials to be conducted at each level for each factor
- The conditions (settings) for each trial

DEFINITIONS

To better explain the concepts, an understanding of the vocabulary is very helpful. The following terms are commonly used in the design of experiments.

Factor: A factor is the variable the experimenter will vary in order to determine its effect on a response variable. A factor is the variable that is set at different levels during an experiment, and results of those changes are observed. It may be time, temperature, an operator, or any other aspect of the system that can be controlled.

Level: A level is the value chosen for the experiment and assigned to change the factor. For instance: Temperature; Level 1: 110° F; Level 2: 150° F.

Controllable factor: When a factor is controllable, it is possible to establish and maintain the particular level throughout the experiment.

Effect: The effect is the result or outcome of the experiment. It is the value of the change in the response variable produced by a change in the factor level(s).

The effect is the change exhibited by the response variable when the factor level is changed.

Response variable: The variable(s) of interest used to describe the reaction of a process to variations in control variables (factors). It is the quality characteristic under study, the variable we want to have an effect on.

Degrees of freedom: At its simplest level, the degrees of freedom in an experiment can be determined by examining the number of levels. For a factor with three levels, L_1 data can be compared with L_2 and L_3 data, but not with itself. Thus a factor with three levels has two degrees of freedom. Extending this to an experiment, the degrees of freedom can be calculated by multiplying the number of treatments by the number of repetitions of each trial and subtracting one ($f = n \times r - 1$).

Interaction: Two or more factors that together produce a result different than what the result of their separate effects would be. Well-designed experiments allow two or more factor interactions to be tested, where other trial and error models only allow the factors to be independently evaluated.

Noise factor: A noise factor is an uncontrollable, but measurable, source of variation in the functional characteristics of a product or process. This error term is used to evaluate the significance of changes in the factor levels.

Treatment: The specific combination of levels for each factor used for a particular run. The number of treatments is based on the number of factors and the levels associated with each factor. For example, if two factors (1, 2) exist and each can be at two levels (A, B), then four treatments are possible. The treatments are factor 1 at level A, factor 1 at level B, factor 2 at level A, factor 2 at level B. A treatment table will show all the levels of each factor for each run.

Run: A run is an experimental trial, the application of one treatment.

Replicate: When an experiment is replicated, it refers to a repeat of the treatment condition. The treatment is begun again from scratch.

Repetition: Repetitions are multiple runs of a particular treatment condition. Repeated measurements are taken from the same setup.

Significance: Significance is a statistical test used to indicate whether a factor or factor combination caused a significant change in the response variable. It shows the importance of a change in a factor in either a statistical sense or in a practical sense.

An Experiment Defined

LEAN SIX SIGMA TOOLS at WORK

A new product development team at JF Inc. has been investigating the parameters surrounding a new part (Figure 21.2). The customer has specified that the part meet very tight tolerances for pierce height. A variety of factors are involved in holding this tolerance. The team would like to determine which of three proposed material suppliers to use. Though all three meet JF's specifications for material yield and tensile strength and for material thickness, there are minute differences in material properties. Because of the interactions with the material, the experiment will also need to determine which machine setting is optimal. For this 35-ton press, the experimenters would like to determine which setting—20, 25, or 30 tons—is appropriate. Several members of the team have worked together to design an appropriate experiment to test their assumptions. To better describe the experiment to other members of their group, they have created the following explanation based on DOE definitions.

Factor: Because factors are variables that are changed during an experiment, two factors in this experiment exist: material supplier and press tonnage. Team members believe that changes to these two factors will affect the response variable, pierce height.

Level: A level is the value assigned to change the factor. In this case, each factor has three levels: supplier (A, B, C) and press tonnage (20, 25, 30).

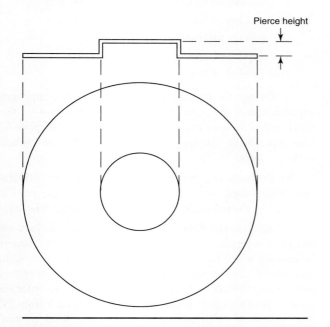

FIGURE 21.2 JF Part

Supplier	Press Tonnage
A	20
B	25
C	30

(Continued)

Controllable factor: When a factor is controllable, it is possible to establish and maintain the particular level throughout the experiment. In this experiment, investigators are able to set the levels for both the material supplier and press tonnage factors.

Effect: The effect is the result or outcome of the experiment. Here the experimenters are interested in consistent pierce height, which is measured as shown in Figure 21.2.

Response variable: The variable used to describe the reaction of a process to variations in the factors. In this example, the response variable is the pierce height.

Degrees of freedom: The degree of freedom in an experiment is the total number of levels for all factors minus 1. Here we have three levels for each factor ($3 \times 3 = 9$), so our degrees of freedom is $9 - 1$ or 8.

Interaction: Because the two factors, supplier and press tonnage, will together produce a result different than what would result if only one or the other were changed, there is an interaction between the two that must be considered in the experiment.

Noise factor: A noise factor is an uncontrollable but measurable source of variation. For this experiment, noise factors may be material thickness, material properties, and press operating temperatures.

Treatment: Because the number of treatments is based on the number of factors and the levels associated with each factor, and both factors have three levels, supplier (A, B, C) and press tonnage (20, 25, 30), the result is $3 \times 3 = 9$ treatments:

Supplier	Press Tonnage
A	20
A	25
A	30
B	20
B	25
B	30
C	20
C	25
C	30

Run: A run is an experimental trial, the application of one treatment to one experimental unit. The supplier A, press tonnage 25 combination is one run.

Replicate: When an experiment is replicated, it refers to a repeat of the treatment condition, a complete redo of the design.

Repetition: Repetitions are multiple results of a treatment condition; for example, the experimenters ran the supplier B, press tonnage 20 combination multiple times.

Significance: The determination of whether the changes to the factor levels had a significant effect on the response variable. In other words, did changing suppliers or press tonnage significantly affect pierce height?

CONDUCTING AN EXPERIMENT: STEPS IN PLANNED EXPERIMENTATION

First and foremost, *plan your experiment!* Successful experiments depend on how well they are planned. Before creating the design of an experiment, those involved must gain a thorough understanding of the process being studied. Achieving a deep understanding of the process will allow those designing the experiment to identify the factors in the process or product that influence the outcome. While planning an experiment, answer the following questions:

What are you investigating?
What is the objective of your experiment?
What are you hoping to learn more about?
What are the critical factors?
Which of the factors can be controlled?
What resources will be used?

The following are typical steps in conducting an experiment:

1. Establish the purpose by defining the problem. Before beginning an experiment design, you should determine the purpose of the experiment. Knowing the objectives and goals of an experiment can help you select the most appropriate experiment design as well as analysis methods.

2. Identify the components of the experiment. When setting up your experiment, be sure to include the following critical information:

The number of factors the design will consider
The number of levels (options) for each factor
The settings for each level
The response factor
The number of trials to be conducted

3. Design the experiment. Determine the appropriate structure of the experiment. Having determined the factors involved in your experiment, select a study template for your experiment. The types of templates will be covered later in this chapter.

4. Perform the experiment. Run your experiment and collect data about the results. Complete the runs as specified by the template at the levels and settings selected.

5. Analyze the data. Perform statistical analyses and analyze the resulting response variables in the experiment. Determine which factors were significant in determining the outcome of the response variables. At this point, rather than performing the calculations by hand, it is easier to enter the results into an analysis program, such as DOEpack from PQ Systems. Such software makes it easier to analyze your experiment. Using statistical tools to analyze your data will enable you to determine the optimal levels for each factor. Statistical analysis methods include analysis of variance,

Process Improvements Based on Design of Experiments

LEAN SIX SIGMA TOOLS at WORK

A process improvement team at JF Inc. studied an assembly process that welded two stamped steel parts together as the foundation of a component. The welds often failed in assembly. The goal of the team was to make the process more robust by eliminating weld failures. Team members consisted of representatives from manufacturing, engineering, setup, production, and quality.

The team turned to design of experiments when it became evident that studying the process using other quality tools and techniques hadn't provided the needed breakthrough to improve the process. An experiment based on the welds was designed and conducted. Careful analysis of the results of the runs is critical to gaining knowledge of the process. As the team members analyzed the results of the experiment, they realized that they were not capturing all of the main process variables. Something outside of the original design created a strong signal. Some unidentified factor that they didn't include in the original experiment was affecting the process.

This situation can sometimes occur with experimental design. Those closest to the process think they know the process inside and out and thus run the risk of overlooking something. Armed with this information, they conducted a second "screening" experiment. The structure of a screening experiment helps identify the true key variables in the process under study. The levels of these key variables can then be optimized in a follow-up experiment. Because they screen out key variables, screening experiments, which fully load the experiment design with variables, are usually the first type of experiment to be run. Is this case, the investigators would have been wiser to begin with such an experiment.

Analysis of the screening experiment enabled them to determine that some parts were cleaner than others. Here, cleaner means that some parts had less lubricant from the stamping process than others. This lubricant on the parts is what generated the noise, the unidentified signal, in the first experiment.

A third experiment was run in which the lubricant from the stamping operation was included as one of the variables. From this experiment, they were able to determine that the lubricant interfered with the ability of the weld to take hold. Excessive lubricant weakened the weld, resulting in the weld failures seen in assembly.

As a result of these experiments, the team decided to eliminate the lubricant variable altogether by changing to a lower viscosity, water-based lubricant. The end result is a welding process that is very consistent, eliminating the weld failures in assembly.

---

analysis of means, regression analysis, pairwise comparison, response plots, and effects plots.

6. Act on the results. Once you have analyzed the information, apply the knowledge you gained from your experiment to the situation under study. Use the information determined about the significant factors to make changes to the process or product.

EXPERIMENT DESIGNS

Characteristics of a Good Experiment Design

There are many characteristics associated with good experiment design. An experiment should be as simple as possible to set up and carry out. It should also be straightforward to analyze and interpret, as well as easy to communicate and explain to others. A good experiment design will include all the factors for which changes are possible. A well-designed experiment should provide unbiased estimates of process variables and treatment effects (factors at different levels). This means that a well-designed experiment will quickly screen out the factors that do not have a pronounced effect on the response variable while also identifying the best levels for the factors that do have a pronounced effect on the response variable. The experiment should plan for the analysis of the results, generating results that are free from ambiguity of interpretation. When analyzed, the experiment should provide the precision necessary to enable the experimenter to detect important differences between significant and insignificant variables. The analysis of the experiment must produce understandable results in a form that can be easily communicated to others interested in the information. The analysis should reliably be able to detect any signals that are present in the data. The experiment should point the experimenter in the direction of improvement.

Single Factor Experiments

A single factor experiment allows for the manipulation of only one factor during an experiment. Designers of these experiments select one factor and vary it while holding the other factors constant. The experiment is run for each variation of each factor, and the results are recorded. The objective in a single factor experiment is to isolate the changes in the response variable as they relate to a single factor. Single factor experiments are simple to analyze because only one thing changes at a time and the experimenter can see what effect that change has on the system or process. Unfortunately, single factor experiments are time consuming due to the need to change only one thing at a time, which results in dozens of repeated experiments. Another drawback of these types of experiments is that interactions between factors are not detectable. These experiments rarely arrive at an optimum setup because a change in one factor frequently requires adjustments to one or more of the other factors in order to achieve the best results. In life, single factor changes rarely occur that are not interrelated to other factors.

> ### Single Factor Experiment Treatment Table
>
> Researchers are studying the effects of tire pressure, vehicle speed, oil type, and gas type on gas mileage. Problem: What combination of factors provides the best gas mileage?
>
> | Factor | Level 1 | Level 2 |
> |---|---|---|
> | Tire pressure | 28 psi | 35 psi |
> | Speed | 55 mph | 65 mph |
> | Oil | 30 weight | 40 weight |
> | Gas | Regular | Premium |
>
> Response variable: Gas mileage
>
> | Tire | Speed | Oil | Gas |
> |---|---|---|---|
> | 28 | 55 | 30 | R |
> | 35 | 55 | 30 | R |
> | 28 | 65 | 30 | R |
> | 28 | 55 | 30 | R |
> | 28 | 55 | 40 | R |
> | 28 | 55 | 30 | R |
> | 28 | 55 | 30 | P |
>
> In each of these treatments, only one factor is changing at a time. The others are reverting to their original settings (Level 1 or Level 2). Either the remaining three will all be at Level 1 or all at Level 2. Note the complexity this creates when trying to study all the different changes in levels. Realize also that this type of experimentation does not allow experimenters to study the interactions that occur when more than one factor is changed at a time within a treatment. To study all the interactions, a full factorial experiment must be conducted.

Full Factorial Experiments

A *full factorial design* consists of all possible combinations of all selected levels of the factors to be investigated. This type of experiment examines every possible combination of all factors at all levels. It is possible to determine the effect that all the factors will have on the output variable as well as what effect their interactions will have. To determine the number of possible combinations or runs, multiply the number of levels for each factor by the number of factors. For example, in an experiment involving six factors at two levels, the total number of combinations will be 2 × 2 × 2 × 2 × 2 × 2, also represented in exponential notation as 2^6 or 64. If the experiment had four factors, two with two levels and two with three levels, a full factorial will have: 2 levels × 2 levels × 3 levels × 3 levels or 36 treatments. A full factorial design allows the most complete analysis because it can determine the following:

Main effects of the factors manipulated on response variables

Effects of factor interactions on response variables

A full factorial design can estimate levels at which to set factors for the best results. This type of experiment design is used when adequate time and resources exist to complete all of the runs necessary. A full factorial experiment design is

> ### Full Factorial Experiment Treatment Table
>
> Researchers are studying the effects of tire pressure, vehicle speed, oil type, and gas type on gas mileage. Problem: What combination of factors provides the best gas mileage?
>
> | Factor | Level 1 | Level 2 |
> |---|---|---|
> | Tire pressure | 28 psi | 35 psi |
> | Speed | 55 mph | 65 mph |
> | Oil | 30 weight | 40 weight |
> | Gas | Regular | Premium |
>
> Response variable: Gas mileage
>
> The complexity of this type of experimentation can be seen in Figure 21.1. Each path that is followed represents one treatment.
>
> | Tire | Speed | Oil | Gas |
> |---|---|---|---|
> | 28 | 55 | 30 | R |
> | 35 | 55 | 30 | R |
> | 28 | 55 | 30 | P |
> | 35 | 55 | 30 | P |
> | 28 | 65 | 30 | R |
> | 35 | 65 | 30 | R |
> | 28 | 65 | 30 | P |
> | 35 | 65 | 30 | P |
> | 28 | 55 | 40 | R |
> | 35 | 55 | 40 | R |
> | 28 | 55 | 40 | P |
> | 35 | 55 | 40 | P |
> | 28 | 65 | 40 | R |
> | 35 | 65 | 40 | R |
> | 28 | 65 | 40 | P |
> | 35 | 65 | 40 | P |

useful when it is important to study all the possible interactions that may exist. Unfortunately, a full factorial experiment design is time consuming and expensive due to the need for numerous runs.

Fractional Factorial Experiments

To reduce the total number of experiments that have to be conducted to a practical level, a limited number of the possibilities shown by a full factorial experiment may be chosen. *A fractional factorial experiment studies only a fraction or subset of all the possible combinations.* A selected and controlled multiple number of factors are adjusted simultaneously. By using this method, the total number of experiments is reduced. Designed correctly, fractional factorial experiments still reveal the complex interactions between the factors, including which factors are more important than others. One must be careful when selecting the fractional or partial group of experiments to be run to ensure that the critical factors and their interactions are studied. Many different experiment designs exist, including Plackett-Burman screening designs and Taguchi designs. When utilizing experiments in industry, reference texts and software programs can provide a wide variety of experiment designs from which an appropriate experiment design can be selected.

Plackett-Burman Screening Designs

Plackett-Burman screening designs are a subset of fractional factorial experiment designs. They provide an effective way to consider a large number of factors with a minimum number of runs. Users select the most appropriate design for their experiment needs. These screening designs are most effectively used when a large number of factors must be studied and time and resources are limited. Screening designs can be selected from preprepared tables from sources such as *Tables of Screening Designs* by Donald Wheeler. Figures 21.3 to 21.6 show several Plackett-Burman designs. The + and − signs show the levels of the factors [Level 1(−), Level 2(+)]. A, B, C, etc., represent the factors. Treatments are labeled 1 to 8.

Taguchi Designs

Dr. Genichi Taguchi (1924–) developed methods that seek to improve quality and consistency, reduce losses, and identify key product and process characteristics before production. Dr. Taguchi's methods emphasize consistency of performance and significantly reduced variation. Dr. Taguchi introduced the concept that the total loss to society generated by a product is an important dimension of the quality of a product. In his "loss function" concept, Dr. Taguchi expressed the

Treatment \ Factors	A	B	C	D	E	F	G
1	−	−	−	−	−	−	−
2	−	−	−	+	+	+	+
3	−	+	+	+	+	−	−
4	−	+	+	−	−	+	+
5	+	+	−	−	+	+	−
6	+	+	−	+	−	−	+
7	+	−	+	+	−	+	−
8	+	−	+	−	+	−	+

FIGURE 21.3 The Basic Eight-Run Plackett-Burman Design

Treatment \ Factors	A	B	C	D	E	F	G	H	I	J	K
1	−	−	−	−	−	−	−	−	−	−	−
2	−	−	−	+	+	+	−	−	+	+	+
3	−	−	+	−	+	+	+	+	−	−	+
4	−	+	+	−	+	−	−	+	+	+	−
5	−	+	+	+	−	−	+	−	+	−	+
6	−	+	−	+	−	+	+	+	−	+	−
7	+	+	−	+	+	−	+	−	−	+	+
8	+	+	−	−	+	+	+	−	+	−	−
9	+	+	+	−	−	+	−	−	−	+	+
10	+	−	+	+	−	+	−	+	+	−	−
11	+	−	+	+	+	−	+	−	−	+	−
12	+	−	−	−	−	−	+	+	+	+	+

FIGURE 21.4 The Basic 12-Run Plackett-Burman Design

Factors / Treatment	A	B	C	D	E	F	G	H	I	J	K	L	M	N	O
1	−	−	−	−	−	−	−	−	−	−	−	−	−	−	−
2	−	−	−	−	−	−	−	+	+	+	+	+	+	+	+
3	−	−	−	+	+	+	+	+	+	+	+	−	−	−	−
4	−	−	−	+	+	+	+	−	−	−	−	+	+	+	+
5	−	+	+	+	+	−	−	−	−	+	+	+	+	−	−
6	−	+	+	+	+	−	−	+	+	−	−	−	−	+	+
7	−	+	+	−	−	+	+	+	+	−	−	+	+	−	−
8	−	+	+	−	−	+	+	−	−	+	+	−	−	+	+
9	+	+	−	−	+	+	−	−	+	+	−	−	+	+	−
10	+	+	−	−	+	+	−	+	−	−	+	+	−	−	+
11	+	+	−	+	−	−	+	+	−	−	+	−	+	+	−
12	+	+	−	+	−	−	+	−	+	+	−	+	−	−	+
13	+	−	+	+	−	+	−	−	+	+	−	+	−	+	−
14	+	−	+	+	−	+	−	+	−	−	+	−	+	−	+
15	+	−	+	−	+	−	+	+	−	+	−	−	+	−	+
16	+	−	+	−	+	−	+	−	+	−	+	+	−	+	−

FIGURE 21.5 The Basic 16-Run Plackett-Burman Design

Factors / Treatment	A	B	C	D
1	+	+	+	+
2	+	+	−	−
3	+	−	−	+
4	+	−	+	−
5	−	−	+	+
6	−	−	−	−
7	−	+	−	+
8	−	+	+	−

FIGURE 21.6 The Eight-Run Reflected Plackett-Burman Design

costs of performance variation (Figure 21.7). Any deviation from target specifications causes loss, he said, even if the variation is within specifications. When the variation is within specifications, the loss may be in the form of poor fit, poor finish, undersize, oversize, or alignment problems. Scrap, rework, warranties, and loss of goodwill are all examples of losses when the variation extends beyond the specifications. Knowing the loss function helps designers set product and manufacturing tolerances. Capital expenditures are more easily justified by relating the cost of deviations from the target value to quality costs. Improving the consistency of performance minimizes losses.

Dr. Taguchi studied experiment design as a method of determining the best way to remove variation from a process. Statistically planned experiments can identify the settings of product and process parameters that reduce performance variation. Dr. Taguchi's methods design the experiment to systematically weed out insignificant elements in a product or process. The focus of experiment efforts is then placed on the significant elements. There are four basic steps:

1. Select the process/product to be studied.
2. Identify the important variables.

Plackett-Burman Screening Designs Experiment Treatment Table

LEAN SIX SIGMA TOOLS at WORK

Researchers are studying the effects of tire pressure, vehicle speed, oil type, and gas type on gas mileage. Problem: What combination of factors provides the best gas mileage?

Factor	Level 1	Level 2
Tire pressure	28 psi	35 psi
Speed	55 mph	65 mph
Oil	30 weight	40 weight
Gas	Regular	Premium

Response variable: Gas mileage

Design: Plackett-Burman Eight-Run Reflected (Figure 21.6)

Tire	Speed	Oil	Gas
35	65	40	P
35	65	30	R
35	55	30	P
35	55	40	R
28	55	40	P
28	55	30	R
28	65	30	P
28	65	40	R

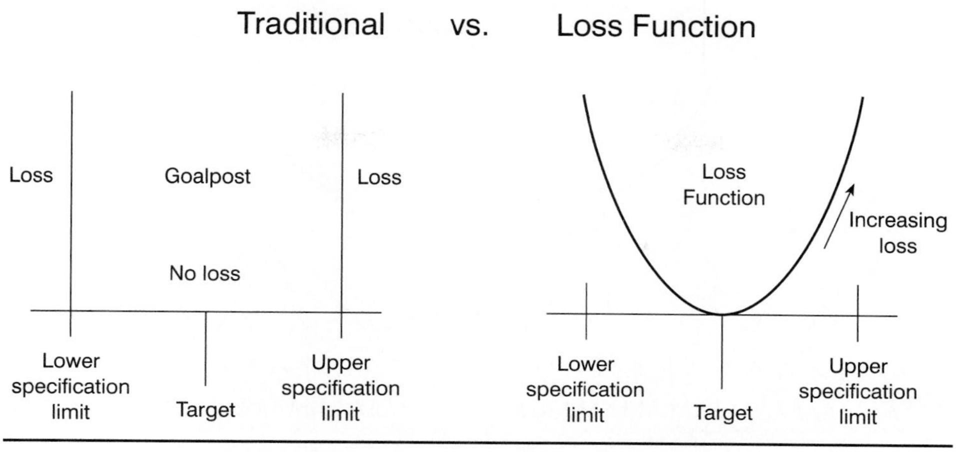

Traditional vs. Loss Function

FIGURE 21.7 Taguchi Loss Function

3. Reduce variation on the important variables through redesign, process improvement, and tolerancing.
4. Open up tolerances on unimportant variables.

The final quality and cost of a manufactured product are determined to a large extent by the engineering designs of the product and its manufacturing process.

Taguchi designs use orthogonal arrays to determine the factors and their levels for the experiments to be conducted. Essentially, Taguchi uses orthogonal arrays that select only a few of the combinations found in a traditional factorial design. This method has the advantage of being very efficient; however, these experiments work best when there is minimal interaction among the factors. Taguchi designs are most effective when the experimenter already has a general feel for the interactions that may be present among the factors. A comparison of the total number of experiments needed using *factorial designs* versus *Taguchi designs* is shown in Table 21.1. Figure 21.8 shows the structure of a traditional factorial experiment and then the reduced Taguchi orthogonal array. The treatments chosen by Taguchi are labeled in the full factorial experiment as T-1, T-2, and so on.

HYPOTHESES AND EXPERIMENT ERRORS

Experimenters approach each experiment with a hypothesis about how changes in various factors will affect the response variable. Often this hypothesis is expressed as follows:

H_0: The change in the factor will have no effect on the response variable.

H_1: The change in the factor will have an effect on the response variable.

Experiments are designed to test hypotheses, yet the experiments themselves are not infallible. This is why the significance of the factors affecting the response variables is studied. Experimenters are also cautious to determine whether errors exist in the experiment. Errors exist in experiments for a variety of reasons, including a lack of uniformity of the material and inherent variability in the experimental technique.

Two types of errors exist:

1. Type I error: The hypothesis is *rejected* when it is *true*. For instance, a Type I error would occur if the experimenter drew the conclusion that a factor does not produce a significant effect on a response variable when, in fact, its effect is meaningful. A Type I error is designated with the symbol alpha (α).

2. Type II error: The hypothesis is *accepted* when it is *false*. For instance, a Type II error would occur if the experimenter drew the conclusion that a factor produces a significant effect on a response variable when, in fact, its effect is negligible (a false alarm). A Type II error is designated with the symbol beta (β).

Table 21.2 shows the relationship of these types of errors.

TABLE 21.1 Comparison of Factorial Design and Taguchi Design

Factors	Level	Total Number of Experiments	
		Factorial Design	Taguchi
2	2	4 (2^2)	4
3	2	8 (2^3)	4
4	2	16 (2^4)	8
7	2	128 (2^7)	8
4	3	81 (3^4)	9

CHAPTER TWENTY ONE

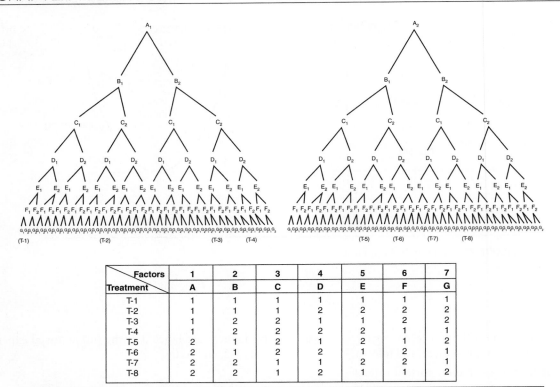

Factors / Treatment	1 A	2 B	3 C	4 D	5 E	6 F	7 G
T-1	1	1	1	1	1	1	1
T-2	1	1	1	2	2	2	2
T-3	1	2	2	1	1	2	2
T-4	1	2	2	2	2	1	1
T-5	2	1	2	1	2	1	2
T-6	2	1	2	2	1	2	1
T-7	2	2	1	1	2	2	1
T-8	2	2	1	2	1	1	2

FIGURE 21.8 Comparing a Full Factorial Experiment with a Taguchi Design

Taguchi Design Experiment Treatment Table

LEAN SIX SIGMA TOOLS at WORK

Researchers are studying the effects of tire pressure, vehicle speed, oil type, and gas type on gas mileage. Problem: What combination of factors provides the best gas mileage?

Factor	Level 1	Level 2
Tire pressure	28 psi	35 psi
Speed	55 mph	65 mph
Oil	30 weight	40 weight
Gas	Regular	Premium

Response variable: Gas mileage

Design: $L_4(2^3)$ Taguchi array (Figure 21.9)

Tire	Speed	Oil	Gas
28	55	30	R
28	55	30	P
28	65	40	R
28	65	40	P
35	55	40	R
35	55	40	P
35	65	30	R
35	65	30	P

Note that a full factorial experiment would have required 16 treatments, whereas a Taguchi experiment required only 8.

Factor No.	1	2	3	4	5	6	7
1	1	1	1	1	1	1	1
2	1	1	1	2	2	2	2
3	1	2	2	1	1	2	2
4	1	2	2	2	2	1	1
5	2	1	2	1	2	1	2
6	2	1	2	2	1	2	1
7	2	2	1	1	2	2	1
8	2	2	1	2	1	1	2

FIGURE 21.9 $L_4(2^3)$ Orthogonal Array

Hypotheses Testing

When a metal is selected for a part that will undergo a metal forming process such as stamping, often the metal is tested to determine its bending failure point. An experiment being conducted at JF Inc. is testing the bending failure of a particular metal that has been provided by two different suppliers. In this experiment, a sample of the metal is bent back and forth until the metal separates at the fold. The experimenters are testing the following hypotheses:

LEAN SIX SIGMA TOOLS at WORK

H_0: There is no difference in the bending failure point (the response variable) with respect to the following factors: suppliers (A and B), or plating (none or plated).

H_1: There is a difference in the bending failure point with respect to supplier or plating.

TABLE 21.2 Error Table

Conclusion from Sample	H_0 True	H_0 False
H_0 True	Correct conclusion	Type II error β
H_0 False	Type I error α	Correct conclusion

EXPERIMENTAL ANALYSIS METHODS

An *analysis of means (ANOM) essentially compares subgroup averages and separates those that represent signals from those that do not.* An ANOM takes the form of a control chart that identifies subgroup averages that are detectably different from the grand average. In this chart, each treatment (experiment) is compared with the grand average. An ANOM is used whenever the experimenter wants to study the differences between the subgroup averages from different treatments of the factors in the experiment. To use an ANOM, you must have more than one observation per subgroup.

Through experiments, researchers hope to identify optimum conditions or settings. In an experiment, many factors may affect optimum performance levels. To sort out the interactions between variables or factors, analysis of variance analyzes the variation attributable to different sources. An *analysis of variance (ANOVA) is a measure of the confidence that can be placed on the results of the experiment.*

This method is used to determine whether changes in factor levels have produced significant effects on a response variable. Variables or factors may be set at any number of levels. In the previous example, the variable or factor of tire pressure has two levels: 28 and 35. An ANOVA analyzes the variability of the data. In this analysis, the variance of the controllable and noise factors is examined. By understanding the source and magnitude of the variance, the best operating conditions can be determined. The response variable identifies the yield of the experiment, in the previous example, the best gas mileage.

When performing an analysis of variance, the samples are assumed to be independent and random from normal populations of equal variances. When conducting an ANOVA, the variance is estimated using two different methods. An ANOVA estimates the variance of the factors in the experiment by using information such as the degrees of freedom present, the sums of squares, and mean squares. If the estimates are similar, then detectable differences between the subgroup averages are unlikely. If the differences are large, then there is a difference between the subgroup averages that is not attributable to background noise alone. ANOVAs are frequently shown as graphs. If an interaction is present, the graph lines will cross (Figure 21.10). If no interaction is present, they will be parallel (Figure 21.11). An ANOVA compares the ratio of the Between Subgroup Variance Estimate with that of the Within Subgroup Variance Estimate. The Between Subgroup Variation Estimate is sensitive to differences between the subgroup averages. The Within Subgroup Variation Estimate is not sensitive to this difference. To learn more about calculating ANOVAs and ANOMs, consult a text on design of experiments.

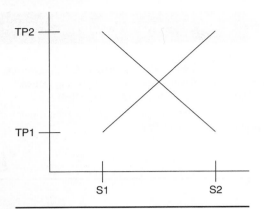

FIGURE 21.10 Interaction Exists

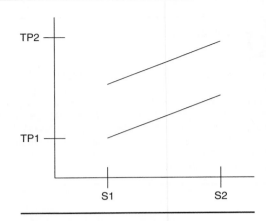

FIGURE 21.11 No Interaction Exists

LEAN SIX SIGMA TOOLS at WORK

Optimizing Order Pick, Pack, and Ship Using Design of Experiments

Order picking, packing, and shipping from a warehouse can be quite complex. Orders are submitted by customers and consist of multiple product types in various quantities. The products on these orders must be picked from the correct storage locations, packaged for safe shipment, and labeled with the appropriate destination. This would be simple if the organization carried only a few items in inventory to be shipped to a few customers. The complexity increases exponentially as the number of items and customers increase.

Doing the job right at every step is critical, not only because the customer wants to receive the correct item, but also because shipping costs are based on the cube of space the package uses up, whether in a truck or the cargo hold of a plane.

At CH Shipping, an automated distribution center, pickers, packers, and shippers handle orders that may contain as many as 50 unique products, numbering up to 5,000 of each unique product. CH Shipping uses "wave" processing, also known as batch picking. This method consolidates orders from several customers into one optimal inventory picking cycle (Figure 21.12). Optimal picking cycles consolidate what has been ordered, the amount ordered, and where the items are stored in the warehouse. As shown in Figure 21.13, once the pick is complete, the wave continues to the packing department, where the orders are automatically sorted using barcodes by the individual customer order and packed. Once packed, the orders proceed to a shipping station to be labeled and staged for shipment via truck or plane. Buffer staging lanes exist between each area. Downtime in a particular area results if orders are not in the system or are held up at an upstream area.

Wave processing is different from "discrete" processing. Discrete processing involves picking one order at a time, packing it, and then readying it for shipment. To manage the wave, order characteristics that optimize flow include how may times the picker visits a particular location in the warehouse, how many items are picked each time, and how much space is taken up by the items picked.

The key departments involved are Pick, Pack, and Ship. Members of the workforce in these departments work together to ensure that customers receive what they need, when they need it. To minimize costs, maximize flexibility, and enhance customer success, the Express Logistics Center Manager has asked the industrial engineers at CH Shipping to optimize order picking, packing, and shipping. In other words, they want to maximize the peaks and valleys in Figure 21.12 through the use of design of experiments. To do so, the team proposed the following experiment design:

Factors

Quantity (Qty): Total number of pieces picked within a given wave

Lines: The different items (products) on the order needing to be picked

Cube: The space needed for the quantity of lines for a particular order (largest outbound parcel box cube or tote cube; a tote is a box without a top)

Weight: The weight of the quantity of lines for a particular order

Level

Max Qty	Max Lines	Max Cube	Max Weight
5000 (+)	50 (+)	Parcel (+)	150 (+)
100 (−)	10 (−)	Tote (−)	25 (−)

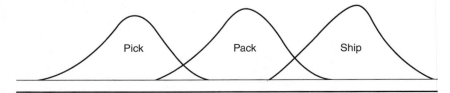

FIGURE 21.12 Department Workload Using Wave Order Processing

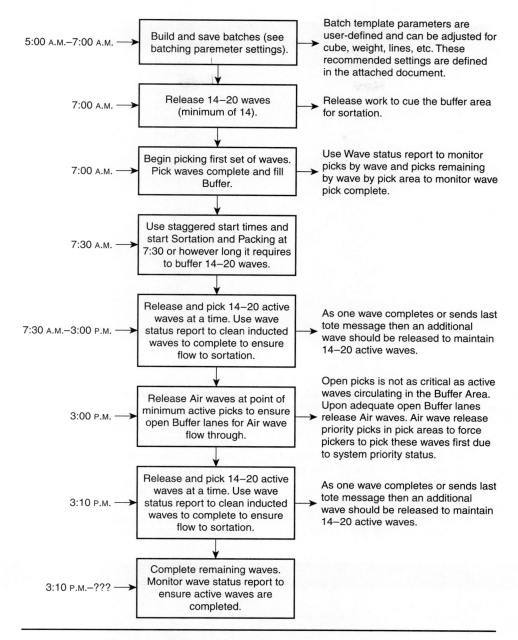

FIGURE 21.13 Flowchart for Order Pick, Pack, and Ship

Controllable Factors
When a factor is controllable, it is possible to establish and maintain the particular level throughout the experiment. In this experiment, investigators are able to set the levels for the quantity, lines, cube, and weight.

Response Variables
These variables are used to describe the reaction of a process to variations in the chosen factors. The response variable for this experiment is Order Per Hour.

Effect
The effect is the result or outcome of the experiment. The investigators are interested in determining the optimal wave parameters. A wave is a batch of customer orders consolidated to optimize warehouse picking activities. Instead of picking one order at a time, many orders are picked.

Treatment
The number of treatments is based on the number of factors and the levels associated with each factor. Based on the number of waves processed per day and the ease of wave building, a full factorial design was chosen (Table 21.3).

Degrees of Freedom
The degree of freedom in an experiment is the total number of levels for all factors minus 1. Here there are two levels for each factor ($4 \times 2 = 8$) so the degrees of freedom equals $8 - 1 = 7$.

Interaction
Because the four factors will work together to produce a result different from what would result if each were only used singly, interactions exist between the four that must be considered in the experiment.

(continued)

TABLE 21.3 Experiment Design for Optimizing Order Pick, Pack, and Ship

Test Wave	Max Qty	Max Lines	Max Cube	Max Weight
1	5,000	50	Parcel	150
2	100	50	Parcel	150
3	5,000	10	Parcel	150
4	100	10	Parcel	150
5	5,000	50	Tote	150
6	100	50	Tote	150
7	5,000	10	Tote	150
8	100	10	Tote	150
9	5,000	50	Parcel	25
10	100	50	Parcel	25
11	5,000	10	Parcel	25
12	100	10	Parcel	25
13	5,000	50	Tote	25
14	100	50	Tote	25
15	5,000	10	Tote	25
16	100	10	Tote	25

Noise Factor

A noise factor is an uncontrollable but measurable source of variation. For this experiment the noise factors are buffer downtime, pick downtime, and shipping downtime.

Run

A run is an experimental trial, the application of one treatment to one experimental unit. A quantity of 100, with 10 lines, using a tote, and weighing 25 pounds is one run.

Replicate

When an experiment is replicated, it refers to a repeat of the treatment condition, a complete rerunning of the design.

Repetition

Repetitions are multiple results of a treatment condition.

Significance

The determination of whether the changes to the factor levels had a significant effect on the response variable. In other words, did changing the quantity, lines, cube, or weight affect the result?

To conduct the experiment, the investigators set up controlled runs based on each of the 16 runs denoted by the full factorial design (Table 21.3). These runs, from actual customer orders, took place on a Saturday, outside of the normal workweek.

Following the experiment, the investigators analyzed the results (Table 21.4) to identify the optimum wave settings to

TABLE 21.4 Experiment Results for Optimizing Pick, Pack, and Ship

Test Wave	Max Qty	Max Lines	Max Cube	Max Weight	Order/Hour
1	5,000	50	Parcel	150	8.2
2	100	50	Parcel	150	8.1
3	5,000	10	Parcel	150	6.1
4	100	10	Parcel	150	6.3
5	5,000	50	Tote	150	7.9
6	100	50	Tote	150	8.2
7	5,000	10	Tote	150	4.5
8	100	10	Tote	150	5.1
9	5,000	50	Parcel	25	8.3
10	100	50	Parcel	25	8
11	5,000	10	Parcel	25	5.9
12	100	10	Parcel	25	6.2
13	5,000	50	Tote	25	7.8
14	100	50	Tote	25	8.1
15	5,000	10	Tote	25	5.1
16	100	10	Tote	25	4.9

maximize flow through Pick, Pack, and Ship. They began their analysis with the most simple of methods: ANOG or analysis of good. This very simple but effective technique ranks the results of the experiment by the most desirable outcome: high orders per hour (Table 21.5). Run 9, with a quantity of 5,000 and 50 lines allows 8.3 orders to be picked per hour.

Further statistical analysis using ANOVA and ANOM revealed that the factor having the most significant impact on Number of Orders Per Hour was Lines Per Wave, the number of products with a unique part number and their quantity. By comparison, the weight and cube size were insignificant. The experiment determined that the combination of unique part numbers that maximizes quantities results in the optimal picking sequence. Now, the company batches its orders based on the combination of unique part numbers that provides the greatest quantity. For instance, 25 unique part numbers whose order quantity totals to 4,925 or 3 unique part numbers whose order quantity totals 4,976 or 1 unique part number whose order quantity totals 5,000. The goal is to have a high number of items in the pick.

TABLE 21.5 Analysis of Good for Optimizing Order Pick, Pack, and Ship

		Analysis of Good			
ANOG Test Wave	Max Qty	Max Lines	Max Cube	Max Weight	Order/Hour
9	5,000	50	Parcel	25	8.3
1	5,000	50	Parcel	150	8.2
6	100	50	Tote	150	8.2
2	100	50	Parcel	150	8.1
14	100	50	Tote	25	8.1
10	100	50	Parcel	25	8
5	5,000	50	Tote	150	7.9
13	5,000	50	Tote	25	7.8
4	100	10	Parcel	150	6.3
12	100	10	Parcel	25	6.2
3	5,000	10	Parcel	150	6.1
11	5,000	10	Parcel	25	5.9
8	100	10	Tote	150	5.1
15	5,000	10	Tote	25	5.1
16	100	10	Tote	25	4.9
7	5,000	10	Tote	150	4.5

SUMMARY

This chapter has provided only a basic introduction to the interesting and valuable technique of design of experiments. Design of experiments enable users to understand more about the products or processes they are studying. Both methods are assets to any problem-solving adventure because they provide in-depth information about the complexities of the products and processes. Because both methods are complex and time consuming, they should be used judiciously.

TAKE AWAY TIPS

1. An experiment design is the plan or layout of an experiment. It shows the treatments to be included and the replication of the treatments.

2. Experimenters can study all the combinations of factors by utilizing a full factorial design.

3. Experimenters can study partial combinations of the factors by using fractional factorial experiment designs such as Plackett-Burman or Taguchi.

4. Design of experiments seeks to investigate the interactions of factors and their effects on the response variable.

CHAPTER QUESTIONS

1. When a metal is selected for a part that will undergo a metal forming process such as stamping, often the metal is tested to determine its bending failure point. An experiment being conducted at JF Inc. is testing the bending failure of a particular metal. In this experiment, a sample of the metal is bent back and forth until the metal separates at the fold. Metal from two different suppliers is currently being studied. Plated metal from each supplier is being compared with nonplated metal. For each piece of metal from each supplier, the metal samples being tested are in two sizes, a 2 in. wide piece and a 3 in. wide piece. If all of the factors are to be tested at each of their levels, create a table showing the factors and their levels for the two sizes, two suppliers, and two surface finishes. Define the components of this experiment.

2. For the metal bending failure experiment, create a matrix showing all the different treatments a full factorial experiment would have run.

3. In plastic injection molding, temperature, time in mold, and injection pressure each play an important role in determining the strength of the part. At JF Inc., experimenters are studying the effects on part strength when the temperature is set at either 250°, 275°, or 300°F; the time in mold is either 5, 7, or 9 seconds; and the injection pressure is either 200, 250, or 300 psi. Create a table showing the factors and their levels for this information.

4. For the information in the above plastic injection molding problem, define the components for this experiment.

5. For the above plastic injection molding problem, how many tests would be necessary to test every possible treatment (a full factorial experiment)? Create a tree similar to those shown in Figure 21.1 showing all possible treatments.

6. Researchers are studying the effects of tire pressure and vehicle speed on gas mileage. Create a hypothesis for the following information:

Factors	Levels
Tire pressure	28 psi, 35 psi
Speed	55 mph, 65 mph

7. Describe the difference between the two types of errors that can occur with experiments.

8. A team is currently working to redesign a compressor assembly cell. The team members have several design changes that they want to test using design of experiments. The ultimate objective of the experiment is to decrease the cell's assembly and packing throughput time. Currently, it takes 45 minutes to assemble and pack a compressor. The team is considering three changes:

    Reduce current 36-inch conveyor width to either 18 inches or 24 inches

    Install new parts presentation shelving, either flat surfaced or angled

    Line form: either straight or curved

    Create a table showing all the factors and their levels that will need to be tested. What is the response variable being investigated?

TWO

FAILURE MODES AND EFFECTS ANALYSIS

Will it fail?
What happens then?

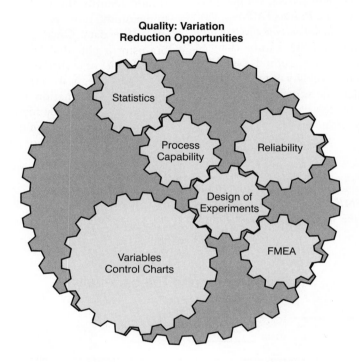

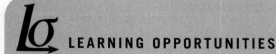

LEARNING OPPORTUNITIES

1. To familiarize the reader with the concept of failure modes and effects analysis
2. To teach the reader to perform a failure modes and effects analysis

FAILURE MODES AND EFFECTS ANALYSIS

Lean Six Sigma practitioners view failure modes and effects analysis (FMEA) as an effective failure prevention technique. **FMEA** *is a systematic approach to identifying both the ways that a product, part, process, or service can fail and the effects of those failures.* Once identified, these potential failure modes are rated by the severity of their effects and the probability the failure will occur. FMEAs generate a list of potential failures, rank the critical characteristics, and generate a list of actions that can be taken to eliminate the causes of failures or at least reduce their rate of occurrence. FMEAs are critical in the design of any system, process, service, or product, especially because they help identify potential failure modes that may adversely affect safety or government regulation compliance. Moreover, they identify these problems before the process or system is used or the product is put into production. Lean Six Sigma organizations utilize FMEA to improve the quality and reliability of their products, services, and processes. Three types of FMEAs exist: system, process, and design.

SYSTEMS FMEA

Systems enable organizations to accomplish tasks on a grand scale. Insurance companies have systems that enable them to process a high volume of claims every day. Manufacturing systems allow users to track customer orders, inventory usage, and production rates. A systems FMEA focuses on the big picture in order to optimize system designs, whether for services or manufacturing industries, determining the possible ways the system can fail. *System FMEAs study the functions of the system and reveal whether design deficiencies exist.* System FMEAs also study the interactions of the system with other systems and how subsystems within the system support and interact with each other.

PROCESS FMEA

Process FMEAs assist in the design or redesign of manufacturing, assembly, or service processes. *Process FMEAs identify the different ways that a process could fail and the effects of those failures.* With this information, processes can be changed, controls can be developed, or detection methods can be put into place that will eliminate the possibility of process failure. Because they focus on potential process deficiencies, lean Six Sigma practitioners often use process FMEAs to identify and rank process improvement opportunities.

DESIGN FMEA

Design FMEAs focus on products. Used during the product development stage, *design FMEAs seek to identify potential product failure modes and the likelihood of those failures occurring.* Design FMEAs assist in evaluating product design requirements and alternatives. Because they identify design deficiencies, they alert manufacturers to potential safety concerns.

CREATING AN FMEA

Though the focus of the three types of FMEAs is different, their construction is similar. To conduct an FMEA, use a form similar to the one shown in Figure 22.1. The form should identify the FMEA; the system, process, or part under study; those responsible; important dates; and any other key data. Once the headings are complete, users brainstorm potential failure modes, causes, effects, and probabilities of occurrence.

FMEAs begin with the study of the system, process, or part. This information is listed by item and function (Figure 22.2). When studying the system, process, service, or part, ask these types of questions to help understand the system, process, service, or part:

Where and how will this system (process, service, part) be used?

In what conditions will this system (process, service, part) operate?

Failure Modes and Effects Analysis

System (Process, Part) _____

Team _____   Date _____

Item	Function	Potential Failure Modes	Potential Causes	Failure Effect	Existing Controls, Countermeasures, Detection Methods	Probability	Severity	Risk Priority Code	Recommended Action, Responsibility, Target Date

FIGURE 22.1 Failure Modes and Effects Analysis Format

What are the interactions with other systems (processes, services, parts)?

What is the ultimate function of this system (process, service, part)?

Potential failure modes are the many different ways in which a system, process, service, or part may fail to perform its intended function (Figure 22.2). For each item, when determining potential failure modes, ask these types of questions:

How can this system (process, service, part) fail to perform its intended function?

How would it be recognized that it didn't perform its intended function?

What could go wrong?

How would it be recognized that something has gone wrong?

Potential failure causes are the different conditions that must be in place for the system, process, service, or part to fail (Figure 22.2). Be sure to find the proximate or true cause of the failure. For each item's potential failure mode, when determining potential failure causes, ask these types of questions:

What might have triggered this reaction in the system (process, service, part)?

What factors need to be in place to cause this failure to occur in the system (process, service, part)?

What would cause this system (process, service, part) to fail in this manner?

Failure Modes and Effects Analysis

System (Process, Service, Part) **Vehicle Exterior Lighting System**

Team **SS 1**   Date **2/14/06**

Item	Function	Potential Failure Modes	Potential Causes	Failure Effect	Existing Controls, Counter-measures, Detection Methods	Probability	Severity	Risk Priority Code	Recommended Action, Responsibility, Target Date
Headlamp System	Provide lighting for vehicle traveling at night	Lightbulb fails to light	Burnt out Not receiving electric current	No light	Visual maintenance check				
		Electric current failure	Wire failure Detached Short	No light	Visual maintenance check				
		Actuation switch failure	Short Broken	No light	Visual maintenance check				
Turn indicator system	Provide turn signals information to nearby vehicles	Lightbulb fails to light	Burnt out No current	No signal lights	Visual maintenance check				
		Electric current failure							
		Actuation switch failure							

FIGURE 22.2 Vehicle Exterior Lighting System FMEA

Effect	Criteria
No Effect	No effect
Very Slight Effect	Very slight effect on performance. Customer not annoyed. Nonvital component failure rarely noticed.
Slight Effect	Slight effect on performance. Customer slightly annoyed. Nonvital component failure noticed occasionally.
Minor Effect	Minor effect on performance. Customer will notice minor effect on system performance. Nonvital component failure always noticed.
Moderate Effect	Moderate effect on performance. Customer experiences some dissatisfaction. Nonvital component requires repair.
Significant Effect	Performance degraded, but operable and safe. Customer experiences discomfort. Nonvital component inoperable.
Major Effect	Performance severely affected but operable and safe. Customer dissatisfied. Subsystem inoperable.
Extreme Effect	Inoperable but safe. Customer very dissatisfied. System inoperable.
Serious Effect	Potentially hazardous effect. Compliance with government regulations in jeopardy.
Hazardous Effect	Hazardous effect. Safety-related. Sudden failure. Noncompliance with government regulations.

FIGURE 22.3 *Severity Rating Examples*

Under what circumstances would this system (process, service, part) fail to perform its intended function?

What can cause this system (process, service, part) to fail to deliver its intended function?

The potential failure effect is the consequence(s) of a system, process, service, or part failing (Figure 22.2). For each item's potential failure mode, when determining potential failure effects, ask these types of questions:

If the system (process, service, part) fails, what will be the consequences on the operation, function, or status of the system (process, service, part)?

If the system (process, service, part) fails, what will be the consequences on the operation, function, or status of the related systems (processes, service, parts)?

If the system (process, service, part) fails, what will be the consequences for the customer?

If the system (process, service, part) fails, what will be the consequence on government regulations?

FMEAs capture the conditions or controls currently in place that work to prevent system, process, service, or part failure. These methods or controls are recorded on the FMEA for each failure mode. These existing countermeasures answer the questions:

How will this cause of failure be recognized?

How can this cause of failure be prevented?

How can this cause of failure be minimized?

How can this cause of failure be mitigated?

The severity of a failure refers to the seriousness of the effect of a potential failure mode. Having completed the sections of the FMEA dealing with failures, causes, and effects (modes), the severity of each effect must be assessed. Severity ratings can range from no effect to hazardous effect. How many severity ratings used in an FMEA is up to those creating it. Figure 22.3 provides examples of severity ratings.

FMEAs also estimate the probability of occurrence or the estimated frequency at which failure may occur during the life of the system, process, or part. In other words, what is the likelihood of this type of failure from this particular cause happening? The probability of occurrence can rank from impossible to almost certain. Figure 22.4 provides examples of occurrences ratings.

Using a risk priority code (RPC), FMEAs rank failures according to their severity and their chance of occurring. Risk priority codes identify and rank the potential design weaknesses so that users can focus their efforts and actions toward eliminating the most probable causes with the most severe results. As shown in Figure 22.5, a risk priority code is the product of the severity rating and the probability of occurrence. Items with RPCs of one are dealt with first.

When FMEAs are complete, lean Six Sigma practitioners develop recommended actions designed to prevent or reduce the occurrence of failure. Starting with items with an RPC equal to one, these actions are designed to reduce the chance of failure occurring as well as the consequences associated with failure. These recommendations will include responsibilities, target completion dates, and reporting requirements for the actions taken.

Occurrence	Criteria
Almost Impossible	Failure unlikely
Improbable	Rare number of failures likely
Remote	Very few failures likely
Possible	Few failures likely
Occasional	Occasional failures likely
Probable	Several failures likely
Moderately High	A moderate number of failures likely
Frequent	A high number of failures likely
Very High	A very high number of failures likely
Almost Certain	Failure almost certain to occur

FIGURE 22.4 *Occurrence Ratings Examples*

Probability of Mishap

Severity of Consequences		F Impossible	E Improbable	D Remote	C Occasional	B Probable	A Frequent
	I Catastrophic			2		1	
	II Critical				2		
	III Marginal			3		2	2
	IV Negligible						

Risk Zones

Code		Action
1		Imperative to suppress
2		Operation requires written guidelines to operate
3		Operation permissible

Note: Personnel must not be exposed to hazards in Risk Zones 1 and 2.

Probability of Mishap

Level	Descriptive Word	Definition
A	Frequent	Likely to occur repeatedly during life cycle of system
B	Probable	Likely to occur several times in life cycle of system
C	Occasional	Likely to occur some time in life cycle of system
D	Remote	Not likely to occur in life cycle of system, but possible
E	Improbable	Probability of occurrences cannot be distinguished from zero
F	Impossible	Physically impossible to occur

FIGURE 22.5 Risk Priority Codes

Failure Modes and Effects Analysis

LEAN SIX SIGMA TOOLS at WORK

RQ Inc.'s management is concerned about overall air quality within the plant, particularly the plating department. Some of the materials used in this room can have an adverse effect on humans. For this reason, people in this room must wear personal protective devices. This area, located on an outside wall, is separated from the rest of the plant by floor-to-ceiling walls. Access to the area is limited to two doors. To verify the effectiveness of the proposed ventilation system upgrade, which consists of two exhaust fans with filters, RQ Inc.'s lean Six Sigma team conducted an FMEA (Figure 22.6).

The lean Six Sigma team members began the FMEA by listing the primary items and their functions in the ventilation system: exhaust fans, employees, and the power source. Their next step involved determining all of the potential failure modes for each of the primary items. Having established the list of potential failure modes, they identified potential causes for each potential failure. They also listed the effects of the failures. Existing controls, failure detection methods, and countermeasures were recorded next. Once a complete list of the potential failures and their causes and effects were identified, the team determined probabilities and severities in order to assign a risk priority code for each potential failure. Having completed the matrix, the team undertook the critical step of recommending specific actions, including target dates and responsibilities, to eliminate or reduce the probability or severity of a failure occurring.

Beginning with the RPCs with a value of one, the team members first need to investigate why no routine maintenance takes place. Their recommended action is to establish a preventive maintenance program for the ventilation system. A second set of RPCs with a value of one are all related to employee procedures. If the employees forget to turn on the ventilation system, fail to check if the fans are operational, or don't understand when to turn on the fans, the ventilation system cannot operate effectively. The lean Six Sigma team members recommend the creation and enforcement of procedures for ventilation system operation. They also recommend investigating the purchase of an air quality monitoring system designed to automatically activate the ventilation fans. Once the RPCs of one have been taken care of, the team can attack those with RPCs of two.

(Continued)

Failure Modes and Effects Analysis

System (Process, Service, Part): Ventilation System

Team: Jim, Norm, Peg

Date: 4/15/06

Item	Function	Potential Failure Modes	Potential Causes	Failure Effect	Existing Controls, Counter-measures, Detection Methods	Probability	Severity	Risk Priority Code	Recommended Action, Responsibility, Target Date
Ventilation system exhaust fans	Maintain quality of air	Electrical: power short	Faulty wiring, crossed wiring, damaged insulation	Unacceptable level of toxins in air causing eye and skin irritation, lung and brain damage	Smoke detector	Occasional	Critical	2	Investigate design of exhaust electrical system
	Maintain quality of air circulation	Electrical: power failure	Power outage	Unacceptable level of toxins in air causing eye and skin irritation, lung and brain damage	None	Occasional	Critical	2	Investigate design of exhaust electrical system
		Broken fan blades	Faulty workmanship, damage by foreign object	Unacceptable level of toxins in air causing eye and skin irritation, lung and brain damage	Visual inspection	Improbable	Critical	3	

Failure Modes and Effects Analysis

Item	Function	Failure Mode	Cause	Effect	Detection Method	Probability	Criticality	Rank	Action
Employees	Maintain quality of air; Maintain air circulation	Interference with fan blade rotation	Dirty, damaged during installation, poor design, damaged by foreign object	Fire, unacceptable level of toxins in air causing eye and skin irritation, lung and brain damage	Visual inspection, noise heard	Remote	Critical	3	
		Corrosion of motor works	Dirty, damaged during installation, poor design, damaged by foreign object	Unacceptable level of toxins in air causing eye and skin irritation, lung and brain damage	Visual inspection	Occasional	Critical	2	Establish preventive maintenance program for ventilation system
		Excessive dirt and grime on blades, unit unable to circulate air effectively	No routine maintenance	Unacceptable level of toxins in air causing eye and skin irritation, lung and brain damage	Visual inspection	Probable	Critical	1	Establish preventive maintenance program for ventilation system
		Power rating not correct or insufficient to circulate air	Not designed or selected correctly	Unacceptable level of toxins in air causing eye and skin irritation, lung and brain damage	Engineering design review	Remote	Critical	3	
		Employees forget to turn on fans	No process in place for operation of fans	Unacceptable level of toxins in air causing eye and skin irritation, lung and brain damage	Visual inspection	Probable	Critical	1	Create procedures for ventilation system operation and enforce

(Continued)

		Potential failure mode	Potential cause	Potential effect	Current controls	Probability	Severity	Risk	Recommended action
		Employees fail to notice if fans are on or not	No process in place for operation of fans	Unacceptable level of toxins in air causing eye and skin irritation, lung and brain damage	Visual inspection	Probable	Critical	1	Create procedures for ventilation system operation and enforce
		Employees don't know when to turn fans on	No process in place for operation of fans	Unacceptable level of toxins in air causing eye and skin irritation, lung and brain damage	None	Frequent	Critical	1	Create procedures for ventilation system operation and enforce
		Blockage of fan or air duct system	No routine maintenance	Unacceptable level of toxins in air causing eye and skin irritation, lung and brain damage	Visual inspection	Occasional	Critical	2	Establish preventive maintenance program for ventilation system
Power source	Provide electricity for fans	Loss of power	Loss of power from company	Unacceptable level of toxins in air causing eye and skin irritation, lung and brain damage	Visual	Occasional	Critical	2	Install emergency backup generators
			Internal failure	Unacceptable level of toxins in air causing eye and skin irritation, lung and brain damage	Visual	Occasional	Critical	2	Install emergency backup generators

FIGURE 22.6 Ventilation System FMEA

SUMMARY

Failure modes and effects analysis is an excellent method for capturing the relationship between the customer requirements and how a system, process, service, or part may fail to meet these requirements. It identifies the causes and effects of the failures within a system, process, service, or part by critically examining a system, process, service, or part. Once identified by the FMEA, weaknesses in the system, process, service, or part can be eliminated or strengthened.

TAKE AWAY TIPS

1. FMEA is a method that seeks to identify failures and keep them from occurring.
2. FMEA divides the system, process, service, or part into manageable segments and records the ways the segment may fail.
3. FMEA uses the risk priority code to rate the degree of hazard associated with failure.
4. When complete, recommendations are made to eliminate the risk priority codes with a value of one.

CHAPTER QUESTIONS

1. Why do lean Six Sigma practitioners use failure modes and effects analysis?
2. Describe the three types of FMEAs discussed in this chapter.
3. Describe the steps involved in creating an FMEA.
4. Create and analyze an FMEA for a refrigerator.
5. Create and analyze an FMEA for a chain saw.
6. Create and analyze an FMEA for a prescription filling process.
7. Create and analyze an FMEA for the operation of a lathe, mill, or drill.
8. Create and analyze an FMEA for a hospital patient check-in procedure.
9. Create and analyze an FMEA for a food preparation area in a restaurant.
10. Create and analyze an FMEA for a product, service, or process that you are familiar with.

CHAPTER TWENTY THREE

LEAN SIX SIGMA

Well begun is half done.
Aristotle

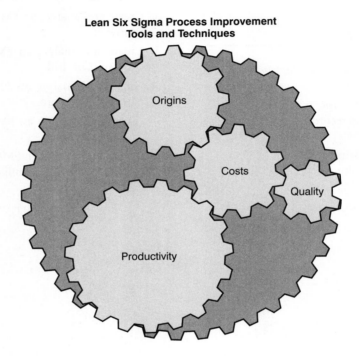

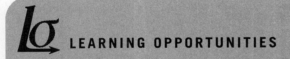

LEARNING OPPORTUNITIES

1. **To summarize the tools and techniques presented in this text**

Organizations that utilize lean Six Sigma tools and techniques have well begun and are half done on their desire to reach their goals of improved productivity, increased quality, and reduced costs. Leaders in lean Six Sigma organizations practice the following effective management techniques:

- Establishing good work procedures and standards (Chapters 10–16)
- Providing sound training in work procedures (Chapters 2 and 3)
- Practicing transparency or visual management (Chapter 14)
- Managing by walking around (Chapter 2)

Initiating lean Six Sigma begins with an understanding of the concepts, tools, and techniques related to these two management philosophies. As we have seen in this text, a wide variety of tools and techniques can be applied from both lean and Six Sigma.

LEAN

Lean focuses on identifying value-added tasks that enable the organization to get the work done the same way every time. Lean tools and techniques identify and remove the seven sources of waste: overproduction, idle time, delivery, waste in the work itself, inventory waste, wasted motion, and waste in rejected parts. Lean also enables the creation of standardized and simplified processes and procedures.

Kaizen (Chapter 13)
Value-added process mapping (Chapter 10)
5S (Chapter 12)
Kanban (pull inventory management) (Chapter 11)

Error proofing (Chapter 13)
Productive maintenance (Chapter 15)
Setup time reduction (Chapter 14)
Reduced batch sizes (Chapter 14)
Line balancing (Chapter 14)
Schedule leveling (Chapter 14)
Standardized work (Chapter 14)
Visual management (Chapter 14)

SIX SIGMA

Six Sigma tools and techniques address the variation present in processes, materials, equipment, products, and services. Six Sigma is instrumental in gathering reliable and consistent data for problem solving and decision making.

DMAIC (PDSA) (Chapter 9)
Value-added process mapping (Chapter 10)
Quality function deployment (Chapter 4)
Failure modes and effects analysis (Chapter 22)
Design of experiments (Chapter 21)
Reliability (Chapter 20)
Teamwork (Chapter 5)
Project management (Chapter 8)
Statistical methods (Chapter 17)
Control charts (Chapter 18)
Process capability analysis (Chapter 19)
Seven tools of quality (Chapter 9)

Lean and Six Sigma methodologies work together so that each step in the process is:

Valuable (customer focused)
Capable (Six Sigma tools and techniques)
Available (lean tools and techniques)
Adequate (Six Sigma and lean tools and techniques)
Flexible (lean tools and techniques)

Lean Six Sigma organizations apply the tools and techniques presented in this text in an integrated fashion in order to reap the benefits of both methodologies to achieve their goals (Figure 23.1). For more information about lean Six Sigma, visit the following websites:

American Society for Quality: www.asq.org

American Society for Quality, Statistics Division: www.asq.org/statistics

American Society for Quality, Quality Audit Division: www.asq.org/audit

American Society for Quality, Reliability Division: www.asq-rd.org

U.S. Commerce Department's National Institute of Standards: www.nist.gov

International Organization for Standardization (ISO 9000 and 14000): www.iso.org

Automotive Industry Action Group: www.aiag.org

Quality Information: www.itl.nist.gov/div898/handbook/pmc/pmc_d.htm

Six Sigma Information: www.sixsigma.com

Quality in Healthcare: www.jcaho.org

Information on Quality: www.qualitydigest.com

Discussion Forum: www.insidequality.com

Juran: www.juran.com

Deming: www.deming.org

European Foundation for Quality Management: www.efqm.org

Quality Standards: http://e-standards.asq.org/perl/catalog.cgi

American Productivity and Quality Center, Benchmarking Studies: www.apqc.org

(*Continued*)

FIGURE 23.1

Benchmarking Exchange: www.benchnet.com
Performance Measures: www.zigonperf.com
Lean Manufacturing: www.nwlean.net
Quality Function Deployment Institute: www.qfdi.org
American Standards Institute (ANSI): www.ansi.org
FMEA Info Centre: www.fmeainfocentre.com

Institute of Industrial Engineers: www.iienet.org
Internal Auditor: www.internal-auditor.com
Institute for Supply Management (ISM): www.ism.ws
Lean SCM: www.leanscm.net
ISO/IED Information Centre: www.standardsinfo.net
Supply-Chain Council: www.supply-chain.org

APPENDIX 1

Normal Curve Areas $P(Z \leq Z_0)$

Z	0.09	0.08	0.07	0.06	0.05	0.04	0.03	0.02	0.01	0.00
−3.8	.0001	.0001	.0001	.0001	.0001	.0001	.0001	.0001	.0001	.0001
−3.7	.0001	.0001	.0001	.0001	.0001	.0001	.0001	.0001	.0001	.0001
−3.6	.0001	.0001	.0001	.0001	.0001	.0001	.0001	.0001	.0002	.0002
−3.5	.0002	.0002	.0002	.0002	.0002	.0002	.0002	.0002	.0002	.0002
−3.4	.0002	.0003	.0003	.0003	.0003	.0003	.0003	.0003	.0003	.0003
−3.3	.0003	.0004	.0004	.0004	.0004	.0004	.0004	.0005	.0005	.0005
−3.2	.0005	.0005	.0005	.0006	.0006	.0006	.0006	.0006	.0007	.0007
−3.1	.0007	.0007	.0008	.0008	.0008	.0008	.0009	.0009	.0009	.0010
−3.0	.0010	.0010	.0011	.0011	.0011	.0012	.0012	.0013	.0013	.0013
−2.9	.0014	.0014	.0015	.0015	.0016	.0016	.0017	.0018	.0018	.0019
−2.8	.0019	.0020	.0021	.0021	.0022	.0023	.0023	.0024	.0025	.0026
−2.7	.0026	.0027	.0028	.0029	.0030	.0031	.0032	.0033	.0034	.0035
−2.6	.0036	.0037	.0038	.0039	.0040	.0041	.0043	.0044	.0045	.0047
−2.5	.0048	.0049	.0051	.0052	.0054	.0055	.0057	.0059	.0060	.0062
−2.4	.0064	.0066	.0068	.0069	.0071	.0073	.0075	.0078	.0080	.0082
−2.3	.0084	.0087	.0089	.0091	.0094	.0096	.0099	.0102	.0104	.0107
−2.2	.0110	.0113	.0116	.0119	.0122	.0125	.0129	.0132	.0136	.0139
−2.1	.0143	.0146	.0150	.0154	.0158	.0162	.0166	.0170	.0174	.0179
−2.0	.0183	.0188	.0192	.0197	.0202	.0207	.0212	.0217	.0222	.0228
−1.9	.0233	.0239	.0244	.0250	.0256	.0262	.0268	.0274	.0281	.0287
−1.8	.0294	.0301	.0307	.0314	.0322	.0329	.0336	.0344	.0351	.0359
−1.7	.0367	.0375	.0384	.0392	.0401	.0409	.0418	.0427	.0436	.0446
−1.6	.0455	.0465	.0475	.0485	.0495	.0505	.0516	.0526	.0537	.0548
−1.5	.0559	.0571	.0582	.0594	.0606	.0618	.0630	.0643	.0655	.0668
−1.4	.0681	.0694	.0708	.0721	.0735	.0749	.0764	.0778	.0793	.0808
−1.3	.0823	.0838	.0853	.0869	.0885	.0901	.0918	.0934	.0951	.0968
−1.2	.0985	.1003	.1020	.1038	.1056	.1075	.1093	.1112	.1131	.1151
−1.1	.1170	.1190	.1210	.1230	.1251	.1271	.1292	.1314	.1335	.1357
−1.0	.1379	.1401	.1423	.1446	.1469	.1492	.1515	.1539	.1562	.1587
−0.9	.1611	.1635	.1660	.1685	.1711	.1736	.1762	.1788	.1814	.1841
−0.8	.1867	.1894	.1922	.1949	.1977	.2005	.2033	.2061	.2090	.2119
−0.7	.2148	.2177	.2206	.2236	.2266	.2296	.2327	.2358	.2389	.2420
−0.6	.2451	.2483	.2514	.2546	.2578	.2611	.2643	.2676	.2709	.2743
−0.5	.2776	.2810	.2843	.2877	.2912	.2946	.2981	.3015	.3050	.3085
−0.4	.3121	.3156	.3192	.3228	.3264	.3300	.3336	.3372	.3409	.3446
−0.3	.3483	.3520	.3557	.3594	.3632	.3669	.3707	.3745	.3783	.3821
−0.2	.3859	.3897	.3936	.3974	.4013	.4052	.4090	.4129	.4168	.4207
−0.1	.4247	.4286	.4325	.4364	.4404	.4443	.4483	.4522	.4562	.4602
−0.0	.4641	.4681	.4721	.4761	.4801	.4840	.4880	.4920	.4960	.5000

(*continued*)

Z	0.00	0.01	0.02	0.03	0.04	0.05	0.06	0.07	0.08	0.09
+0.0	.5000	.5040	.5080	.5120	.5160	.5199	.5239	.5279	.5319	.5359
+0.1	.5398	.5438	.5478	.5517	.5557	.5596	.5636	.5675	.5714	.5753
+0.2	.5793	.5832	.5871	.5910	.5948	.5987	.6026	.6064	.6103	.6141
+0.3	.6179	.6217	.6255	.6293	.6331	.6368	.6406	.6443	.6480	.6517
+0.4	.6554	.6591	.6628	.6664	.6700	.6736	.6772	.6808	.6844	.6879
+0.5	.6915	.6950	.6985	.7019	.7054	.7088	.7123	.7157	.7190	.7224
+0.6	.7257	.7291	.7324	.7357	.7389	.7422	.7454	.7486	.7517	.7549
+0.7	.7580	.7611	.7642	.7673	.7704	.7734	.7764	.7794	.7823	.7852
+0.8	.7881	.7910	.7939	.7967	.7995	.8023	.8051	.8078	.8106	.8133
+0.9	.8159	.8186	.8212	.8238	.8264	.8289	.8315	.8340	.8365	.8389
+1.0	.8413	.8438	.8461	.8485	.8508	.8531	.8554	.8577	.8599	.8621
+1.1	.8643	.8665	.8686	.8708	.8729	.8749	.8770	.8790	.8810	.8830
+1.2	.8849	.8869	.8888	.8907	.8925	.8944	.8962	.8980	.8997	.9015
+1.3	.9032	.9049	.9066	.9082	.9099	.9115	.9131	.9147	.9162	.9177
+1.4	.9192	.9207	.9222	.9236	.9251	.9265	.9279	.9292	.9306	.9319
+1.5	.9332	.9345	.9357	.9370	.9382	.9394	.9406	.9418	.9429	.9441
+1.6	.9452	.9463	.9474	.9484	.9495	.9505	.9515	.9525	.9535	.9545
+1.7	.9554	.9564	.9573	.9582	.9591	.9599	.9608	.9616	.9625	.9633
+1.8	.9641	.9649	.9656	.9664	.9671	.9678	.9686	.9693	.9699	.9706
+1.9	.9713	.9719	.9726	.9732	.9738	.9744	.9750	.9756	.9761	.9767
+2.0	.9772	.9778	.9783	.9788	.9793	.9798	.9803	.9808	.9812	.9817
+2.1	.9821	.9826	.9830	.9834	.9838	.9842	.9846	.9850	.9854	.9857
+2.2	.9861	.9864	.9868	.9871	.9875	.9878	.9881	.9884	.9887	.9890
+2.3	.9893	.9896	.9898	.9901	.9904	.9906	.9909	.9911	.9913	.9916
+2.4	.9918	.9920	.9922	.9925	.9927	.9929	.9931	.9932	.9934	.9936
+2.5	.9938	.9940	.9941	.9943	.9945	.9946	.9948	.9949	.9951	.9952
+2.6	.9953	.9955	.9956	.9957	.9959	.9960	.9961	.9962	.9963	.9964
+2.7	.9965	.9966	.9967	.9968	.9969	.9970	.9971	.9972	.9973	.9974
+2.8	.9974	.9975	.9976	.9977	.9977	.9978	.9979	.9979	.9980	.9981
+2.9	.9981	.9982	.9982	.9983	.9984	.9984	.9985	.9985	.9986	.9986
+3.0	.9987	.9987	.9987	.9988	.9988	.9989	.9989	.9989	.9990	.9990
+3.1	.9990	.9991	.9991	.9991	.9992	.9992	.9992	.9992	.9993	.9993
+3.2	.9993	.9993	.9994	.9994	.9994	.9994	.9994	.9995	.9995	.9995
+3.3	.9995	.9995	.9995	.9996	.9996	.9996	.9996	.9996	.9996	.9997
+3.4	.9997	.9997	.9997	.9997	.9997	.9997	.9997	.9997	.9997	.9998
+3.5	.9998	.9998	.9998	.9998	.9998	.9998	.9998	.9998	.9998	.9998
+3.6	.9998	.9998	.9999	.9999	.9999	.9999	.9999	.9999	.9999	.9999
+3.7	.9999	.9999	.9999	.9999	.9999	.9999	.9999	.9999	.9999	.9999
+3.8	.9999	.9999	.9999	.9999	.9999	.9999	.9999	.9999	.9999	.9999

APPENDIX 2

Factors for Computing Central Lines and 3σ Control Limits for $\bar{X}$, s, and R Charts

Observations in Sample, n	Chart for Averages			Chart for Ranges							Chart for Standard Deviations					
	Factors for Control Limits			Factor for Central Line		Factors for Control Limits					Factor for Central Line	Factors for Control Limits				
	A	A_2	A_3	d_2	d_1	D_1	D_2	D_3	D_4		c_4	B_3	B_4	B_5	B_6	
2	2.121	1.880	2.659	1.128	0.853	0	3.686	0	3.267		0.7979	0	3.267	0	2.606	
3	1.732	1.023	1.954	1.693	0.888	0	4.358	0	2.574		0.8862	0	2.568	0	2.276	
4	1.500	0.729	1.628	2.059	0.880	0	4.698	0	2.282		0.9213	0	2.266	0	2.088	
5	1.342	0.577	1.427	2.326	0.864	0	4.918	0	2.114		0.9400	0	2.089	0	1.964	
6	1.225	0.483	1.287	2.534	0.848	0	5.078	0	2.004		0.9515	0.030	1.970	0.029	1.874	
7	1.134	0.419	1.182	2.704	0.833	0.204	5.204	0.076	1.924		0.9594	0.118	1.882	0.113	1.806	
8	1.061	0.373	1.099	2.847	0.820	0.388	5.306	0.136	1.864		0.9650	0.185	1.815	0.179	1.751	
9	1.000	0.337	1.032	2.970	0.808	0.547	5.393	0.184	1.816		0.9693	0.239	1.761	0.232	1.707	
10	0.949	0.308	0.975	3.078	0.797	0.687	5.469	0.223	1.777		0.9727	0.284	1.716	0.276	1.669	
11	0.905	0.285	0.927	3.173	0.787	0.811	5.535	0.256	1.744		0.9754	0.321	1.679	0.313	1.637	
12	0.866	0.266	0.886	3.258	0.778	0.922	5.594	0.283	1.717		0.9776	0.354	1.646	0.346	1.610	
13	0.832	0.249	0.850	3.336	0.770	1.025	5.647	0.307	1.693		0.9794	0.382	1.618	0.374	1.585	
14	0.802	0.235	0.817	3.407	0.763	1.118	5.696	0.328	1.672		0.9810	0.406	1.594	0.399	1.563	
15	0.775	0.223	0.789	3.472	0.756	1.203	5.741	0.347	1.653		0.9823	0.428	1.572	0.421	1.544	
16	0.750	0.212	0.763	3.532	0.750	1.282	5.782	0.363	1.637		0.9835	0.448	1.552	0.440	1.526	
17	0.728	0.203	0.739	3.588	0.744	1.356	5.820	0.378	1.622		0.9845	0.466	1.534	0.458	1.511	
18	0.707	0.194	0.718	3.640	0.739	1.424	5.856	0.391	1.608		0.9854	0.482	1.518	0.475	1.496	
19	0.688	0.187	0.698	3.689	0.734	1.487	5.891	0.403	1.597		0.9862	0.497	1.503	0.490	1.483	
20	0.671	0.180	0.680	3.735	0.729	1.549	5.921	0.415	1.585		0.9869	0.510	1.490	0.504	1.470	

APPENDIX 3
Values of t Distribution

Values of t Distribution

df	$t_{0.10}$	$t_{0.05}$	$t_{0.025}$	$t_{0.01}$	$t_{0.005}$	df
1	3.078	6.314	12.706	31.821	63.656	1
2	1.886	2.920	4.303	6.965	9.925	2
3	1.638	2.353	3.182	4.541	5.841	3
4	1.533	2.132	2.776	3.747	4.604	4
5	1.476	2.015	2.571	3.365	4.032	5
6	1.440	1.943	2.447	3.143	3.707	6
7	1.415	1.895	2.365	2.998	3.499	7
8	1.397	1.860	2.306	2.896	3.355	8
9	1.383	1.833	2.262	2.821	3.250	9
10	1.372	1.812	2.228	2.764	3.169	10
11	1.363	1.796	2.201	2.718	3.106	11
12	1.356	1.782	2.179	2.681	3.055	12
13	1.350	1.771	2.160	2.650	3.012	13
14	1.345	1.761	2.145	2.624	2.977	14
15	1.341	1.753	2.131	2.602	2.947	15
16	1.337	1.746	2.120	2.583	2.921	16
17	1.333	1.740	2.110	2.567	2.898	17
18	1.330	1.734	2.101	2.552	2.878	18
19	1.328	1.729	2.093	2.539	2.861	19
20	1.325	1.725	2.086	2.528	2.845	20
21	1.323	1.721	2.080	2.518	2.831	21
22	1.321	1.717	2.074	2.508	2.819	22
23	1.319	1.714	2.069	2.500	2.807	23
24	1.318	1.711	2.064	2.492	2.797	24
25	1.316	1.708	2.060	2.485	2.787	25
26	1.315	1.706	2.056	2.479	2.779	26
27	1.314	1.703	2.052	2.473	2.771	27
28	1.313	1.701	2.048	2.467	2.763	28
29	1.311	1.699	2.045	2.462	2.756	29
30	1.310	1.697	2.042	2.457	2.750	30
31	1.309	1.696	2.040	2.453	2.744	31
32	1.309	1.694	2.037	2.449	2.738	32
33	1.308	1.692	2.035	2.445	2.733	33

Values of t Distribution (*continued*)

df	$t_{0.10}$	$t_{0.05}$	$t_{0.025}$	$t_{0.01}$	$t_{0.005}$	df
34	1.307	1.691	2.032	2.441	2.728	34
35	1.306	1.690	2.030	2.438	2.724	35
40	1.303	1.684	2.021	2.423	2.704	40
45	1.301	1.679	2.014	2.412	2.690	45
50	1.299	1.676	2.009	2.403	2.678	50
55	1.297	1.673	2.004	2.396	2.668	55
60	1.296	1.671	2.000	2.390	2.660	60
70	1.294	1.667	1.994	2.381	2.648	70
80	1.292	1.664	1.990	2.374	2.639	80
90	1.291	1.662	1.987	2.368	2.632	90
100	1.290	1.660	1.984	2.364	2.626	100
200	1.286	1.653	1.972	2.345	2.601	200
400	1.284	1.649	1.966	2.336	2.588	400
600	1.283	1.647	1.964	2.333	2.584	600
800	1.283	1.647	1.963	2.331	2.582	800
999	1.282	1.646	1.962	2.330	2.581	999

ANSWERS TO SELECTED PROBLEMS

Chapter 17

17.10 Mean = 39
       Mode = 34
       Median = 38
17.11 Mean = 1.123
       Mode = 1.122
       Median = 1.123
17.13 Mean = 119.8
       Mode = 119.8
       Median = 119.8
       Range = 0.8
       Std. Dev = 0.4
17.14 Area = 0.8413
17.15 Area = 0.0047 or 0.47% of the parts are above 0.93 mm
17.16 Area = 0.9525
17.19 Mean = 0.0015
       Mode = 0.0025, 0.0004
       Median = 0.0014
       $\sigma = 0.0008$
       R = 0.0028
       73.41% of the parts will meet spec.

Chapter 18

18.2 $\overline{\overline{X}} = 16, UCL_x = 20, LCL_x = 12,$
      $\overline{R} = 7, UCL_R = 15, LCL_R = 0$
18.3 $\overline{\overline{X}} = 50.2, UCL_x = 50.7, LCL_x = 49.7,$
      $\overline{R} = 0.7, UCL_R = 1.6, LCL_R = 0$
18.4 $\overline{\overline{X}} = 349, UCL_x = 357, LCL_x = 341,$
      $\overline{R} = 14, UCL_R = 30, LCL_R = 0$
18.7 $\overline{\overline{X}} = 0.0627, UCL_x = 0.0629, LCL_x = 0.0625,$
      $\overline{R} = 0.0003, UCL_R = 0.0006, LCL_R = 0$

Chapter 19

19.3 $\sigma = 3, 6\sigma = 18, C_p = 0.44, C_{pk} = 0.33$
19.4 $\sigma = 0.3, 6\sigma = 1.8, C_p = 0.6, C_{pk} = 0.3$
19.8 $\sigma = 0.04, 6\sigma = 0.24, C_p = 0.42, C_{pk} = 0.25$

Chapter 20

20.3 $\lambda = 0.00035 \; \theta = 2886.25$ hours
20.4 $\lambda = 0.00197 \; \theta = 508$ hours
20.7 Reliability of the system = 0.9768
20.8 Reliability of the system = 0.9561
20.9 Reliability of the system = 0.9859

BIBLIOGRAPHY

Abarca, D. "Making the Most of Internal Audits." *Quality Digest*, February 1999, pp. 26–28.

Adam, P., and R. VandeWater. "Benchmarking and the Bottom Line: Translating BPR into Bottom-Line Results." *Industrial Engineering*, February 1995, pp. 24–26.

Adcock, S. "FAA Orders Fix on Older 737s." *Newsday*, May 8, 1998.

Adrian, N. "Incredible Journey." *Quality Progress*, June 2008, pp. 36–41.

Adrian, N. "It All Ties Together." *Quality Progress*, May 2008, pp. 20–27.

Adrian, N. "Small Business, Big Feat." *Quality Progress*, July 2008, pp. 40–43.

Aeppel, T. "Still Made in the USA." *The Wall Street Journal*, July 8, 2004.

Aeppel, T. "Tire Recalls Show Flaws in the System." *The Wall Street Journal*, November 1, 2007.

Albert, M. "Five Steps to Stellar Customer Service." *Quality Digest*, August 2007, pp. 49–53.

Alsup, F., and R. Watson. *Practical Statistical Process Control*. New York: Van Nostrand Reinhold, 1993.

Alukal, G. "Keeping Lean Alive." *Quality Progress*, October 2006, pp. 67–69.

AMA Management Briefing. *World Class Quality*. New York: AMA Publications Division, 1990.

Amari, D., and D. James. "ISO 9001 Takes on a New Role—Crime Fighter." *Quality Progress*, May 2004, pp. 57–61.

American Society for Quality Control, P.O. Box 3005, Milwaukee, WI 53201-3005.

Anand, G. "The Henry Ford of Heart Surgery." *The Wall Street Journal*, November 21, 2009.

Arthur, J. "Statistical Process Control for Health Care." *Quality Digest*, June 2008, pp. 34–38.

Azeredo, M., S. Silva, and K. Rekab. "Improve Molded Part Quality." *Quality Progress*, July 2003, pp. 72–76.

Bacus, H. "Liability: Trying Times." *Nation's Business*, February 1986, pp. 22–28.

Bala, S. "A Practical Approach to Lean Six Sigma." *Quality Digest*, July 2008, pp. 30–34.

Balestracci, D. "2007 Resolutions: Weight and . . . Budget? (Pt. 2)" *Quality Digest*, February 2007, p. 20.

Balestracci, D. "Why Deming Was Such a Curmudgeon," *Quality Digest*, October 2006, p. 19.

Bamford, J. "Order in the Court." *Forbes*, January 27, 1986, pp. 46–47.

Barcellos, P., and A. Mueller "The Right Move." *Quality Progress*, August 2008, pp. 28–35.

Baranzelli, J. "The Road to Improvement." *Quality Progress*, August 2008, pp. 28–35.

Barrett, J. "Whirlpool Cleans Up Its Delivery Act." *The Wall Street Journal*, September 23, 2009.

Bernowski, K., and B. Stratton. "How Do People Use the Baldrige Award Criteria?" *Quality Progress*, May 1995, pp. 43–47.

Berry, T. *Managing the Total Quality Transformation*. Milwaukee, WI: ASQC Quality Press, 1991.

Besterfield, D. *Quality Control*, 4th ed. Englewood Cliffs, NJ: Prentice Hall, 1994.

Biesada, A. "Strategic Benchmarking." *Financial World*, September 29, 1992, pp. 30–36.

Bilke, T., and Sinn, J. "Bright Idea." *Quality Progress*, August 2008, pp. 20–27.

Bisgaard, S. "Geared Toward Innovation." *Quality Progress*, September 2008, pp. 20–25.

Bishara, R., and M. Wyrick. "A Systematic Approach to Quality Assurance Auditing." *Quality Progress*, December 1994, pp. 67–69.

Block, M. "The White House Manages Green." *Quality Progress*, July 2003, pp. 90–91.

Borsai, T., B. Ludovico, and G. Dzialas. "ISO 13485: A Path to the Global Market." *Quality Digest*, July 2007, pp. 35–39.

Bossert, J. "Lean and Six Sigma—Synergy Made in Heaven." *Quality Progress*, July 2003, pp. 31–32.

Bovet, S. F. "Use TQM, Benchmarking to Improve Productivity." *Public Relations Journal*, January 1994, p. 7.

Breyfogle, F., and B. Meadows. "Bottom-Line Success with Six Sigma." *Quality Progress*, May 2001, pp. 101–104.

Brocka, B., and M. Brocka. *Quality Management: Implementing the Best Ideas of the Masters.* Homewood, IL: Business One Irwin, 1992.

Bruder, K. "Public Benchmarking: A Practical Approach." *Public Press,* September 1994, pp. 9–14.

Brumm, E. "Managing Records for ISO 9000 Compliance." *Quality Progress,* January 1995, pp. 73–77.

Butz, H. "Strategic Planning: The Missing Link in TQM." *Quality Progress,* May 1995, pp. 105–108.

Byrnes, D. "Exploring the World of ISO 9000." *Quality,* October 1992, pp. 19–31.

Byron, E. "Beauty, Prestige and Worry Lines," *The Wall Street Journal,* August 20, 2007.

Campanella, J., Ed. *Principles of Quality Costs.* Milwaukee, WI: ASQC Quality Press, 1990.

Camperi, J. A. "Vendor Approval and Audits in Total Quality Management." *Food Technology,* September 1994, pp. 160–162.

Carey, S. "Racing to Improve." *The Wall Street Journal,* March 24, 2006.

Carson, P. P. "Deming Versus Traditional Management Theorists on Goal Setting: Can Both Be Right?" *Business Horizons,* September 1993, pp. 79–84.

Chapman, C. D. "Clean House with Lean 55," *Quality Progress,* June 2005, pp. 27–32.

Chowdhury, S. *The Ice Cream Maker.* New York: Currency Doubleday, 2005.

Chircop, J. "On the Same Page." *Quality Progress,* November 2008, pp. 40–46.

Cochran, C. "The Ten Biggest Quality Mistakes." *Quality Digest,* February 2007, pp. 28–32.

Columbus, L. "Quality Partnerships with Your Customers" *Quality Digest,* August 2007, pp. 44–48.

Conner, G. *Lean Manufacturing for the Small Shop.* Dearborn, MI: Society of Manufacturing Engineers, 2001.

Cook, B. M. "Quality: The Pioneers Survey the Landscape." *Industry Week,* October 21, 1991, pp. 68–73.

Crosby, P. B. *The Eternally Successful Organization: The Art of Corporate Wellness.* New York: New American Library, 1988.

Crosby, P. B. *Quality Is Free.* New York: Penguin Books, 1979.

Crosby, P. B. *Quality Is Free: The Art of Making Quality Certain.* New York: McGraw-Hill, 1979.

Crosby, P. B. *Quality without Tears: The Art of Hassle-Free Management.* New York: McGraw-Hill, 1979.

Cross, C. "Everything But the Kitchen Sink." *Industrial Engineer,* April 2007, pp. 32–37.

Cross, C. "Fighting for Air." *Industrial Engineer,* February 2007, pp. 30–35.

Crownover, D. "Baldrige: It's Easy, Free, and It Works." *Quality Progress,* July 2003, pp. 37–41.

Crownover, D. *Take it to the Next Level.* Dallas: NextLevel Press, 1999.

Cullen, C. "Short Run SPC Re-emerges." *Quality,* April 1995, p. 44.

"Customer Satisfaction Hits Nine-Year High." *Quality Progress,* April 2004, p. 16.

Dalpino, N. "Outsourcing the Quality Function." *Quality Digest,* July 2008, pp. 43–46.

Daniels, S. "From One-Man Show to Baldrige Recipient." *Quality Progress,* July 2007, pp. 50–55.

Darden, W., W. Babin, M. Griffin, and R. Coulter. "Investigation of Products Liability Attitudes and Opinions: A Consumer Perspective." *Journal of Consumer Affairs,* June 22, 1994.

Davis, P. M. "New Emphasis on Product Warnings." *Design News,* August 6, 1990, p. 150.

Day, C. R. "Benchmarking's First Law: Know Thyself." *Industry Week,* February 17, 1992, p. 70.

Day, R. G. *Quality Function Deployment.* Milwaukee, WI: ASQ Quality Press, 1993.

Dean, M., and C. Tomovic. "Does Baldrige Make a Business Case for Quality?" *Quality Progress,* April 2004, pp. 40–45.

Dearing, J. "ISO 9001: Could It Be Better?" *Quality Progress,* February 2007, pp. 23–27.

DeFoe, J. A. "The Tip of the Iceberg." *Quality Progress,* May 2001, pp. 29–37.

Deming, W. E. *The New Economics.* Cambridge, MA: MIT CAES, 1993.

Deming, W. E. *Out of the Crisis.* Cambridge, MA: MIT Press, 1986.

Dennis, P. *Lean Production Simplified: A Plain Language Guide to the World's Most Powerful Production System.* New York: Productivity Press, 2002.

Denny, W. "While We Are Sleeping." *Quality Digest,* July 2008, pp. 38–42.

DeToro, I. "The Ten Pitfalls of Benchmarking." *Quality Progress,* January 1995, pp. 61–63.

DeVor, R., T. Chang, and J. Sutherland. *Statistical Quality Design and Control.* New York: Macmillan, 1992.

Disney, L. "Wave of Relief." *Industrial Engineer,* February 2007, pp. 24–29.

Dobyns, L., and C. Crawford-Mason. *Quality or Else: The Revolution in World Business.* Boston: Houghton Mifflin, 1991.

Dusharme, M. "RFID Tunes Into Supply Chain Management." *Quality Digest*, October 2006, pp. 37–41.

Eaton, B. "Cessna's Approach to Internal Quality Audits." *IIE Solutions, Industrial Engineering*, June 1995, pp. 12–16.

Eckert, C. "Eliminating Defects in the Vertical Supply Chain." *Quality Digest*, December 2008, pp. 24–27.

Engle, P. "When Inventory Goes Missing." *Industrial Engineer*, November 2006, p. 22.

Ericson, J. "Lean Inspection Through Supplier Partnership." *Quality Progress*, November 2006, pp. 36–42.

Eureka, W. E., and N. E. Ryan. *The Customer Driven Company*. Dearborn, MI: ASI Press, 1988.

Farahmand, K., R. Becerra, and J. Greene. "ISO 9000 Certification: Johnson Controls' Inside Story." *Industrial Engineering*, September 1994, pp. 22–23.

Feigenbaum, A. V. "Changing Concepts and Management of Quality Worldwide." *Quality Progress*, December 1997, pp. 43–47.

Feigenbaum, A. V. *Total Quality Control*. New York: McGraw-Hill, 1983.

Feigenbaum, A. V. "The Future of Quality Management." *Quality Digest*, May 1998, pp. 24–30.

Feigenbaum, A. V. "How to Manage for Quality in Today's Economy." *Quality Progress*, May 2001, pp. 26–27.

Feigenbaum, A.V. "The International Growth of Quality." *Quality Progress*, February 2007, pp. 36–40.

Feigenbaum, A. V. "Raising the Bar." *Quality Progress*, July 2008, pp. 22–27.

Feigenbaum, A. V. "Spring Into Action" *Quality Progress*, November 2009, pp. 18–22.

Fenn, D. *Alpha Dogs: How Your Small Business Can Become a Leader of the Pack*. New York: Collins, 2005.

Foster, S. T. *Managing Quality: Integrating the Supply Chain*. Upper Saddle River, NJ: Prentice Hall, 2007.

Franco, V. R. "Adopting Six Sigma." *Quality Digest*, June 2001, pp. 28–32.

Frum, D. "Crash!" *Forbes*, November 8, 1993, p. 62.

Fuhrmans, V. "Insurers, FDA Team Up to Find Problem Drugs." *The Wall Street Journal*, April 15, 2008.

Galpin, D., R. Dooley, J. Parker, and R. Bell. "Assess Remaining Component Life with Three Level Approach." *Power*, August 1990, pp. 69–72.

Galpin, T. *The Human Side of Change*. San Francisco, CA: Jossey-Bass Publishers, 1996.

Gardner, R. A. "Resolving the Process Paradox." *Quality Progress*, March 2001, pp. 51–59.

Gates, R. "Deployment: Start Off on the Right Foot." *Quality Progress*, August 2007, pp. 51–57.

Gerling, A. "How Jury Decided How Much the Coffee Spill Was Worth." *The Wall Street Journal*, September 4, 1994.

George, M., D. Rowlands, M. Price, and J. Maxey. *Lean Six Sigma Pocket Toolbook*. New York: McGraw-Hill, 2005.

Gest, T. "Product Paranoia." *U.S. News & World Report*, February 24, 1992, pp. 67–69.

Geyelin, M. "Product Liability Suits Fare Worse Now." *The Wall Street Journal*, July 12, 1994.

Ghattas, R. G., and S. L. McKee. *Practical Project Management*. Upper Saddle River, NJ: Prentice Hall, 2001.

Gitlow, H. S. *Planning for Quality, Productivity, and Competitive Position*. Homewood, IL: Business One Irwin, 1990.

Gitlow, H. S., and S. J. Gitlow. *The Deming Guide to Quality and Competitive Position*. Englewood Cliffs, NJ: Prentice Hall, 1987.

Goetsch, D. L. *Effective Supervision*. Upper Saddle River, NJ: Prentice Hall, 2002.

Goetsch, D. L., and S. B. Davis. *ISO 14000 Environmental Management*. Upper Saddle River, NJ: Prentice Hall, 2001.

Goetsch, D. L., and S. B. Davis. *Understanding and Implementing ISO 9000 and ISO Standards*. Upper Saddle River, NJ: Prentice Hall, 1998.

Goodden, R. "Better Safe Than Sorry." *Quality Progress*, May 2008, pp. 28–35.

Goodden, R. "How a Good Quality Management System Can Limit Lawsuits." *Quality Progress*, June 2001, pp. 55–59.

Goodden, R. L. *Product Liability Prevention: A Strategic Guide*. Milwaukee, WI: ASQ Quality Press, 2000.

Goodden, R. "Product Reliability Considerations Empower Quality Professionals." *Quality*, April 1995, p. 108.

Goodden, R. "Reduce the Impact of Product Liability on Your Organization." *Quality Progress*, January 1995, pp. 85–88.

Grahn, D. "The Five Drivers of Total Quality." *Quality Progress*, January 1995, pp. 65–70.

Grant, E., and T. Lang. "Why Product-Liability and Medical Malpractice Lawsuits Are So Numerous in the United States." *Quality Progress*, December 1994, pp. 63–65.

Grant, E., and R. Leavenworth. *Statistical Quality Control*. New York: McGraw-Hill, 1988.

Greasley, A. *Operations Management*. West Sussex, England: John Wiley & Sons, 2006.

Grenny, J. "Five Crucial Conversations for Successful Projects." *Quality Digest*, August 2007, pp. 26–31.

Groover, M. *Work Systems*. Upper Saddle River, NJ: Pearson Education, 2007.

Hale, R., and D. Kubiak. "Waste's Final Foothold." *Industrial Engineering*, August 2007, pp. 36–38.

Hare, L. "SPC: From Chaos to Wiping the Floor." *Quality Progress*, July 2003, pp. 58–63.

Hare, L., R. Hoerl, J. Hromi, and R. Snee. "The Role of Statistical Thinking in Management." *Quality Progress*, February 1995, pp. 53–59.

Harrington, H. "Harrington's Wheel of Fortune." *Quality Digest*, February 2007, p. 16.

Harrington, H. "Hit the Nail, Not the Thumb." *Quality Digest*, October 2006, p. 16.

Harry, M., and R. Schroeder. *Six Sigma: The Breakthrough Management Strategy Revolutionizing the World's Top Corporations*. New York: Doubleday, 2000.

Haupt, H. "Getting Credit for Service." *Quality Progress*, November 2006, pp. 51–55.

Hayes, B. "The True Test of Loyalty." *Quality Progress*, June 2008, pp. 20–27.

Hayward, S. *Churchill on Leadership*. New York: Forum, 1997.

Himelstein, L. "Monkey See, Monkey Sue." *Business Week*, February 7, 1994, pp. 112–113.

Hockman, K., R. Grenville, and S. Jackson. "Road Map to ISO 9000 Registration." *Quality Progress*, May 1994, pp. 39–42.

Hoisington, S., and E. Menzer. "Learn to Talk Money." *Quality Progress*, May 2004, pp. 44–49.

Houghton, P. "Improving Pharmacy Service." *Quality Digest*, October 2006, pp. 49–54.

Hoyer, R. W., and B. B. Hoyer. "What Is Quality?" *Quality Progress*, July 2001, pp. 52–62.

Hutchins, D. *Just-in-Time*. England: Gower Technical Press, 1989.

Hutchins, G. "The State of Quality Auditing." *Quality Progress*, March 2001, pp. 25–29.

Hutchinson, E. E. "The Road to TL 9000: From the Bell Breakup to Today." *Quality Progress*, January 2001, pp. 33–37.

Hutton, D. W. *From Baldrige to the Bottom Line*. Milwaukee, WI: ASQ Quality Press, 2000.

Iannello, P. "Unplugged . . . Untangled . . . and Informed." *Quality Digest*, February 2007, pp. 54–59.

Imai, M. *Gemba Kaizen: A Commonsense, Low Cost Approach to Management*. New York: McGraw-Hill, 1997.

Imai, M. *Kaizen: The Key to Japan's Competitive Success*. New York: McGraw-Hill/Irwin, 1986.

Ireson, W., and C. Coombs. *Handbook of Reliability Engineering and Management*. New York: McGraw-Hill, 1988.

"ISO Says 8 of 10 Cars to 'Run' on ISO 9001:2000." *Quality Progress*, July 2003, p. 12.

Ishikawa, K. *Guide to Quality Control*, rev. ed. White Plains, NY: Kraus International Publications, 1982.

Ishikawa, K. *What Is Total Quality Control? The Japanese Way*. Englewood Cliffs, NJ: Prentice Hall, 1985.

Iyer, S. "Don't Forget the People." *Quality Progress*, October 2006, pp. 60–66.

Jaffrey, S. "ISO 9001 Made Easy." *Quality Progress*, May 2004, p. 104.

Japan Human Relations Association. *Kaizen Teian 1*. Portland, OR: Productivity Press, 1992.

Jargon, J. "Latest Starbucks Buzzword: 'Lean' Japanese Techniques." *The Wall Street Journal*, August 4, 2009.

Johnson, K. "Print Perfect." *Quality Progress*, July 2003, pp. 48–56.

Johnson, W. "AS9100: On Course and Gaining." *Quality Digest*, February 2007, pp. 43–48.

Jones, D. M. "Merging Quality Cultures in Contract Manufacturing." *Quality Progress*, February 2007, pp. 41–46.

Juran, J. "A Close Shave." *Quality Progress*, May 2004, pp. 41–43.

Juran, J. M. *A History of Managing for Quality*. Milwaukee, WI: ASQC Quality Press, 1995.

Juran, J. M. *Juran on Leadership for Quality: An Executive Handbook*. New York: Free Press, 1989.

Juran, J. M. *Juran on Planning for Quality*. New York: Free Press, 1988.

Juran, J. M. *Juran on Quality by Design: The New Steps for Planning Quality into Goods and Services*. New York: Free Press, 1992.

Juran, J. M. "The Quality Trilogy." *Quality Progress*, August 1986, pp. 19–24.

Juran, J. M., and F. M. Gryna. *Quality Planning and Analysis: From Product Development through Usage*. New York: McGraw-Hill, 1970.

Kackar, R. "Taguchi's Quality Philosophy: Analysis and Commentary." *Quality Progress*, December 1986, pp. 21–29.

Kanholm, J. "New and Improved ISO 9000:2000." *Quality Digest*, October 1999, pp. 28–32.

Kausek, J. "10 Auditing Rules." *Quality Progress*, July 2008, pp. 44–49.

Kececioglu, D. *Reliability and Life Testing Handbook*. Englewood Cliffs, NJ: Prentice Hall, 1993.

Keller, C. "QOS—A Simple Method for Big or Small." *Quality Progress*, July 2003, pp. 28–31.

Ketola, J., and K. Roberts. *ISO 9001:2000 in a Nutshell*, 2d ed. Chico, CA: Paton Press, 2001.

Ketola, J., and K. Roberts. "Transition Planning for ISO 9001:2000." *Quality Digest*, March 2001, pp. 24–28.

Kirscht, R. "Quality and Outsourcing." *Quality Digest*, July 2007, p. 40–43.

Kolarik, W. J. *Creating Quality: Concepts, Systems, Strategies and Tools.* New York: McGraw-Hill, 1995.

Krzykowski, B. "K'NEX Success." *Quality Progress*, May 2008, pp. 36–43.

Krzykowski, B. "Now Hear This." *Quality Progress*, June 2008, pp. 28–35.

Kubiak, T. "An Integrated Approach System." *Quality Progress*, July 2003, pp. 41–45.

Lancaster, J. *Making Time: Lillian Moller Gilbreth.* Boston, MA: Northeastern University Press, 2004.

Lareau, W. *Office Kaizen*. Milwaukee, WI: ASQ Quality Press, 2003.

Lazalier, M. "Coax, Don't Squeeze." *Industrial Engineer*, April 2007, pp. 26–31.

Leibfried, K. H. *Benchmarking: A Tool for Continuous Improvement.* New York: Harper Business, 1992.

Levinson, W. A. "ISO 9000 at the Front Line." *Quality Progress*, March 2001, pp. 33–36.

Lindborg, H. "Get Rid of Clutter." *Quality Progress*, February 2007, p. 50.

Lunsford, J., and D. Michaels. "After Four Years in the Rear, Boeing Is Set to Jet Past Airbus." *The Wall Street Journal*, June 10, 2005.

Mader, D. "DFSS and Your Current Design Process." *Quality Progress*, July 2003, pp. 88–89.

Mader, D. "How to Identify and Select Lean Six Sigma Projects." *Quality Progress*, July 2007, pp. 58–60.

Manos, A. "The Benefits of Kaizen and Kaizen Events." *Quality Progress*, February 2007, p. 47.

Mascitelli, R. "Design's Due Process." *Industrial Engineer*, April 2007, pp. 38–43.

Mason, R., and J. Young. "Multivariate Thinking." *Quality Progress*, April 2004, pp. 89–91.

Matthews, C. "Linking the Supply Chain to TQM." *Quality Progress*, November 2006, pp. 29–34.

May, M. "Lean Thinking for Knowledge Work." *Quality Progress*, June 2005, pp. 33–39.

McCartney, S. "Airlines Address Long Waits—Sort Of." *The Wall Street Journal*, April 29, 2008.

McCartney, S. "The Truth About Flight Delays." *The Wall Street Journal*, April 21, 2004.

McElroy, A., and I. Fruchtman. "Use Statistical Analysis to Predict Equipment Reliability." *Power*, October 1992, pp. 39–46.

McManus, K. "No Time For Projects." *Industrial Engineer*, November 2006, p. 20.

McManus, K. "The Last Great Fad." *Industrial Engineer*, December 2006, p. 18.

McManus, K. "The Pull of Lean." *Industrial Engineer*, April 2007, p. 20.

Malcolm Baldrige National Quality Award, U.S. Department of Commerce, Technology Administration, National Institute of Standards and Technology, Gaithersburg, MD.

Marcus, A. "Limits on Personal-Injury Suits Urged." *The Wall Street Journal*, April 23, 1991.

Mathews, J. "The Cost of Quality." *Newsweek*, September 7, 1992, pp. 48–49.

Matthews, C. "Linking the Supply Chain to TQM." *Quality Progress*, November 2006, pp. 29–35.

Mehta, M., and K. Rampura. "Squeezing Out Extra Value." *Industrial Engineer*, December 2006, pp. 29–35.

Meier, B. "Court Rejects Coupon Settlement in Suit Over G.M. Pickup Trucks." *New York Times*, April 18, 1995.

Merrill, P. "In the Lead." *Quality Progress*, September 2008, pp. 26–32.

Michaels, D. "Kiosks to Help Trace Baggage." *The Wall Street Journal*, July 1, 2009.

Michaels, D., and I. Lunsford. "Boeing, Airbus Look to Auto Companies for Production Tips." *The Wall Street Journal*, April 1, 2005.

Milas, G. "How to Develop a Meaningful Employee Recognition Program." *Quality Progress*, May 1995, pp. 139–142.

Miller, I., and J. Freund. *Probability and Statistics for Engineers.* Englewood Cliffs, NJ: Prentice Hall, 1977.

Miller, J. R., and J. S. Morris. "Is Quality Free or Profitable?" *Quality Progress*, January 2000, pp. 50–53.

Miscikowski, D., and E. Stein. "Empowering Employees to Pull the Quality Trigger." *Quality Progress*, October 2006, pp. 43–48.

Moen, R., T. Nolan, and L. Provost. *Improving Quality through Planned Experimentation.* New York: McGraw-Hill, 1991.

Montgomery, D. *Introduction to Statistical Quality Control.* New York: John Wiley & Sons, Inc., 2001.

Moran, J. W., and P. C. La Londe. "ASQ Certification Program Gains Wider Recognition." *Quality Progress*, April 2000, pp. 29–41.

Mukherjee, S. "A Dose of DMAIC." *Quality Progress*, August 2008, pp. 44–51.

Mullenhour, P., and J. Flinchbaugh. "Bringing Lean Systems Thinking to Six Sigma." *Quality Digest*, March 2005, pp. 38–41.

Munoz, J., and C. Nielsen. "SPC: What Data Should I Collect? What Charts Should I Use?" *Quality Progress*, January 1991, pp. 50–52.

Munro, R. A. "Linking Six Sigma with QS-9000." *Quality Progress*, May 2000, pp. 47–53.

Munro, S. "The Last Bastion of Inefficiency." *Industrial Engineer*, November 2006, pp. 34–39.

Murphy, J. "Honda CEO Vies for Green Mantle." *The Wall Street Journal*, June 16, 2008.

Nakhai, B., and J. Neves. "The Deming, Baldrige, and European Quality Awards." *Quality Progress*, April 1994, pp. 33–37.

Nash, M., S. Poling, and S. Ward. "Six Sigma Speed." *Industrial Engineer*, November 2006, pp. 40–44.

Neave, H. *The Deming Dimension*. Knoxville, TN: SPC Press, 1990.

"Needed: A Backup for Ma Bell." *U.S. News & World Report*, September 30, 1991, p. 22.

Neenan, R. "Who Is Keeping Score?" *Quality Progress*, June 2008, pp. 48–52.

Nelsen, D. "Your Gateway to Quality Knowledge." *Quality Progress*, April 2004, pp. 26–34.

Nesbitt, T. "Flowcharting Business Processes." *Quality*, March 1993, pp. 34–38.

Neuscheler-Fritsch, D., and R. Norris. "Capturing Financial Benefits From Six Sigma." *Quality Progress*, May 2001, pp. 39–44.

Nicholas, J. *Competitive Manufacturing Management*. New York: Irwin McGraw-Hill, 1998.

Ohno, T. *Toyota Production System: Beyond Large Scale Production*. New York: Productivity Press, 1988.

Okes, D. "Complexity Theory Simplifies Choices." *Quality Progress*, July 2003, pp. 35–37.

Okes, D. "Driven by Metrics." *Quality Progress*, September 2008, pp. 48–53.

Orsini, J. "What's Up Down Under?" *Quality Progress*, January 1995, pp. 57–59.

Palmer, B. "Overcoming Resistance to Change." *Quality Progress*, April 2004, pp. 35–39.

Palmes, P. "How to Fail the ISO 9001 Driver's Test." *Quality Progress*, October 2006, pp. 33–36.

Parker, D., and A. M. Parker. "Turning the Frown Upside Down." *Quality Progress*, October 2006, p. 88.

Parsons, C. "The Big Spill: Hot Java and Life in the Fast Lane." *Gannett News Services*, October 25, 1994.

Passariello, C. "Logistics Are in Vogue with Designers." *The Wall Street Journal*, June 27, 2008.

Pasztor, A. "Study Cites Maintenance Risk." *The Wall Street Journal*, April 21, 2004.

Pearson, T. A. "Measure for Six Sigma Success." *Quality Progress*, February 2001, pp. 35–40.

Perry, M. "The Fish(bone) Tale." *Quality Progress*, November 2006, p. 88.

Pesansky, G. "Process Mapping for Knowledge Transfer." *Quality Digest*, November 2008, pp. 31–34.

Phillips-Donaldson, D. "100 Years of Juran." *Quality Progress*, May 2004, pp. 25–39.

Pietras, A. "Supply Chain Management." *Quality Digest*, December 2008, pp. 40–43.

Pittle, D. "Product Safety: There's No Substitute for Safer Design." *Trial*, October 1991, pp. 110–114.

Pond, R. *Fundamentals of Statistical Quality Control*. New York: Merrill, 1994.

PQ Systems. *Applying Design of Experiments Using DOEpack*. Dayton, OH: PQ Systems, 2001.

Press, A., G. Carroll, and S. Waldman. "Are Lawyers Burning America?" *Newsweek*, March 20, 1995, pp. 30–35.

Prevette, S. "Systems Thinking—An Uncommon Answer." *Quality Progress*, July 2003, pp. 32–35.

Pritts, B. A. "Industry-wide Shakeout." *Quality Progress*, January 2001, pp. 61–64.

Pyzdek, T. "Managing Metric Madness." *Quality Digest*, February 2007, p. 22.

Pyzdek, T., and P. Keller. *The Six Sigma Handbook*. New York: McGraw-Hill, 2007.

Radziwill, N., D. Olson, A. Vollmar, T. Leppert, T. Mattis, K. Van Dewark, and J. Sinn. "Starting from Scratch." *Quality Progress*, September 2008, pp. 40–47.

Ramberg, J. S. "Six Sigma: Fad or Fundamental." *Quality Digest*, May 2000, pp. 28–32.

Ramu, G. "In the Know." *Quality Progress*, August 2008, pp. 36–43.

Redmond, M. "60 Minutes to a Solution." *Quality Progress*, February 2007, p. 80.

Reid, R. "Developing the Voluntary Healthcare Standard." *Quality Progress*, November 2006, pp. 68–71.

Reid, R. D. "From Deming to ISO 9001:2000." *Quality Progress*, June 2001, pp. 66–70.

Reid, R. D. "Tips for Automotive Auditors." *Quality Progress*, May 2004, pp. 72–76.

Reid, R. D. "Why QS 9000 Was Developed and What's in Its Future." *Quality Progress*, April 2000, pp. 115–117.

Reidenbach, R., and R. Goeke. "Six Sigma, Value and Competitive Strategy." *Quality Progress*, July 2007, pp. 45–49.

RFID Study Group. "Challenges in RFID Enabled Supply Chain Management." *Quality Progress*, November 2006, pp. 23–28.

Rienzo, T. F. "Planning Deming Management for Service Organizations." *Business Horizons*, May 1993, pp. 19–29.

Robitaille, D. "The Basics of Internal Auditing." *Quality Digest*, June 2007, pp. 42–46.

Rooney, S., and J. Rooney. "Lean Glossary." *Quality Progress*, June 2005, pp. 41–47.

Roy, R. *A Primer on the Taguchi Method.* New York: Van Norstrand Reinhold, 1990.

Roy, R. K. "Sixteen Steps to Improvement." *Quality Digest*, June 2001, pp. 24–27.

Russell, J. *Quality Management Benchmark Assessment.* Milwaukee, WI: ASQC Press, 1991.

Russell, J. "Quality Management Benchmark Assessment." *Quality Progress*, May 1995, pp. 57–61.

Russell, J. P. "Auditing ISO 9001:2000." *Quality Progress*, July 2001, pp. 147–148.

Russell, J. P. "Know and Follow ISO 19011's Auditing Principles." *Quality Progress*, February 2007, pp. 29–34.

Salot, S. "Controlling Heavy Metals (and Other Nasty Stuff)." *Quality Digest*, October 2006, pp. 32–36.

Santos, J., R. Wysk, and J. Torres *Improving Production with Lean Thinking.* Hoboken, NJ: John Wiley & Sons, 2006.

Savell, L. "Who's Liable When the Product Is Information?" *Editor and Publisher*, August 28, 1993, pp. 35–36.

Saxena, R. "The Changing Chain." *Industrial Engineer*, April 2007, p. 24.

Schatz, A. "Regulators Push for Smarter Tires." *The Wall Street Journal*, July 14, 2004.

Schonberger, R. "Make Cells Work for You." *Quality Progress*, April 2004, pp. 58–63.

Schooley, J. "No Longer Waiting for Answers." *Quality Progress*, November 2008, pp. 34–39.

Schwinn, D. R. "Six Sigma and More: On Not Losing Sight of the Big Picture." *Quality E-line*, May 2, 2001, www.pqsystems.com.

Scovronek, J. "Reliability Sample Testing, A Case History." *Quality Progress*, February 2001, pp. 43–45.

Sedlock, R. "Under One Roof." *Quality Progress*, July 2007, p. 80.

Sharrock, R. "A Well-Appointed Establishment." *Industrial Engineer*, April 2007, pp. 44–48.

Shingo, S. *A Revolution in Manufacturing: The SMED System.* Nashville, TN: Productivity Press, 1985.

Shipley, D. "ISO 9000 Makes Integrated Systems User Friendly." *Quality Progress*, July 2003, pp. 25–28.

Shiu, M., and M. Tu. "QFD's Evolution in Japan and the West." *Quality Progress*, July 2007, pp. 30–38.

"Short-Run SPC Re-emerges." QEI Speaker Interview. *Quality*, April 1995, p. 44.

Smith, G. *Statistical Process Control and Quality Improvement.* New York: Merrill, 1991.

Smith, L. "Would You Like Standardization with That?" *Quality Digest*, February 2007, pp. 10–11.

Smith, R. "Faults Still Plague Electric System as Peak Summertime Use Nears." *The Wall Street Journal*, April 13, 2004.

Smith, R. "The Benchmarking Boom." *Human Resources Focus*, April 1994, pp. 1–6.

Snee, R. "Eight Essential Tools." *Quality Progress*, December 2003, pp. 86–88.

Snee, R. D. "Dealing with the Achilles' Heel of Six Sigma Initiatives." *Quality Progress*, March 2001, pp. 66–72.

Spearman, M. "Realities of Risk: Supply Chains Must Scrutinize the Unknown." *Industrial Engineer*, February 2007, pp. 36–41.

Spendolini, M. *The Benchmarking Book.* New York: Amcom, 1992.

Spigener, J. B., and P. A. Angelo. "What Would Deming Say?" *Quality Progress*, March 2001, pp. 61–64.

Srikanth, M., and S. Robertson. *Measurements for Effective Decision Making.* Wallingford, CT: Spectrum Publishing Co., 1995.

Stamatis, D. H. "Who Needs Six Sigma, Anyway?" *Quality Digest*, May 2000, pp. 33–38.

Stein, P. "By Their Measures Shall Ye Know Them." *Quality Progress*, May 2001, pp. 72–74.

Stevens, T. "Dr. Deming: Management Today Does Not Know What Its Job Is." *Industry Week*, January 17, 1994, pp. 20–28.

Stevenson, J., and A. Kashef. "Newer, Better, Faster." *Quality Progress*, September 2008, pp. 34–39.

Stevenson, W. *Operations Management*, 7th ed. New York: McGraw-Hill Irwin, 2002.

Surak, J. G. "Quality in Commercial Food Processing." *Quality Progress*, February 1999, pp. 25–29.

Suzaki, K. *The New Manufacturing Challenge*, New York: The Free Press, 1987.

Taguchi, G. *Introduction to Quality Engineering.* Dearborn, MI: ASI Press, 1986.

Till, D. *The Recipe for Simple Business Improvement.* Milwaukee, WI: ASQ Quality Press, 2004.

Tobias, R. *Applied Reliability.* New York: Van Nostrand Reinhold, 1986.

Torok, J. "The Where and Why: A 1-2-3 Model for Project Success." *Quality Progress*, April 2004, pp. 46–50.

Toyoda, E. *Toyota: Fifty Years in Motion*. Kodansha International, 1987.

Travalini, M. "The Evolution of a Quality Culture." *Quality Progress*, May 2001, pp. 105–108.

Traver, R. "Nine-Step Process Solves Product Variability Problems." *Quality*, April 1995, p. 94.

"U.S. FAA: FAA Orders Immediate Inspection for High-time Boeing 737s, Extends Inspection Order." *M2 Press Wire*, May 11, 1998.

Van Patten, J. "A Second Look at 5S." *Quality Progress*, October 2006, pp. 55–59.

Van Weerdenburg, N. "Can We Improve Continuous Improvement?" *Quality Digest*, February 2009, pp. 40–43.

Vardeman, S. *Statistics for Engineering Problem Solving*. Boston: PWS Publishing, 1994.

Vardeman, S., and J. Jobe. *Statistical Quality Assurance Methods for Engineers*. New York: John Wiley & Sons, Inc., 1999.

Vermani, S. "Capability Analysis of Complex Parts." *Quality Progress*, July 2003, pp. 65–71.

Verseput, R. "Digging Into DOE." *Quality Digest*, June 2001, pp. 33–36.

Voelkel, J. "What Is 3.4 per Million?" *Quality Progress*, May 2004, pp. 63–65.

Voelkel, J., and C. Chapman. "Value Stream Mapping." *Quality Progress*, May 2003, pp. 65–69.

Walker, H. "The Innovation Process and Quality Tools." *Quality Progress*, July 2007, pp. 18–22.

Wallace, R. "ISO/TS 16949 Pitfalls." *Quality Digest*, June 2008, pp. 40–43.

Walpole, R., and R. Myers. *Probability and Statistics for Engineers and Scientists*. New York: Macmillan, 1989.

Walters, J. "The Benchmarking Craze." *Governing*, April 1994, pp. 33–37.

Walton, M. *Deming Management at Work*. New York: Pedigree Books, 1991.

Walton, M. *The Deming Management Method*. New York: Putnam, 1986.

Watson, G. "Digital Hammers and Electronic Nails—Tools of the Next Generation." *Quality Progress*, July 1998, pp. 21–26.

Watson, G. "The Legacy of Ishikawa." *Quality Progress*, April 2004, pp. 54–57.

Watson, G. "Total Quality, Total Commitment." *Quality Progress*, November 2008, pp. 20–27.

Watson, R. "Modified Pre-control." *Quality*, October 1992, p. 61.

Weiler, G. "What Do CEOs Think About Quality?" *Quality Progress*, May 2004, pp. 52–56.

Weimer, G. A. "Benchmarking Maps the Route to Quality." *Industry Week*, July 20, 1992, pp. 54–55.

Welborn, C. "Using FMEA to Assess Outsourcing Risk." *Quality Progress*, August 2007, pp. 17–21.

West, J. E. "Implementing ISO 9001:2000, Early Feedback Indicates Six Areas of Challenge." *Quality Progress*, May 2001, pp. 65–70.

Westcott, R. "Continual Innovation and Reinvention." *Quality Progress*, July 2007, pp. 56–57.

Westcott, R. "Managing Expectations." *Quality Progress*, November 2009, pp. 24–29.

Westcott, R. "Maximize the Use of Your Abilities." *Quality Progress*, October 2006, pp. 70–71.

Wheeler, D. *Advanced Topics in Statistical Process Control*. Knoxville, TN: SPC Press, 1995.

Wheeler, D. *Short Run SPC*. Knoxville, TN: SPC Press, 1991.

Wheeler, D. *Understanding Industrial Experimentation*. Knoxville, TN: SPC Press, Inc., 1990.

Wheeler, D. *Understanding Variation: The Key to Managing Chaos*. Knoxville, TN: SPC Press, 1993.

Wheeler, D., and D. Chambers. *Understanding Statistical Process Control*. Knoxville, TN: SPC Press, 1992.

White, J. "How Reliable Is Your Car?" *The Wall Street Journal*, June 30, 2004.

Wiesendanger, B. "Benchmarking for Beginners." *Sales and Marketing Management*, November 1992, pp. 59–64.

Williams, B. "Calibration Management in the ISO/IEC 17025 Accredited Facility." *Quality Digest*, June 2007, pp. 37–41.

Wilson, L. *Eight-Step Process to Successful ISO 9000 Implementation*. Milwaukee, WI: ASQC Press, 1996.

Wilson, L. *How to Implement Lean Manufacturing*. New York: McGraw Hill, 2009.

Winslow, R. "Hospitals' Weak Systems Hurt Patients, Study Says." *The Wall Street Journal*, July 5, 1995.

Womack, J., and D. Jones. *Lean Thinking: Banish Waste and Create Wealth in Your Corporation*. New York: Simon and Schuster, 1996.

Womack, J., D. Jones, and D. Roos. *The Machine That Changed the World, The Story of Lean Production*. New York: Harper Perennial, 1991.

Wood, D. "Blurred Vision." *Quality Progress*, July 2008, pp. 28–33.

Zaciewski, R. "Attribute Charts Are Alive and Kicking." *Quality*, March 1992, pp. 8–10.

INDEX

Accuracy, 180–181
Acronyms of Six Sigma, 7
Airbus, 125
Aircraft landing gear life cycle management, 239–240
Airline accidents, 238
Airline costs for mishandled luggage, 46–47
Airlines, lean, 3
Air traffic control systems, 238
Akao, Yoji, 26
American Society for Quality, 4–5
Analysis of means (ANOM), 257
Analysis of variance (ANOVA), 257
Analytical data analysis, 187–193
    kurtosis, 192–193
    measures of central tendency (mean, median, mode), 187–189
    range, 188, 190
    skewness, 192–193
    standard deviation, 190–192
Analyze (in DMAIC), 87–94
    technique: brainstorming, 87–88
    technique: cause-and-effect diagrams, 89–90
    technique: run charts, 92, 93–94
    technique: scatter diagrams, 91
    technique: WHY-WHY diagrams, 90
Andons, 122–123, 153, 154
ANOM. *See* Analysis of means
ANOVA. *See* Analysis of variance
Appraisal costs, 50
Assembly line
    first modern (Henry Maudslay), 146–147
    Henry Ford's, 135
Assignable causes, 204
Attribute data, 178
Autonomation (jidoka), 122–123
Autonomous maintenance, 162
Availability, 240–241
Average chart formulas, 222
Average life, 241
Averages, compared with individual values, 225–226

Backup components, 243
Balanced Scorecard method (Kaplan and Norton)
    Customer Focus, 62–64
    Financial Analysis, 62–64
    Internal Processes, 62–64
    Learning and Growth, 62–64
Banking, 148
Basic training, 162
Bell Laboratories, 203–204
Bimodal distribution, 186–187

Black Belt Certification, 4–5, 7
    compared to Certified Quality Engineer (CQE), 7, 8
Block and tackle system, 146–147
Boeing, 152
Borthwick, J. S., 224
Brainstorming, 87–88
British Petroleum (BP), 161
Budgets, for projects, 76–77
Bulletin board, 153–154
Burger King, 2

Capability index, 232
    See also Process capability
Capability ratio, 232–234
    See also Process capability
Carroll, Lewis, *Alice's Adventures in Wonderland*, 19
Case studies. *See* Lean Six Sigma Tools at Work (case studies)
Cause-and-effect diagrams, 37, 38, 83, 89–90, 159, 208
    for quality costs, 54
    See also Process improvement teams
Centerline, 207, 209, 212
Central limit theorem, 193–195
Central tendency (mean, median, mode), 187–189
Certified Quality Engineer (CQE), 7
    compared to Black Belt certification, 7, 8
Chain reaction (Deming), 135
Challenger (space shuttle) disaster, 238
Change, 40–42
    change cycle, 40
    change matrix, 41
    control charts and, 214, 216
    supporting, 42
Change control systems, 77–78
"Changing Concepts and Management of Quality Worldwide" (Feigenbaum), 25
Check sheets, 38, 82–83, 85
Child, Julia, 124–125
    Mastering the Art of Cooking Volumes I and II, 124
Child, Paul, 124
Cleanliness, 128–129, 130
Combination system, reliability in, 243–244
Combine, 134
Communication, 41–42
    clear, 42
    defined, 41
    for reliability, 238
Competitive global economy, 1
Computer technology transition, 70, 72, 73, 74, 78
Concluding stage of teams, 43
Confidence intervals, 198–200
Conformance to requirements, 45–46

Consultative style of leadership, 17
Contingency plans, for projects, 77–78
Continuous data, 178
Continuous improvement, kaizen and, 133
Continuous improvement/quality management, compared to Malcolm Baldrige Award criteria, 8
Control (in DMAIC), 94–95
Control, for projects, 77–78
Control charts, 17, 38, 83, 160
    See also Variable control charts
Controllable factor, defined, 248
Control limits, 209, 211–213
    vs. specification limits, 227–228
    See also Variable control charts
Corrective action request form, 84
Costs
    appraisal, 50, 51
    failure, 50, 51
    intangible, 50, 51
    prevention, 49–50, 51
    projects and, 69
    total quality costs, 50–52
    See also Costs of quality
Costs of quality, 45–58
    appraisal costs, 50, 51
    categories of, 51
    cost of poor quality, 49
    defining, 47–49
    examples of, 48
    failure costs, 50, 51
    iceberg of, 48
    intangible costs, 50, 51
    measurement system for, 52–56
    prevention costs, 49–50, 51
    quality definition: conformance to requirements, 45–46
    quality measurement: costs of quality, 45–46
    quality performance standard; zero defects, 45–46
    quality system: prevention of defects, 45–46
    reduced by problem-solving process, 53–56
    total quality costs, 50–52
    utilizing for decision making, 57
    vs. profit, 51
    zero defects, 45–46
CPM. *See* Critical path method
CQE. *See* Certified Quality Engineer
Critical path method (CPM), 74, 76–77
Crosby, Philip, 45–46
Culture
    creating a lean Six Sigma, 12–15
    defined, 12
Customer, in just-in-time (JIT) system, 119
Customer Focus (Balanced Scorecard method), 62–64
Customer focus
    customer value perceptions, 24–26
    quality function deployment (QFD), 26–36
Customer knowledge, 6
Customer perceived value, 69
Customer results, 60
Customer satisfaction
    creating value and, 99–100
    defined, 25
    using a process map to improve, 113–115
    See also Value-added process mapping

Cycles
    change, 40
    recurring, 215, 217
Cycle time, 3, 61

Data
    attribute, 178
    continuous, 178
    discrete, 178
    grouped, 178–179
    subgroups, 208–210
    ungrouped, 178
    variables, 178
Data analysis. *See* Analytical data analysis; Graphical data analysis
Data collection, 178–179, 209, 211
    See also Analytical data analysis; Graphical data analysis; Statistics
Decision making, utilizing quality costs for, 57
Deductive statistics, 178
Defects
    prevention of, 45–46
    zero, 45–46
Defects waste, 137
Define (in DMAIC), 82
Define, measure, analyze, improve, control (DMAIC)
    problem-solving process, 6, 81–97, 100, 159
        analyze step, 87–94
        control step, 94–96
        define step, 82
        improve step, 94
        kaizen and, 133
        measure step, 82–87
        Plan-Do-Study-Act (PDSA) cycle and, 82
        problem solving, 81–82
        See also Plan-Do-Study-Act (PDSA) problem solving; Six Sigma
Definitions of Six Sigma, 7
Degrees of freedom, 249
Delegating style of leadership, 17–18
Deming, W. Edwards, 4, 7, 11–12, 59, 82, 100, 132
    14 points, 133
    change and, 40
    continuous improvement and, 133
    economic chain reaction, 12, 135
    on motivating employees, 40–41
    Out of the Crisis, 11, 12, 153
    on training and education, 39
Deploying strategic plans, 21–23
Deployment process map, 108
Design failure modes and effects analysis, 264
    See also Failure modes and effects analysis (FMEA)
Design of experiments (DOE), 7, 247–262
    characteristics of a good DOE, 251
    conducting experiments, 250–251
    defined, 247, 249–250
    definitions of terms, 248–249
    experimental analysis methods, 257
    fractional factorial experiments, 253
    full factorial experiments, 252–253
    hypotheses and experiment errors, 255, 257
    Plackett-Burman screening designs, 253–254
    single factor experiments, 251–252
    Taguchi designs, 253–255, 256
    trial and error experiments, 247–248

Directing style of leadership, 17
Disaster prevention, 161, 238
Discrete data, 178
Dispersion, 188–190, 192–193
Distribution, 168–169
    See also Supply chain management
DMAIC problem-solving process. *See* Define, measure, analyze, improve, control (DMAIC) problem-solving process
DOE. *See* Design of experiments
Drug safety, 50
Dunkin' Donuts, 2

E-commerce, supply chains and, 168
Economic chain reaction (Deming), 12
Economic Control of Quality of Manufactured Product (Shewhart), 204
Education, 38–40
Effect, defined, 248–249
Eisenhower, Dwight, 13
Eliminate, 134
Emergency room performance measures, 65
Equipment losses, 157–160
    See also Productive maintenance
Error proofing (poka-yoke), 142–144
Errors, types I and II, 255, 257
Estimation, of population sigma, 226–228
Experimental analysis methods, 257
Experiments. *See* Design of experiments
External failure costs, 50, 53–54

Factor, defined, 248
Factorial experiments
    fractional, 253
    full, 252–253
Fail-safe features, 237, 238
Failure costs, 50
    increase in, 49
    See also Costs of quality
Failure modes and effects analysis (FMEA), 7, 263–271
    analysis of, 267–270
    creating, 264–267
    defined, 263
    design, 264
    format, 265
    process, 264
    systems, 263–264
Failure rates, 240–241
Farm and lawn equipment quality, 53–56
Feigenbaum, Armand, 24–25, 236
    "Changing Concepts and Management of Quality Worldwide", 25
    Total Quality Control, 24–25
Final customers, 166, 167
Financial Analysis (Balanced Scorecard method), 62–64
Financial strength of organizations, project selection and, 69
Fishbone diagram (cause-and-effect diagram), 37
    See also Process improvement teams
5S, 119, 124–131
    kaizen and, 133
    seiketsu (standardize), 129, 130
    seiri (separate or sort), 125, 129
    seiso (shine, sweep, cleanliness), 128–129, 130
    seiton (set in order, simplify), 126–128, 129–130
    shitsuke (sustain, self-discipline), 129, 130
    tool board, 153
Five Ws and two Hs (questions), 107, 134
Flowcharts, 38, 83, 100, 103–108, 111, 259
    See also Process mapping; Value-added process mapping
Flow kaizen events, 134
Force field analysis, 94, 95
Ford, Henry, 1, 2, 43, 99, 118, 132
    assembly line, 2, 135
    Model T, 135
    Today and Tomorrow, 134
    waste and, 134
    workforce of, 135
Ford Motor Company, 2, 37, 43
Formation stage of teams, 43
Formulas
    control charts, 222
    process capability, 234–235
    reliability systems, 245
    statistics, 200–201
Fractional factorial experiments, 253
Frequency diagrams, 182–183, 225
Full factorial experiments, 252–253
Functional managers, vs. project managers, 78–79
Future predictions, 224–225
    See also Process capability

Galvin, Bob, 4
Gantt charts, 74–75
Gap analysis, 64–65
Gas mileage, 252, 254, 256–257
Genchi genbutsu, 14
Gilbreth, Frank and Lillian, 126–128
Global business environment, 24
Goals and objectives, for projects, 70–72
Graphical data analysis, 182–187
    frequency diagrams, 182–183
    histograms, 184–187
Grau, D. Scotty, 132
Green Belt Certification, 4–5, 7
Grouped data, 178–179

Hirano, Hiroyuki, 125
Histograms, 38, 83, 184–187, 205, 206–207
Homogeneous subgroups, 208–209
Horizontal component of quality function deployment, 26
Hospitals, process improvement at, 99
Humans in the system, reliability and, 237, 238
Hypotheses, 255, 257

If You Haven't Got the Time to Do It Right, When Will You Find the Time to Do It Over? (Mayer), 137
Imai, Masaaki, 132
    Kaizen, The Key to Japan's Competitive Success, 133
Improve (in DMAIC), 94
    technique: force field analysis, 94, 95
Inductive statistics, 178
Information sharing, supply chains and, 168
Intangible costs, 50
Interaction, defined, 249
Internal failure costs, 50

Internal Processes (Balanced Scorecard method), 62–64
Inventory management, 167–168
Ishikawa diagram (cause-and-effect diagram or fishbone diagram), 37
  See also Process improvement teams
Ishikawa, Kaoru, 37
ISO 9000, compared to Malcolm Baldrige Award and continuous improvement/quality management, 8

Japanese quality paradigm, 38
Jidoka (autonomation), 118, 122–123
JIT. *See* Just-in-time (JIT) system
Juran, Joseph, 83, 85
Just-in-time (JIT) system, 118–120
  defined, 118–119
  problem removal, 120
  summary of, 120
  Takt time, 120
  See also Kanban system

Kaizen projects, 69, 119, 132–142
  continuous improvement and Deming's 14 points and, 133
  defined, 132
  DMAIC and PDSA and, 133
  example of, 138–141
  5S and, 133
  five Ws and two Hs (questions), 134
  guiding words for (combine, simplify, eliminate), 134
  length of, 133
  waste elimination and, 134, 137–138
Kaizen, The Key to Japan's Competitive Success (Imai), 133
Kanban system (pull inventory management), 120–122
  See also Just-in-time (JIT) system
Key process input variables (KPIV), 6, 61–62, 82
Key process output variables (KPOV), 6, 61–62, 82
KPIV. *See* Key process input variables
KPOV. *See* Key process output variables
Kurtosis, 185–186, 192–193

Lawn mower defect, 49
LCL. *See* Lower control limit
Leadership, 11–18
  to achieve business results, 13
  lean Six Sigma culture, 12–15
  managing by fact and with a knowledge of variation, 16–17
  roles, 13
  styles of, 17–18
  traits, 11–12
  See also Managers
Lead time, 3, 61
Lean enterprises
  defined, 2–3
  error proofing (poka-yoke), 142–144
  5S, 3, 119, 124–131
  kaizen projects, 69, 119, 132–142
  kanban system (pull inventory management), 120–122
  lean tools, 3
  line balancing, 148–149
  measures of, 3
  origins of, 2–3
  reduced lot sizes (single piece flow), 151–152
  schedule leveling, 152
  setup time reduction, 149–151
  standardized work, 152–153
  summary, 272–273
  value stream process mapping, 108–112, 134
  visual management, 153–155
  See also Productive maintenance
Lean measures, 3
Lean Six Sigma, 7, 9
  organizations and websites, 273–274
  process improvement, 2
  summary, 272–273
Lean Six Sigma projects, 68
  See also Project management
Lean Six Sigma teams, 42–43
  See also Process improvement teams
Lean Six Sigma Tools at Work (case studies)
  accuracy and precision, 181
  airlines, 3, 46–47, 125, 238, 239–240
  brainstorming, 88
  bricklaying, 127
  calculating range values, 190
  cause-and-effect diagrams, 89, 208
  central limit theorem, 194–195
  change, 41
  check sheets, 85
  collecting data, 211
  communication, 42
  computer technology transition, 70, 72, 73, 74, 78
  confidence intervals, 198–200
  cost of poor quality, 49
  costs of quality, 46–47, 52–56
  CPM (critical path method), 77
  data, 178–179, 211
  define, measure, analyze, improve, control (DMAIC) problem solving, 83–85, 86–87, 88, 89, 91–92, 93, 94–96
  design of experiments, 248, 249–250, 251, 252, 254, 256–257, 258–261
  determining the mean, 187
  determining the median, 188
  determining the mode, 188
  emergency room performance measures, 65
  error proofing (poka-yoke), 142
  evaluating the solution, 95–96
  failure modes and effects analysis (FMEA), 267–270
  5S, 125, 127, 129–130
  force field analysis, 95
  frequency diagrams, 182–183
  histograms, 184–185, 186
  improving flow, 136–137
  indicators or measures of performance, 63
  justifying a project, 61
  kaizen events, 136–137, 138–142
  kanban system, 122
  leadership, 14, 15, 16
  logistics and a tsunami, 169
  managing by fact and with a knowledge of variation, 16
  measuring performance, 66
  Motorola and Six Sigma, 4
  negotiations and supply chains, 173–174
  organizational evolution, 9
  Pareto charts, 86

patient care measures, 64
PERT (program evaluation and review technique) network, 76
prevention costs in drug development, 50
problem identification, 83–84
process capability, 229, 231, 232, 234
process improvement, 43, 99
process mapping steps, 103
process performance measures, 61, 63, 64, 65, 66
productive maintenance, 157–158, 159–160, 161, 163
project management, 70, 72, 73, 74, 76, 77, 78
quality function deployment (QFD), 27–35
quoting process (process mapping), 106–108
range values, 190
relationship of mean, mode, median, 189
reliability, 238, 239–240, 241, 242, 243, 244
run charts, 93
sample taking, 177
scatter diagrams, 91
setup time reduction, 149, 150–151
single piece flow, 152
standard deviation, 191
standardized work through well-designed work instructions, 153
standard normal probability distribution, 197
Starbucks, 2, 43, 142
statistics, 177, 178–179, 181, 182–183, 184–185, 186, 187–190, 191–192, 194–195, 197, 198–200
strategic planning, 22–23
subgroup sample size, 209–210
supply chain management, 172, 173–174
takt time, 120, 148
Toyota Suggestions System, 39
value-added process mapping, 99, 103, 106–108, 111–112, 113–115
variable control charts, 207, 208, 209–211, 212, 219–221
variation, 207
work optimization, 148–151, 152, 153
Lean thinking, 2–3
Lean Thinking (Womack), 2
Lean tools, 3
Learning and Growth (Balanced Scorecard method), 62–64
Leptokurtic distribution, 186
Level, defined, 248
Levels
    change, jump, or shift in, 214, 216
    trends or steady changes in, 214, 215
Line balancing, 148–149
Logistics, supply chain management and, 168–169
Long-term maintenance planning, 161
Lower control limit (LCL), 207
Luggage, mishandled, 46–47

Machine That Changed the World, The (Womack), 2
Maintenance
    of a system for reliability, 237, 238
    autonomous, 162
    of performance records for reliability, 237–238
    predictive, 162
    preventive, 160–162
    See also Productive maintenance
Maintenance intervals, 161

Maintenance prevention, 160–162
    See also Productive maintenance
Maintenance records, 161–162
Malcolm Baldrige National Quality Award, 4
    compared to ISO 9000 and continuous improvement/quality management, 8
Management
    visual, 153–155
    See also Supply chain management
Managers
    functional, 78
    for projects, 78–79
    See also Leadership; Project management
Managing, by fact, 16–17
Manufacturing methods, 237
Mapping symbols, 102, 109
Market-in quality, 38
Master Black Belts, 7
Mastering the Art of Cooking Volumes I and II (Child), 124
Matrices
    for change, 41
    quality function deployment, 26–27
Maudslay, Henry, 146–147
Mayer, Jeffrey J., *If You Haven't Got the Time to Do It Right, When Will You Find the Time to Do It Over?*, 137
McDonald's, 2
Mean, 187–189, 191–192
Mean life, 240–241
Mean time between failures (MTBF), 241
Mean time to failure (MTTF), 241
Measure (in DMAIC), 82–87
    technique: check sheets, 82–83
    technique: Pareto analysis, 83, 85–87
Measurement error, 180–182
Measurement system, of quality costs, 52–56
Measures
    defined, 59
    See also Process performance measures
Median, 187–189, 191–192
Metal bending failure point, 257
Milk production, 157–158, 159–160
Minnich, Jay E., 132
Mission statements, 14, 15
Mode, 185, 187–189, 191–192
Model T, 134
Motion economy, 126–128
Motion waste, 138
Motorola, 4
MTBF. *See* Mean time between failures
MTTF. *See* Mean time to failure
Multimodal distribution, 186
Mura (unevenness), 119
Muri (overburden), 119

Negotiations, supply chain management and, 172–174
Noise factor, defined, 249
Non-faulty systems, 4
Normal curves, 225
Normal frequency distribution, 195–198

Occurrence ratings examples, 266
OEE. *See* Overall equipment effectiveness

Ohno, Taiichi, 118–119
    sources of waste, 135, 137–138
One-piece-at-a-time production, 119
    See also Just-in-time (JIT) system
Operational necessities, for projects, 69
Optimizing order: pick, pack, and ship, 258–261
Organizational results, 60–61
Organizations, for lean Six Sigma, 273–274
Oscillating trend, 214, 215
Out of the Crisis (Deming), 11, 12, 153
Overall equipment effectiveness (OEE), 157
Overproduction waste, 135

Packaging problems, 61, 138–141
Parallel system, reliability in, 242–243
Pareto charts, 38, 83, 85–87, 159
Pareto, Vilfredo, 85
Participative style of leadership, 17
Patient care measures, 64
Performance, defined, 69
Performance graphs, 16
Performance records, 237–238
    See also Record-keeping procedures
Performing stage of teams, 43
Personal commitment to quality, 39
PERT. *See* Program evaluation and review technique
Peterson, Donald, 37
Plackett-Burman screening designs, 253–254
Plan-Do-Study-Act (PDSA) problem solving, 6–7, 7, 14, 82, 100
    kaizen and, 133
    See also Define, measure, analyze, improve, control (DMAIC) problem-solving process
Planning, for projects, 72–73
Platykurtic distribution, 186
Poka-yoke (error proofing), 142–144
Poor quality
    costs of, 49
    See also Costs of quality
Populations, 177–178
    two, 217, 219
Population sigma, estimation of, 226–227
Power outages, 238
Precision, 180–181
Predictive maintenance, 162
Prevention costs, 49–50
Prevention of defects, 46
Preventive maintenance, 160–162
Price, vs. quality, 25
Problems
    defining, 83–84, 208
    See also Define, measure, analyze, improve, control (DMAIC) problem-solving process
Problem solving, 81–82
    quality costs reduced by, 53–56
    See also Define, measure, analyze, improve, control (DMAIC) problem-solving process; Plan-Do-Study-Act (PDSA) problem solving
Problem-solving tools, 3, 6, 38, 53–56, 83
    See also Lean enterprises; Lean Six Sigma Tools at Work
Process capability, 224–235
    calculating, 230–234
    capability index, 232
    capability ratio, 232–234
    control limits vs. specification limits, 227–228
    defined, 224
    estimation of population sigma from sample data, 226–227
    formulas, 234–235
    future predictions, 224–225
    individual values compared with averages, 225–226
    Six Sigma spread vs. specification limits, 228–230
Process cycle efficiency, 138
Processes
    defining, 98–99
    identifying key processes, 99
    See also Value-added process mapping
Process failure modes and effects analysis, 264
    See also Failure modes and effects analysis (FMEA)
Process flowcharts. *See* Process mapping; Value-added process mapping
Process improvement, 100–101
    lean thinking key steps, 3
Process improvement teams (quality circles), 7, 37–44
    1. formation stage, 43
    2. stormy stage, 43
    3. performing stage, 43
    4. concluding stage, 43
    background, 37
    cause-and-effect diagram (Ishikawa diagram/fishbone diagram), 37
    change cycle, 40
    change matrix, 41
    communication and, 41–42
    education and training, 39–40
    enhancing team performance, 40–42
    focus areas of, 38–39
    fundamentals of, 38
    lean Six Sigma teams, 42–43
    seven tools of quality, 38
    supporting change, 42
Process mapping, 56, 101–108, 139, 144
    defined, 101
    example of, 102
    symbols for, 102, 109
    See also Value-added process mapping
Process owners, 7
Process performance measures, 59–67
    Balanced Scorecard method (Customer Focus, Internal Processes, Learning and Growth, Financial Analysis), 62–64
    customer results: measures related to products and services, 60
    cycle time and, 61
    defining, 59–60
    gap analysis and, 64–65
    justifying a project, 61
    key process input variables (KPIV), 61–62
    key process output variables (KPOV), 61–62
    lead time and, 61
    measures of patient care, 64
    measuring patient room performance, 65
    organizational results: measures focused on strategic intent, 60
    strategic planning diagram, 65
    traditional and overall organization measures, 61
    value-creation time and, 61

Process sequencing, 151
Process waste, 137–138
Product dimensions, 54
Productive maintenance, 156–164
    autonomous maintenance, 162
    defined, 156–157
    disaster prevention, 161
    elimination of equipment losses, 157–160
    long-term maintenance planning, 161
    maintenance intervals, 161
    maintenance records, 161–162
    overall equipment effectiveness (OEE), 157
    predictive maintenance, 162
    preventive maintenance, 160–162
    spare parts control, 161
    speed losses, 159
Productivity, 1
Product life cycle curve, 239–240
Profit vs. cost of quality, 51
Program evaluation and review technique (PERT), 74, 75–76
Project management, 68–80
    contingency plans and change control systems, 77–78
    critical path method (CPM), 74, 76–77
    Gantt charts, 74–75
    lean Six Sigma projects, 68
    program evaluation and review technique (PERT), 74, 75–76
    project budgets, 76–77
    project characteristics (performance, cost, time), 69
    project goals and objectives, 70–72
    project managers, 78–79
    project plans, 72–73
    project proposals, 70, 71
    project schedules, 73–76
    project selection, 69–70
Projects. *See* Project management
Proposals, for projects, 70, 71
Pull production systems, 119–122
    See also Kanban system
Purchasing, 169–174, 237
    See also Supply chain management
Purchasing cycle, 170
Purchasing triangle, 170

QFD. *See* Quality function deployment
Quality, 1
    defined, 4, 7, 25
    education and, 38–40
    seven tools of, 38, 83
    websites, 273–274
    See also Costs of quality
Quality characteristic, 208
Quality circles, 37
    See also Process improvement teams
Quality costs. *See* Costs of quality
Quality Digest, 25
Quality function deployment (QFD), 7, 26–35
    background, 26
    creating a, 27–35
    defined, 26
    horizontal and vertical components of, 26
    matrix summary, 26–27
    process of, 26

Quality Progress, 25, 37
Quoting process, improving, 106–108

Range, 188, 190, 191–192
R charts, 204, 205, 211–222, 226–228
    formulas for, 222
    overlaying on the $\overline{X}$ chart, 229
    See also Variable control charts
Record-keeping procedures, 161–162, 237–238
Reduced lot sizes (single piece flow), 3, 151–152
Redundant systems, reliability in, 237, 238, 243
Reliability, 7, 236–246
    of a combination system, 244
    calculating system reliability, 241–244
    defined, 236–237, 241
    failure rate, mean life, and availability, 240–241
    formulas, 245
    measures of, 240–244
    in parallel systems, 242–243
    product life cycle curve, 239
    programs for, 237–238
    in redundant systems and backup components, 243
    in series, 242
Repeatability of measurements, 180
Replication, defined, 249
Reproducibility of measurements, 180
Response variable, defined, 249
Run, defined, 249
Run charts, 92, 93–94
Runs, control charts and, 215–217

Samples, 177–178
    estimation of population sigma from sample data, 226–228
    homogeneous subgroups, 208–209
Scatter diagrams, 38, 83, 91, 93
Schedule leveling, 152
Schedules, for projects, 73–76
Second Order of the Sacred Treasure (Japan), 11
Seiketsu (standardize), 129, 130
Seiri (separate or sort), 125, 129
Seiso (shine, sweep, cleanliness), 128–129, 130
Seiton (set in order, simplify), 126–128, 129–130
Selecting projects, 69–70
Series, reliability in, 242
Setup time reduction, 149–151
Seven tools of quality, 38, 83
Severity rating examples, 266
Shewhart, Walter, 7, 17, 82, 100, 203–205
    Economic Control of Quality of Manufactured Product, 204
Shingo, Shigeo, 118, 119
Shitsuke (sustain, self-discipline), 129, 130
Significance, defined, 249
Significant figures, 180, 182
Simplicity of design, 237, 238
Simplification, 126–128, 129–130, 134
Single factor experiments, 251–252
Single minute exchange of dies (SMED), 149–150
Single piece flow (reduced batch sizes), 151–152
Six Sigma
    acronyms and definitions, 7
    calculating, 5–6
    defined, 3–4, 5

eight phases of, 6
measures of, 6
origins of, 3–5
problem-solving steps (summary), 82
summary, 273
tools for, 6–7
vs. specification limits, 228–230
websites, 273–274
See also Define, measure, analyze, improve, control (DMAIC) problem-solving process
Six Sigma Breakthrough Strategy, 4
six sigma formula for process capability, 5
Six Sigma Report on Business Magazine, 203
Skewness, 185–186, 192–193
Smith, Bill, 4
Solutions, evaluating, 95–96
Sort, 128, 129
Spare parts control, 161
Special causes, 20
Specification limits vs. control limits, 227–228
Speed losses, 159
Sports drink bottles design (quality function deployment), 27–35
Standard deviation, 190–192, 204
See also Variable control charts
Standardization, 129, 130, 152–153
Standard normal probability distribution: Z tables, 196–198
Starbucks, 2, 43, 142
kaizen and error proofing at, 142
Statistics, 176–202
accuracy, precision, and measurement error, 180–182
central limit theorem, 193–195
confidence intervals, 198–200
data analysis: analytical, 187–193
data analysis: graphical, 182–187
data collection, 178–179
deductive, 178
defined, 176–177
formulas, 200–201
inductive, 178
normal frequency distribution, 195–198
populations vs. samples, 177–178
See also Analytical data analysis; Graphical data analysis; Variable control charts
Stormy stage of teams, 43
Strategic intent, 60
Strategic planning, 19–23
components of, 20–21
defined, 19–20
diagram, 65
elements needed, 20
generic plan, 21
issues to address, 20
Subgroup samples, 208–210
Supply chain management, 165–175
benefits of, 166
challenges of, 174
complex supply chain, 167
design questions, 166
e-commerce, 168
information sharing, 168
inventory management, 167–168
logistics, 168–169
negotiations, 172–174
participants in, 166
purchasing, 169–174
selecting suppliers, 171–172
single and multiple suppliers, 171–172
supplier partnerships, 171
supply chain defined, 165–166
Sustain, 129, 130
Symmetry, 185
Systemization, 129, 130
Systems failure modes and effects analysis, 263–264

Taguchi designs, 253–255, 256
Taguchi, Genichi, 253
Takt time calculations, 120, 148, 154
Taylor, Charles E., 239
Taylor, Frederick Winslow, 124
Teams
stages of formation, 43
types of lean Six Sigma, 42–43
See also Process improvement teams
Teamwork. *See* Process improvement teams (quality circles)
Therbligs, 127–128
Time frame for projects, 69
Time-terminated tests, 240–241
Tool board, 153
Tools, of quality, 38, 83
Total Quality Control (Feigenbaum), 24–25
Total quality costs, 50–52
See also Costs of quality
Total quality movement, 24–25
Toyada, Ohno, 118
Toyota Motor Company, 118–119
See also Just-in-time (JIT) system
Toyota Production System: Beyond Large Scale Production (Ohno, T.), 119
Toyota Suggestions System, 39
Traditional performance measures, 61
Training, 39–40
basic, 162
Transparency, 153
Transportation waste, 135, 137
Treatment, defined, 249
Trends, 214
Trial and error experiments, 247–248
See also Design of experiments
Trial control limits, 209, 211–213
Tsuruoka, Tadamitsu, 3
Type I errors, 255, 257
Type II errors, 255, 257

UCL. *See* Upper control limit
Uncontrolled variation, 204
Ungrouped data, 178
Upper control limit (UCL), 207

Value
customers' perception of, 24–26
defined, 25
Value-added process mapping, 98–117
creating value and generating customer satisfaction, 99–100

defining processes, 98–99
identifying key processes, 99
to improve work flow and increase customer satisfaction, 113–115
process improvement, 100–101
process mapping, 101–108
separating the non-value-added activities from the value-added activities, 112–113
value stream process mapping, 108–112
Value-creation time, 3, 61
Value stream process mapping, 108–112, 134
Variable control charts, 203–223
centerline, 207, 209, 212–213
change, jump, or shift in level, 214–215, 216
controlled variation (chance or common causes), 204
as decision-making tools, 205
formulas, 222
interpretation of, 213–221
introduction, 203–206
patterns, 213–214
as problem-solving tools, 206
R charts, 204, 205, 211–222, 226–229
recurring cycles, 215, 217
revising, 217–221
runs, 215–217
trends or steady changes in level, 214
two populations, 217, 219
uncontrolled variation (special or assignable causes), 204
for variables, 207
variation, 207
Walter Shewhart and, 203–205
$\bar{X}$ charts, 204, 205, 207–222, 226–229
Variables
for control charts, 207
defined, 207
Variables data, 178
Variation, 16–17
control charts for, 207
controlled, 204
defined, 207
uncontrolled, 204
Ventilation system upgrade, 267–270
Vertical component of quality function deployment, 26
Vision statements, 14
Visual management, 153–155

Wait time waste, 138
Warranty claims, 53
Warranty investigation report, 56
Waste
defects waste, 137
eliminating, 2
Henry Ford and, 134
motion waste, 138
overproduction waste, 135
process waste, 137–138
sources of, 3, 119
transportation waste, 135, 137
wait time waste, 138
Websites for lean Six Sigma organizations, 273–274
White board with takt time, 153, 154
WHY-WHY diagrams, 55, 90, 139, 159
Womack, Jim
Lean Thinking, 2
Machine That Changed the World, The, 2
Worker involvement, 38
Work flow, using a process map to improve, 113–115
Workforce, importance of an excellent, 135
Work optimization, 146–155
block and tackle system, 146–147
line balancing, 148–149
schedule leveling, 152
setup time reduction, 149–151
single piece flow (reduced batch sizes), 151–152
standardized work, 152–153
visual management, 153–155
Work standards, 152–153
Workstation layout, 143
Wright Brothers, 239

$\bar{X}$ charts, 204, 205, 207–222
estimation of population sigma from sample data, 226–228
formulas for, 222
overlaying the R chart pattern on, 229

Yasuda, Yuzo, 39

Zero defects, 45–46
Z tables, 196–198